Medicinal and Aromatic Plants of Himachal Pradesh

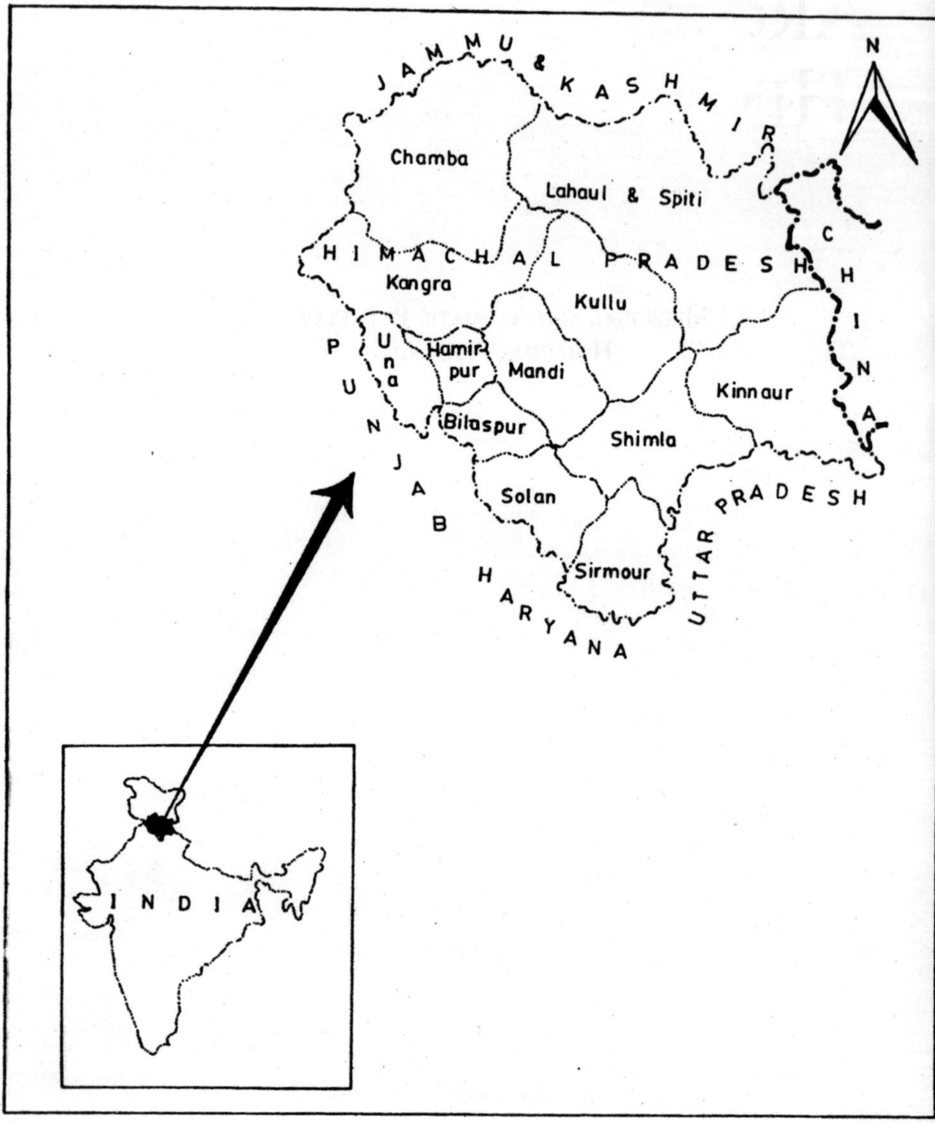

JAMMU & KASHMIR
N
Chamba
Lahaul & Spiti
HIMACHAL PRADESH
CHINA
Kangra
Kullu
PUNJAB
Una
Hamir-pur
Mandi
Kinnaur
Bilaspur
Shimla
Solan
UTTAR PRADESH
Sirmour
HARYANA
INDIA

Medicinal and Aromatic Plants of Himachal Pradesh

SECOND EDITION

NARAIN SINGH CHAUHAN

INDUS
PUBLISHING COMPANY

Published by
INDUS PUBLISHING COMPANY
FS-5, Tagore Garden, New Delhi-110027
Tel.: 42133022, 42133622, 25151333
indus@indus-publishing.com
www.indus-publishing.com

Second Edition 2006

ISBN 81-7387-098-5

Printed at
Chaman Enterprises, New Delhi

Foreword

India has a rich heritage of use of plants as medicines. Indian systems of medicines utilize 80 per cent of the material derived out of plants. There are over 2500 plant species in India having documented medicinal value, majority of them growing in a wild state; whereas only a few are cultivated. Out of these about 750 plant species form the ingredients of 14,000 published recipes of Ayurveda, Sidha and Unani medicines. The therapeutic value of 48 plant species has been established by modern investigations and are extensively used by the modern medical practitioners in one form or the other.

In recent years, a few developments in the drug industry have brought Indian Medicinal Wealth and their Ecosystem into world focus. In particular, the clinical reports on many native plants like *Rauvolfia, Tylophora, Guggulu, Chirayita, Cinchona, Valeriana, Taxus, Arjuna* etc. have supported their claim of containing specific pharmacological activity for which they are used in the Indian System of Medicines. Traditional systems of medicines continue to be widely practiced. Global estimates indicate that 80 per cent of population can not afford the products of the western pharmaceutical industry and have to rely upon the use of traditional medicines mainly derived from plants. An inventory of over 20,000 species has been compiled for the entire world. According to a survey report by WHO, about 25 per cent of prescribed human medicines are derived from plants and 80 per cent people still depend on traditional systems of medicines.

Due to heavy demand from pharmaceutical industry leading to over-exploitation, destruction of habitats and unsustainable harvesting practices sans conservation and cultivation support many valuable herbs especially in the Himalayan region are facing the danger to their survival in their natural abode. The knowledge of status of existing resources particularly their biosystematics and utilization potential are the pre-requisites for conservation in its broad sense. Much is desired to formulate a comprehensive strategy to cover identification, collection, storage, preservation, sustainable utilization, supply, restoration of the ecological balance and regeneration aspects. An insight into ethno-botanical aspects provides an added fillip to the efforts, unraveling the nature's mysterious and miraculous generosities. This grassroot approach of local inventory will enrich and strengthen the macro approach of national and global biodiversity enumerations.

It gives me immense pleasure to appreciate and compliment the tireless efforts of the author, Dr. Narain Singh Chauhan, a scientist of international repute, spreading over three decades to document this most valuable and multifaceted knowledge under the title, '*Medicinal and Aromatic Plants of Himachal Pradesh*'. This book is based on the author's personal explorations made in all parts on the State supported by collections, identification, compilation of information on traditional knowledge gathered during surveys, covering status of existing resources, utilization potential, conservation needs, cultivation efforts in herbal gardens and establishment of Herbaria at IRISM, Jogindernagar and in Dr. Y.S. Parmar University of Horticulture & Forestry. Important plants include the name, description, distribution in H.P.; parts used, chemical constituents, medicinal uses, market prices, cultivation notes supported by data on commercial importance, business potential, listing of endangered species, scope for pharmaceutical industries, scope of cultivation in different agroclimatic zones, potent substitutes and formulation in Indian systems of medicines. This book will be of immense use to all those concerned wholly or enthusiastically with the medicinal and aromatic plants, especially pharmaceutical industry, students and research scientists in Ayurveda, Unani, traditional healers, botany, pharmacognosy, phytochemistry, ethnobotany, forestry, agriculture/horticulture, naturalists and conservationalists. The drawings and coloured photographs have made the book a most desirable asset not only for the Western Himalaya, but also for the entire Hindu Kush Himalayan region. The multifaceted information brought close together at one place is an added advantage, the book will give to the readers of any walk of life.

PROF. L.R. VERMA
Vice-Chancellor
Dr. Y.S. Parmar University of Horticulture and Forestry,
Nauni, Solan, Himachal Pradesh 173 230 India

PREFACE

The use of plants as sources of medicines and human sustenance has been in vogue since antiquity. Ancient Indian literature like Rigveda (4000-1500 BC), Atharveda (1500 BC), Upnishada (1000-600 BC), Mahabharata and Puranas (700-400 BC), Charak Samhita and Sushruta Samhitas (1000-600 BC) include medicinal plants used for drugs, essences, worships, food, fuels, poisons, agricultural tools and other purposes. Charaka and Sushruta Samhitas describe the uses of about 1,200 plant drugs. The epic Ramayana has reference to Sanjivani, Sandhankarni, Vishalyakarni etc. In other parts of the world notable records like Papyrus Ebers (1550 BC), works of Hypocrites (460-370 BC), Disocoroides (78 AD) and Galen (131-200 AD) are worthy of mention. Chinese, Summerian, Babylonian, Greek and other civilizations of the world have earliest records of use of plants as medicine. It can be concluded that herbs and herbal products occupied a prime place in all ages and in all civilizations on earth totally unconcerned with the development of synthetic and chemotherapeutic agents. The Indian System of Medicines, Ayurveda, passing through several phases of turmoils resulting from invasion of Islam and the Britishers, is still the mainstay for about 90 per cent of our population. Indian medicinal plants got a boost during British regime as various botanists, physicians, travelers, navigators etc. highlighted and documented a lot of information on plant-based drugs of our country. The contributions of Garcia D'Orta, Van Rheede (*Hortus malabaricus*), Roxburgh (*Flora indica*), Ainslie (*Materia indica*) and George Watt (Dictionary of Economic Products in India) are worth mentioning. WHO (World Health organisation) has listed about 20,000 plant species in the world yielding drugs. In India over 2,500 species are credited with medicinal value. Recently the Ministry of Environment through a co-ordinated research project on ethnobotany has succeeded in getting the local uses of about 7,500 species documented. About 1,100 plant species are frequently used in Indian Systems of Medicines. Out of these, 500 plants, are commonly used in the preparation of Ayurvedic, Unani and Homoeopathic drugs. There are around 7,000 units engaged in manufacturing plant-based drugs in India covering around 14,000 recipes.

Plant-based drugs have created a sweeping resurgence in the west in their favour and the demand for these has reached a scale never witnessed before. Newer and newer sources of drugs are being screened to combat the menace of dreadful diseases like cancer and AIDS. The focus has now shifted from chemical screening to biological screening of plants. Every species growing on earth, even the so-called obnoxious weeds are subjected to biological screening to find out its activity which has brought into focus various new sources of alkaloids, anti-malarial, anti-protozoal, anti-arthritic, anti-asthmatic, anti-ulcer, anti-diabetic, anti-neuplastic and anti-implantation agents. More than two-thirds of the deaths result from heart attacks and it is a number one killer in the world. Plants afford the cure for cardiovascular diseases. In the area of hepatoprotective drugs too, the plant kingdom has proved to be a fruitful ground for the discovery of effective remedies, which the modern system of medicine can not provide.

The researches in herbal drugs and more so, the process of discovering new and valuable drugs from plants, brings together various disciplines like Taxonomy, Pharmacognosy (including biopharmaceutics, pharmacokinetics, phytopharmacy, chemotaxonomy etc.), Phytochemistry, Pharmacology, Toxicology and Clinical screening. It is widely accepted that the major initial problem in a plant drug research programme is the availability of the raw material and selection of the plant or plant parts. The credibility of any system of medicine depends on the use of genuine and authenticated material. The Indian Systems of Medicines suffers from the confusion and controversies due to prevalent synonyms. Many names are assigned to a plant species and the same name is applicable to may different species. To facilitate the physician and the pharmaceutical expert in selecting the genuine drug, there is a prime need to minimize, if not the altogether removal of controversies and confusion, by identifying the genuine material, adulterants and substitutes and establishing standards that could be adopted uniformly throughout the country.

Till date, the maximum reliance for drug plants to meet the requirement of raw material for drugs, pharmaceuticals, perfumery, cosmetics and related industries is on the plant wealth growing in forests, valleys, wastelands, temperate forests, alpine meadows and cold deserts and only a few are cultivated. The excessive exploitation from the nature has resulted in shrinkage in population of many; while a few others are facing threat of extinction. Increasing population pressure and unprecedented search for finding newer and newer sources of drugs coupled with growing popularity of traditional herbal lores have necessitated identification of our genetic resources, inventorisation, qualitative and quantitative estimation, conservation and commercial cultivation leading to their sustainable utilization. The knowledge of natural resources, their enumeration alongwith evaluation of their resource potential and life-style of people are pre-requisite to standardized their systematic exploitation. This grass-root approach of local inventory will enrich and strengthen the macro approach of national and global biodiversity enumerations.

Keeping in view the recent, growing popularity of herbal remedies, the Indian medicinal plants, more so the Himalayan herbs have received unprecedented attention, the world over. The literature on medicinal plants in quite vast, scattered and not available at one place even in big libraries. A great necessity was felt to have relevant literature providing multi-disciplinary information on such resources. The present book will serve the cause of the readers from diverse categories, who have the curiousity to know the plants and make use of them. It covers around 700 species of medicinal and aromatic plants of Himachal Pradesh, representative of not only Western Himalaya, but also of the entire Hindu Kush Himalayan region. Vegetation of the state has been discussed under each climatic zone elucidating with relevant examples. The plants described in details in the text cover Latin name (following latest nomenclature), family, common and local names, distribution in Himachal Pradesh (voucher numbers indicated in parenthesis alongwith the name of the collector), description of the plant, parts used, market rate, chemical constituents, formulations, uses and cultivation notes. Sketches and colour photographs provided will facilitate identity and comprehension by the readers. Appendices include information on commercial plants available for trade and drug industry, annual outturn of raw material during the past five years; endangered species, plants for phytochemical industry, dhoop and incense, tans and dyes, essential oils, plants causing health hazards, having ethnomedical importance and the species suitable for incorporation in cropping systems in different climatic zones of the state. Separate chapters have been devoted to collection, cultivation and policy issues. This is only the beginning and efforts are on to explore the herbal wealth extensively at micro level which will add much to the present knowledge in the coming years. This book will serve as a useful companion to the students, researchers, teachers, industrialists, collectors, cultivators, foresters, planners, nature lovers, those interested in biodiversity of the Himalayan regions and above all the students and practitioners of Indian Systems of Medicines, Homoeopathy and ethnomedicines.

NARAIN SINGH CHAUHAN

ACKNOWLEDGEMENTS

I feel a great pleasure in acknowledging the unflinching support, field assistance, companionship, and encouragement received from all those who have been associated with me at any level during the course of my explorations of the vast treasure of herbal wealth in this Himalayan state and bringing the information together in the present book form. The field and assisting staff at the Ayurvedic Research Institute, Jogindernagar (Mandi) and at the Dr. Y.S. Parmar University of Horticulture and Forestry, Nauni-Solan (HP) deserve appreciation for their consistent support.

The affection and encouragement received from Dr. M.R. Thakur (former Vice-Chancellor, Dr. YSPUHF, Solan), late Dr. G.C. Negi (former Vice-Chancellor, HPKV, Palampur), late Dr. Pushkar Nath, (Vice-Chairman, HP Science, Technology and Environment Council, Himachal Pradesh), will always guide and strengthen my spirits to unravel more and more of the mysteries of nature. Professor L.R. Verma (present Vice-Chancellor, Dr. YSPUHF, Solan) has always been encouraging the developmental activities on medicinal and aromatic plants and I am thankful to him for all the facilities and for writing the Foreword for this book.

The Indian Systems of Medicines have become popular all over the world due to the sincere and dedicated efforts of few industrial houses, practitioners and philanthropists. Dr. S.P. Kaushal and Dr. Manju Chawla of Chinese Medicine Clinic and Cancer Research Clinic, New Delhi are such pioneers in the country. Their love for herbs can be gauged from their offer of subsidy for printing the book on their very first visit to my office. The constant support, encouragement and perseverance of my wife, daughters and son, Maneesh has always been a driving force behind me.

The help sought from available literature to supplement the information is gratefully acknowledged. I am thankful to Shri Joginder Bansal for assisting in preparing drawings, sketches and maps and typing the manuscript. I am also thankful to M/s Indus Publishing Company for their painstaking efforts in printing the book.

NARAIN SINGH CHAUHAN

CONTENTS

Foreword 5

Preface 7

Acknowledgements 11

List of Medicinal & Aromatic Plants 15

List of Sketches 21

List of Plates 25

INTRODUCTION 29

- Geographical Position
- Geology, Rock and Soil
- Climate and Vegetation
- Resources of Medicinal and Aromatic Plants

DESCRIPTIVE PROFILE OF PLANTS 47

COLLECTION AND MARKETING OF PLANTS 449

APPENDICES

1. Commercial plants for trade and drug industry 479
2. Annual outturn of medicinal and aromatic plants for trade and commerce 498
3. Endangered species of medicinal and aromatic plants 500
4. Plant species having fluctuating demand 510
5. Unexploited rare species of medicinal and aromatic plants 516
6. Demand of state-run Ayurvedic pharmacies 562
7. Raw material for dhoop and incense industry 566
8. Raw material for manufacturing of phytochemicals for the drug industry 568

9. Essential oil-yielding sources 570
10. Plants for insecticide and pesticide industry 574
11. Tannin-yielding plants 577
12. Medicinal and aromatic plants to fit in different cropping systems 579
13. Domestication hints for medicinal and aromatic plants 588
14. Aromatic plants responsible for health hazards 594
15. Potent substitutes for indigenous and exotic drug materials 596

Bibliography 599

Index of Botanical Names 611

Index of Local Names 619

List of Medicinal & Aromatic Plants

1. *Abies spectabilis* (D. Don.) Spach 47
2. *Abrus precatorius* Linn. 49
3. *Acacia catechu* Willd. 52
4. *Acacia nilotica* Delile 56
5. *Achillea millefolium* Linn. 59
6. *Achyranthes aspera* Linn. 62
7. *Aconitum deinorrhizum* Stapf. 65
8. *Aconitum heterophyllum* Wall. 69
9. *Acorus calamus* Linn. 72
10. *Adiantum venustum* G. Don. 76
11. *Aegle marmelos* (L.) Corr. 77
12. *Aesculus indica* Colebr. 80
13. *Ajuga bracteosa* Wall. ex Benth. 83
14. *Albizia lebbek* Benth. 85
15. *Allium sativum* Linn. 88
16. *Aloe barbadensis* Mill. 90
17. *Angelica glauca* Edgew. 91
18. *Arctium lappa* Linn. 94
19. *Arnebia benthami* (Wall. ex G. Don.) Johnston 96
20. *Arnebia euchroma* (Royle) Johnston 98
21. *Artemisia brevifolia* Wall. 100
22. *Asparagus abscendens* Roxb. 102
23. *Atropa acuminata* Royle 104
24. *Azadirachta indica* A. Juss. 107
25. *Baliospermum montanum* Muell-Arg. 110

26. *Bauhinia variegata* Linn. 111
27. *Berberis aristata* DC. 114
28. *Berberis lycium* Royle 117
29. *Bergenia ciliata* (How.) Stern. 120
30. *Bergenia stracheyi* (Hook f. & Thoms) Engl. 123
31. *Betula utilis* D. Don. 125
32. *Bombax ceiba* L. 128
33. *Bunium persicum* (Boiss.) Fedt. 132
34. *Butea monosperma* (Lamk.) Taub. 133
35. *Calotropis gigantea* (Linn.) Ait. f. 137
36. *Cannabis sativa* Linn. 139
37. *Capsicum annuum* Linn. 143
38. *Carum carvi* Linn. 144
39. *Cassia absus* Linn. 147
40. *Cassia fistula* Linn. 148
41. *Cassia occidentalis* Linn. 150
42. *Cassia tora* Linn. 151
43. *Cedrus deodara* (Roxb.) Loud. 153
44. *Centella asiatica* (Linn.) Urban 154
45. *Chenopodium ambrosioides* Linn. 156
46. *Cichorium intybus* Linn. 157
47. *Cinnamomum tamala* Nces & Ebesm. 160
48. *Cissampelos pareira* Linn. 162
49. *Clerodendrum serratum* Spreng. 164
50. *Coriandrum sativum* Linn. 165
51. *Corydalis govaniana* Wall. 167
52. *Costus speciosus* Smith 169
53. *Crocus sativus* Linn. 172
54. *Cuscuta reflexa* Roxb. 174
55. *Cynodon dactylon* (Linn.) Pers. 176
56. *Cyperus rotundus* Linn. 177
57. *Dactylorhiza hatagirea* (D. Don.) Soo 180
58. *Datura fastuosa* Linn. 183

59. *Datura metel* Linn. 184
60. *Datura stramonium* Linn. 187
61. *Dioscorea deltoidea* Wall. ex Kunth. 190
62. *Dolichos uniflorus* Lank. 194
63. *Drocera peltata* Sm. 195
64. *Eclipta prostrata* Roxb. 196
65. *Emblica officinalis* Gaertn. 199
66. *Ephedra gerardiana* Wall. 202
67. *Fagopyrum esculentum* Moench 205
68. *Fumaria parviflora* Lam. 206
69. *Gentiana kurroo* Royle 207
70. *Geranium nepalense* Sweet 210
71. *Ginkgo biloba* Linn. 211
72. *Gloriosa superba* Linn. 212
73. *Hedychium acuminatum* Rosc. 215
74. *Helicteres isora* Linn. 218
75. *Heracleum lanatum* Michx. 221
76. *Hippophae rhamnoides* Linn. 223
77. *Hippophae salicifolia* D. Don. 225
78. *Holarrhena antidysenterica* Wall. ex DC. 225
79. *Humulus lupulus* Linn. 229
80. *Hyoscyamus niger* Linn. 231
81. *Hypericum perforatum* Linn. 234
82. *Hyssopus officinalis* Linn. 237
83. *Inula racemosa* Hook f. 240
84. *Jasminum humile* Linn. 242
85. *Juglans regia* Linn. 246
86. *Juniperus communis* Linn. 250
87. *Juniperus macropoda* (Boiss.) 253
88. *Juniperus recurva* Buch. Ham. 255
89. *Jurinea dolomiaea* Boiss. 256
90. *Justicia adhatoda* Linn. 259
91. *Leucas cephalotes* (Roth) Spreng. 260

92. *Mallotus philippinensis* Muell. Arg. 262
93. *Mentha longifolia* (L.) Huds. 265
94. *Morchella esculenta* (L.) Pers. 268
95. *Moringa oleifera* Lamk. 269
96. *Mucuna pruriens* (L.) DC. 272
97. *Murraya koenigii* Spreng. 274
98. *Myrica esculenta* Buch. Ham. 276
99. *Myrsine africana* Linn. 279
100. *Nardostachys grandiflora* DC. 280
101. *Nerium indicum* Mill. 283
102. *Nyctanthes arbor-tristis* Linn. 285
103. *Ocimum sanctum* Linn. 288
104. *Onosma bracteatum* Wall. 291
105. *Origanum vulgare* Linn. 293
106. *Oroxylum indicum* Vent. 296
107. *Paeonia emodi* Wall. 298
108. *Parmelia perlata* (Huds.) Ach. 300
109. *Physochlaina praealta* (G. Don.) Miers. 301
110. *Phytolacca acinosa* Roxb. 304
111. *Picrorhiza kurooa* Royle ex Benth. 305
112. *Pinus roxburghii* Sarg. 308
113. *Pinus wallichiana* A.B. Jackson 309
114. *Pistacia khinjuk* Stew. ex Brandis 311
115. *Plumbago zeylanica* Linn. 314
116. *Podophyllum hexandrum* Royle 317
117. *Polygonum aviculare* Linn. 320
118. *Pongamia pinnata* (Linn.) Merr. 323
119. *Potentilla fulgens* Wall. 324
120. *Prinsepia utilis* Royle 325
121. *Prunella vulgaris* Linn. 327
122. *Prunus cerasoides* D. Don. 329
123. *Pueraria tuberosa* DC. 332
124. *Punica granatum* Linn. 333

125. *Putranjiva roxburghii* Wall. 337
126. *Pyracantha crenulata* (D. Don.) M. Roemer 338
127. *Randia dumetorum* (Retz.) Poir 340
128. *Rauvolfia serpentina* Benth. ex Kutz. 342
129. *Rheum australe* D. Don. 346
130. *Rheum moorcroftianum* Royle 348
131. *Rhododendron anthopogon* D. Don. 351
132. *Rhododendron arboreum* Sm. 353
133. *Rhododendron campanulatum* D. Don. 355
134. *Ricinius communis* Linn. 357
135. *Roscoea alpina* Royle 360
136. *Rubia manjith* Roxb. ex Fleming 363
137. *Salvia moorcroftiana* Wall. ex Benth. 366
138. *Sapindus mukorossi* Gaertn. 368
139. *Saussurea costus* (Falc.) Lipsch. 370
140. *Selinum vaginatum* C.B. Clarke 373
141. *Selinum wallichianum* (DC.) Raizada & Saxena ex Raizada 374
142. *Sesamum indicum* Linn. 376
143. *Sida cordifolia* Linn. 377
144. *Sida rhombifolia* Linn. 380
145. *Solanum khasianum* C.B. Clarke 381
146. *Solanum nigrum* Linn. 383
147. *Solanum surattense* Burm. f. 386
148. *Spilanthes acmella* Linn. 388
149. *Swertia chirayita* Karst. 389
150. *Symplocos paniculata* Wall. 394
151. *Tagetes minuta* Linn. 396
152. *Taraxacum officinale* Wigg. 398
153. *Taxus wallichiana* Zucc. 401
154. *Terminalia arjuna* W. & A. 404
155. *Terminalia bellirica* Roxb. 407
156. *Terminalia chebula* Retz. 410

157. *Thalictrum foliolosum* DC. 412
158. *Thevetia neriifolia* Juss. 415
159. *Thymus serpyllum* Linn. 416
160. *Tinospora cordifolia* (Willd.) Miers. 418
161. *Trillidium govanianum* (D. Don.) Kunth. 420
162. *Uraria picta* Desv. 422
163. *Urginea indica* (Roxb.) Kunth. 423
164. *Valeriana hardwickii* Wall. 425
165. *Valeriana jatamansi* Jones. 426
166. *Vanda tessellata* Lodd. ex Loud. 428
167. *Vetiveria zizanioides* (Linn.) Nash. 430
168. *Viola serpens* Wall. 431
169. *Viscum album* Linn. 434
170. *Vitex negundo* Linn. 436
171. *Withania somnifera* Dunal. 439
172. *Woodfordia fruticosa* Kurtz. 442
173. *Xanthium strumarium* Linn. 444
174. *Zanthoxylum armatum* DC. 446

List of Sketches

1. Map of Himachal Pradesh 2
2. Distribution map of important medicinal and aromatic plants 41
3. *Abies spectabilis* 48
4. *Abrus precatorius* 51
5. *Acacia catechu* 54
6. *Acacia nilotica* 57
7. *Achillea millefolium* 60
8. *Achyranthes aspera* 64
9. *Aconitum deinorrhizum* 67
10. *Aconitum heterophyllum* 71
11. *Acorus calamus* 74
12. *Aegle marmelos* 79
13. *Aesculus indica* 81
14. *Ajuga bracteosa* 84
15. *Albizia lebbek* 87
16. *Angelica glauca* 93
17. *Arnebia benthami* 97
18. *Arnebia euchroma* 99
19. *Artemisia brevifolia* 101
20. *Atropa belladonna* 106
21. *Azadirachta indica* 108
22. *Bauhinia variegata* 112
23. *Berberis aristata* 116
24. *Berberis chitria* 501
25. *Berberis lycium* 118
26. *Bergenia ciliata* 121

27. *Bergenia stracheyi* 124
28. *Betula utilis* 126
29. *Bombax ceiba* 130
30. *Butea monosperma* 135
31. *Calotropis gigantea* 138
32. *Cannabis sativa* 141
33. *Carum carvi* 145
34. *Cassia fistula* 149
35. *Centratherum anthelminticum* 527
36. *Cichorium intybus* 159
37. *Cinnamomum tamala* 161
38. *Cissampelos pareira* 163
39. *Corydalis govaniana* 168
40. *Costus speciosus* 170
41. *Cuscuta reflexa* 175
42. *Cyperus rotundus* 178
43. *Dactylorhiza hatagirea* 181
44. *Datura metel* 186
45. *Datura stramonium* 188
46. *Dioscorea deltoidea* 192
47. *Eclipta prostrata* 197
48. *Emblica officinalis* 200
49. *Ephedra gerardiana* 203
50. *Euphorbia pilosa* 534
51. *Fritillaria oxypetala* 504
52. *Gentiana kurroo* 208
53. *Gloriosa superba* 214
54. *Habenaria pectinata* 539
55. *Hedychium acuminatum* 217
56. *Helicteres isora* 219
57. *Heracleum candicans* 222
58. *Holarrhena antidysenterica* 227
59. *Humulus lupulus* 230
60. *Hyoscyamus niger* 232
61. *Hypericum perforatum* 235

62. *Hyssopus officinalis* 238
63. *Inula racemosa* 241
64. *Jasminum humile* 244
65. *Jasminum officinale* 245
66. *Juglans regia* 247
67. *Juniperus communis* 251
68. *Juniperus macropoda* 254
69. *Jurinea dolomiaea* 257
70. *Leucas cephalotes* 261
71. *Mallotus philippinensis* 264
72. *Malva verticilata* 542
73. *Mentha longifolia* 267
74. *Moringa oleifera* 271
75. *Murraya koenigii* 275
76. *Myrica esculenta* 277
77. *Nardostachys grandiflora* 281
78. *Nerium indicum* 284
79. *Nyctanthes arbor-tristis* 287
80. *Ocimum sanctum* 289
81. *Onosma bracteatum* 292
82. *Origanum vulgare* 294
83. *Oroxylum indicum* 297
84. *Physochlaina praealta* 302
85. *Picrorhiza kurrooa* 306
86. *Pistacia khinjuk* 312
87. *Plumbago zeylanica* 315
88. *Podophyllum hexandrum* 318
89. *Polygonatum multiflorum* 505
90. *Polygonum aviculare* 321
91. *Potentilla nepalensis* 550
92. *Prinsepia utilis* 326
93. *Prunella vulgaris* 328
94. *Prunus cerasoides* 331
95. *Punica granatum* 335
96. *Randia dumetorum* 341

97. *Rauvolfia serpentina* 344
98. *Rheum australe* 347
99. *Rheum moorcroftianum* 349
100. *Rhododendron anthopogon* 352
101. *Rhododendron arboreum* 354
102. *Rhododendron campanulatum* 356
103. *Ricinius communis* 359
104. *Roscoea alpina* 361
105. *Rubia manjith* 364
106. *Salvia moorcroftiana* 367
107. *Sapindus mukorossi* 369
108. *Saussurea costus* 371
109. *Sida cordifolia* 378
110. *Skimmia laureola* 556
111. *Solanum nigrum* 384
112. *Solanum surattense* 387
113. *Swertia angustifolia* 390
114. *Swertia chirayita* 391
115. *Symplocos paniculata* 395
116. *Taraxacum officinale* 399
117. *Taxus wallichiana* 402
118. *Terminalia arjuna* 405
119. *Terminalia bellirica* 408
120. *Terminalia chebula* 411
121. *Thalictrum foliolosum* 413
122. *Thymus serpyllum* 417
123. *Trillidium govanianum* 421
124. *Valeriana jatamansi* 427
125. *Viola serpens* 432
126. *Viscum album* 435
127. *Vitex negundo* 437
128. *Withania somnifera* 440
129. *Woodfordia fruticosa* 443
130. *Zanthoxylum armatum* 447

List of Plates

1. *Achillea millefolium*
2. *Aconogonum rumicifolium*
3. *Allium carolinianum*
4. *Aloe barbadensis*
5. *Androsace rotundifolia*
6. *Angelica glauca*
7. *Arctium lappa*
8. *Arnebia euchroma*
9. *Arnebia guttata*
10. *Artocarpus lakoocha*
11. *Asparagus adsendens*
12. *Atropa belladonna*
13. *Bauhinia variegata*
14. *Berberis asiatica*
15. *Berberis lycium*
16. *Bergenia ciliata*
17. *Bergenia stracheyi*
18. *Boerhaavia diffusa*
19. *Bombax ceiba*
20. *Bryonopsis laciniosa*
21. *Bunium persicum* (complete plant showing bulb to flowers)
22. *Butea monosperma*
23. *Calotropis gigantea*
24. *Capparis spinosa* (flowers)
25. *Capparis spinosa* (fruits)

26. *Celastrus paniculatus*
27. *Centella asiatica*
28. *Centratherum anthelminticum*
29. *Codonopsis ovata*
30. *Codonopsis ovata* (showing fleshy, milky roots)
31. *Colchicum luteum*
32. *Corydalis govaniana*
33. *Costus speciosus*
34. *Crocus sativus*
35. *Dactylorhiza hatagirea*
36. *Datura fastuosa*
37. *Digitalis lanata*
38. *Digitalis purpurea*
39. *Dioscorea deltoidea*
40. *Dioscorea deltoidea* (specimen showing rhizomes)
41. *Eclipta alba*
42. *Emblica officinalis*
43. *Ephedra gerardiana* (shrub in fruits)
44. *Evolvulus alsinoides*
45. *Fraxinus xanthoxyloides*
46. *Gentianna Kurroo*
47. *Geum elatum*
48. *Ginkgo biloba*
49. *Gloriosa superba*
50. *Heracleum candicans*
51. *Heracleum candicans* (root stock)
52. *Hippophae rhamnoides* (red fruited type)
53. *Hippophae rhamnoides* (orange fruited type)
54. *Hippophae salicifolia*
55. *Holarrhena antidysenterica*
56. *Humulus lupulus*
57. *Hyssopus officinalis*

58. *Jasminum humile*
59. *Juniperus communis*
60. *Juniperus macropoda*
61. *Jurinea dolomiaea*
62. *Malaxis muscifera*
63. *Malaxis acuminata*
64. *Matricaria chamomilla*
65. *Meconopsis aculeata* (in front and cobra plant in the background)
66. *Morina longifolia*
67. *Mucuna pruriens*
68. *Myrica esculenta*
69. *Nardostachys grandiflora*
70. *Ocimum sanctum*
71. *Oroxylum indicum*
72. *Paeonia emodi*
73. *Physochlaina praealta*
74. *Picrorhiza kurrooa*
75. *Pistacia integerrima* (showing fruits and galls)
76. *Plumbago zeylanica*
77. *Podophyllum hexandrum*
78. *Polygonatum verticillatum*
79. *Prinsepia utilis*
80. *Pyracantha crenulata*
81. *Rauvolfia serpentina*
82. *Rheum australe*
83. *Rheum australe* (showing roots)
84. *Rheum moorcroftianum*
85. *Rhododendron arboreum*
86. *Rhododendron campanulatum*
87. *Roscoea alpina*
88. *Salvia moorcroftiana*

89. *Saussurea costus*
90. *Saussurea gossypiphora*
91. *Saussurea obvallata*
92. *Semecarpus anacardium*
93. *Solanum khasianum*
94. *Solanum surattense*
95. *Spilanthes acmella*
96. *Taraxacum officinale*
97. *Taxus wallichiana*
98. *Thermopsis inflata*
99. *Thymus serpyllum*
100. *Trillidium govanianum*
101. *Typhonium diversifolium*
102. *Valeriana jatamansi*
103. *Viola serpens*
104. *Viscum album*
105. *Waldhemia glabra*
106. *Woodfordia fruticosa*

Introduction

Himachal Pradesh, one of the pioneer Himalayan States, is a rich repository of medicinal wealth, which occupies an important place in the Vedic treatise. In ancient times, it had been the abode of *rishies* and *munies*, who pursued their meditational and scholarly endeavours here. In this land of gods, goddesses and religious resorts, lived Pandavas, demons, Yakshas, Kinners and various *maharishies*. Rishi Vyasa, Agnivesha, Bhardwaja, Punarvasu Atreya and Kali Dass wrote about this hill abode including treatises on herbs and herbal lores. To cure the sufferings and ailments of the people, the first ever seminar is reported to have been held in some parts of Himachal Pradesh, at a place situated some where in Shivalik range (*Charak Samhita*, 19/3). Punarvasu Atreya, a great scholar in Ayurveda, is believed to have lived in Chandra-Bhaga river catchment in Lahaul and Spiti. Indra, the son of Kashyapa, is reported to have had *soma* in order to immortalise and rejuvenate himself here. References to Kailash, Kinner Pradesh (Kinnaur) and Gandharva-desha are found in ancient literature.

Nagarjuna, the scholar in medicinal chemistry, learned and worked in Jwalamukhi (Kangra) and later shifted his research activities to Triloknath in Lahaul and Spiti. In *Valmiki Ramayana*, there is a reference to the existence of rare drugs like Sanjivani, Savarnyakarni, Sandhankarni and Vishalyakarni in Kanchan hills and Kailash mountains. These were the drugs carried by Hanumanji to Lanka to cure Lakshmana and the wounded soldiers. Hanumanji is said to have rested at Jakhu (Shimla) on his way and enquired about the herbs and the route from a *yogi* staying in the temple. Several *yogies* like Gorakhnath, Bhairavnath, Matsyendranath and Charpatnath lived here and benefitted the people through their exceptional knowledge in herbal treatments (Anon, 1977).

It is quite evident from the above account that Ayurveda, the ancient science of medicine, has its origin in Himachal Pradesh.

In this very sacred land, away from the noisy city life, under the nature's all-embracing paradise, calm, peaceful and tranquil atmosphere, these *rishies* propagated the physical, spiritual as well as the social values that form an integral part of the religion and culture prevalent in this state. Because of its geographical conditions, the traditions, myths, legends and folklores of ancient India are still in existence in lores and mores, temples and historical places in Himachal Pradesh.

Vedic literature followed by the writings of Charaka, Sushruta, Dhanwantri, Nagarjuna, Prashara and various other scholars bear testimony to it. Alexander the great was also much influenced by the Himalayan herb science and made a great publicity of it in Yunan and Rome during the middle ages. In the Western Himalayan region, the sources of nearly 80 per cent of Ayurvedic, 46 per cent of Unani and 33 per cent of allopathic medicines are existing, which contribute a major part in the national economy by feeding indigenous industries to earn foreign exchange through the export of these drugs in crude or purified form (Nautiyal & Nautiyal, 1983).

The researches and practices of *rishies* and *munies* on herbs proved a panacea for the people of the region. It created awareness and curiosity among the people to seek real knowledge in close association with these ascetics and scholars. The knowledge acquired by a few dedicated local inhabitants, descended down from generation to generation still forms the traditional practice of herbal cure. *Jantras, mantras* and the uses of herbs are still practised to cure various diseases including snake bite, scorpion sting, rabbies, bone fractures, eczema, wounds, gangrene and also to effect birth control. There are many verses, songs, folk-tales, myths and legends in existence in different parts of the state about plants. The evaluation of this information is likely to yield amazing results and will add many new potent sources to the drug industry.

Owing to illiteracy and lack of proper written records, the practice of herbal cure, which mostly, often, literally depended on the phenomenally good memory of the practitioner, had to suffer a lot during the course of time. The person who knew the art of healing, usually did not reveal his knowledge to any one for fear of being misused. As a result, much of the valuable knowledge of the practitioners used to be burnt in the funeral pyre with the death of such a person. This situation is even existent in most of the cases. Despite a rich heritage and enormous herbal wealth existing in the state, it is disheartening to point out that the knowledge acquired and disseminated by the great saints and scholars, is facing extinction in the very land of its origin.

It is estimated that even today the herbal drugs contribute about 80 per cent material in Indian medicines. Some of these drugs have withstood the test of time and have shown their remarkable curative properties, even after being subjected to vigorous testing by the modern scientific techniques. Many others have, however, reached the list of endangered species or have become extinct with the passage of time due to human ignorance or greed for collecting more quantity to earn instant money. In India, there are over 2000 plants credited with medicinal properties and only a few of these are cultivated, while others grow in a wild state (Kirtikar and Basu, 1933; Chopra et al. 1958). The biological activity of about 3500 plants has been screened at Central Drug Research Institute, Lucknow (Dhar et al. 1968; Bhakuni et al. 1969 & 1971; Dhar et al. 1973, 1974, Dhawan et al. 1976, 1979 & Aswal et al. 1984) and Regional Research Laboratories of CSIR (Sharma et al. 1973). The National Chemical Laboratory Pune, the Central Council for Research in Ayurveda and Sidh; various university departments and medical institutions are also engaged in chemical and bio-chemical screening. There are still many more herbs growing in nature which have yet to be recognized, identified and properly evaluated. It is likely that with extensive survey and screening of vegetal wealth, many newer sources of alkaloids, saponins, tannins, antimalarial, antineuplastic drugs etc. are available for the services of mankind. The folklore medicines of the primitive tribes and aboriginals are also the reservoirs of information in this direction (cf. Mehra et al. 1969; Nautiyal & Nautiyal, 1983 and Chauhan, 1984, 1988, 1989).

The fascinating medicines, as practised in Indian epics/ancient India to cure serious diseases have aroused our inquisitiveness to recall and identify them. According to Charaka, a great propounder of Ayurveda, each and every plant growing on this earth has medicinal value and the Himalayan resources excel all other parts of the world. According to him, the nature has produced the cure for all diseases, but it depends upon the skill and intellect of the physician to utilise it. In many cases, the properties of many herbs remained localised due to the biophysical environment (Chakrapani, 1941—Charaksutra, 26:12; 1:51). It is also reported that drugs growing in a particular country or region always suit best to the inhabitants of that country/ locality.

The unsystematic exploitation of the important drug-yielding plants to support pharmaceutical industry has become a suicidal trap. Their regeneration, protection and preservation in nature is today a challenge to restore our biological heritage. The rich repositories of the past have been butchered with ruthless greed and several species of drug plants like *Dioscorea, Jurinea, Orchis, Aconitum, Polygonatum, Angelica, Nardostachys, Valeriana, Arnebia* and even *Berberis aristata* are facing threat to their survival. It is only recently that the Department of Environment, Government of India; Ministry of Health and the Ministry of Science and Technology have shown their concern and advocated the need for legislation to control and regularise the exploitation of herbal wealth of the Himalaya. National Forest Policy, 1988 has also emphasised the need for achieving nature conservation and maintenance of ecological balance on one hand and meeting the essential requirement of the people on the other. This will enable human beings to live in complete harmony and peace with nature. The Himalayan ecosystem is very fragile and calls for serious efforts in protecting it. It is a veritable emporium of valuable resources and the saviour of our granaries in the plains. The cultivation, protection and conservation of high-yielding strains of important drug species is, thus, a necessity to provide sustainable supply of crude drugs and also to ease pressure on natural resources. The knowledge of natural resources, their enumeration along with evaluation of their resource potential and the life styles of the people are pre-requisites to standardise their systematic exploitation. Santapau (1958) emphasised the need for local floras illustrating the medicinal value of plants contained therein.

Keeping the above factors in view, the present book has been prepared highlighting the most important medicinal and aromatic plant resources of the state. More than 700 species of medicinal and aromatic plants have been covered in the book out of which only 175 medicinal and aromatic plants have been selected for detailed description covering Latin name (based on current nomenclature), local/common name(s), family, distribution in the state specifying the actual locality supported by voucher numbers, description of the plant with special emphasis on official parts, preparations in Ayurvedic medicines, main chemical ingredients, uses, retail market price (wherever possible), cultivation notes and harvesting instructions. Rest of the species find their place in appendices and in text elsewhere. This shall serve as an important reference source for the people engaged in indigenous drugs, pharmaceutical industries, farmers, collectors, scholars, teachers, students and the common man, by providing a variety of information. To make it all the more useful, local/vernacular name(s), sketches/drawings coloured photographs alongwith authentic distributional data etc. will come in handy for the reader enabling him to make judicious use of the resources available in his area. The spectrum of vegetation has been discussed in order to highlight the diversity of floristic elements existing in the state. The reference has also been made to the existing collection procedures, acts and laws. Scientific methods, indicating appropriate collection season have been suggested for official parts obtained from different species. The voucher specimens of the drug plants, included in the present work are housed in the Herbarium of College of Forestry, Deptt. of Forest Products; Dr. Y.S. Parmar University of Horticulture and Forestry, Nauni-Solan; R.R.C. (Ay.) Mandi (a unit of CCRAS) and the Herbarium of the Institute of Research in Indian System of Medicine, Jogindernagar, district Mandi (H.P.), collected by the author from all parts of the state for the last more than 29 years.

Geographical Position

The State of Himachal Pradesh lies between 30^0 22' N to 33^0 13' N latitude and 75^0 23'24" E to 79^0 00' 50" E longitude having a maximum east-west stretch of about 315 km and north-south expansion of about 310 km covering an area of 55,673 sq km. The area is bounded in the north by Jammu & Kashmir; south-east by Tehri Garhwal and Dehradun districts of Uttar Pradesh; in the south, by Ambala district of Haryana; on the west by Ropar, Hoshiarpur and Gurdaspur districts of Punjab; whereas in the east, it has a common international boundary with Tibet. Overall, Himachal

Pradesh is mostly a mountainous state with altitude ranging from 247 m to 7000 m above m.s.l. The general elevation increases from south to north and west to east. There are many parallel ranges and valleys with numerous rivers and rivulets draining the region. The southern and outer Himalayan hills are low and composed of unconsolidated deposits, which are subject to severe erosional activity. South of the outer hills, lie the valleys (Duns) viz. Paonta valley in Sirmour, Balh valley in Mandi and Kangra valley. North of the outer Himalayan zone constitutes the lesser part of Himalaya, which rises in elevation towards north. This region contains several parallel valleys with an exception of Kullu valley. Dhauladhar and Pir Panjal are the main ranges in this zone. The Great Himalayan zone lies further north of the lesser Himalaya. This area is glaciated, highly eroded country and consists of high mountains and deep gorges. The range traverses the state from north-west to south-east at an elevation of about 5000-6000 m with passes like Kangla (5,248 m), Baralacha (4,512 m) separating the drainage of Spiti and Beas river. Zanskar range makes the eastern boundary with Tibet and is cut through by river Sutlej at Ship-ki-La.

The main rivers which drain the area are: Yamuna (Pabbar, Tons, Giri) merging in the Bay of Bengal; Sutlej, Beas, Ravi, Chenab (Chandra-Bhaga) flowing towards west into the Arabian sea; while the Malung drains towards north to join river Indus.

Geology, Rock and Soil

The entire region of the state mostly comprises of tertiary rocks. In the southern part from west to east, there is a great thickness of detrital rocks—clay and conglomerates. North of this belt has mainly granite and other crystalline rocks having unfossiliferous sediments. In southern Spiti and upper Sutlej region, rocks are mainly metamorphic with intermittent granite. Northern part of the state has a complete sequence of the bare laying strata of fossiliferous palaeozoic and mesozoic rocks. The giumal sandstone with yellow brown sandstone and quartzite overlie in the north Spiti valley. The grey limestone with soft grey calcareous shales occur towards north and north-east of the area and this part is rich in fossil and has many well exposed sections.

The soils in the state are young with considerable depth in the valleys and its depth depends upon the nature of the gradient. In general, the soils in Himachal Pradesh vary according to altitude and climate.

Himachal Pradesh is rich in mineral wealth. There are deposits of limestone, gypsum, dolomite, pyrites, clays, mica, building stone, slate and lead in various parts of the state. Black salt mines exist at Gumma and Drang in Mandi. Uranium deposits have been reported near Jari in Kullu district.

Climate and Vegetation

Owing to the mountainous terrain and varying altitudinal zonations, the diversified climatic conditions are characterised by the short mid summer and considerably long severe winters. In general, the climate is humid subtropical in southern low areas, temperate in middle altitudinal area, cold alpine (semi-arctic) and glacial in north and north-eastern higher mountainous area. In winters, snowfall takes place down to an elevation of 1500 m, whereas above 4500 m, there is perpetual snow. Total precipitation varies from 900 mm in Shivalik zone to 3,550 mm at the foothills of Dhauladhar range, beyond which it decreases to 750 mm on the southern side of the Great Himalayan range. It further falls to about 500 mm in Lahaul and Kinnaur and to about 200 mm in Spiti region (Ghosh, 1977).

The wide range of altitude, topography and climatic conditions have endowed Himachal Pradesh with a rich and diversified flora. The vegetation is broadly known through the accounts of Collet (1902), Puri (1960), Champion and Seth (1968), M.A. Rau (1975), N.C. Nair (1977), S.K. Jain & A.R.K. Sastry (1978), A.S. Rao (1978), Choudhry and Wadhwa (1984) and from the reports of various workers like Shabnam (1956), Gupta (1962, 1964), Uniyal and Chauhan (1971, 1972, 1973) and Chauhan (1974, 1976, 1984, 1988, 1989 and 1997). As a result of these studies, more than 3500 plant species of higher plants have been identified in the state flora. There are a large number of ferns, mosses, fungi and asco-lichens occurring in various parts of the state. Broadly speaking, the vegetation of the state encompassing different plant resources, can be classified into the following main types.

1. Thorny scrub of *Acacia* and *Zizyphus*

These occur in the foothills along Punjab and Haryana borders. The dominating species are *Acacia modesta, A. nilotica, A. catechu, Aegle marmelos, Flacourtia indica, Randia dumentorum, Dodonaea viscosa, Vitex negundo, Justicia adhatoda, Woodfordia fruticosa, Agave americana, Carissa opaca, Zizyphus mauritiana, Z. oxycantha, Z. nummularia, Calotropis procera* and the grasses like *Eulaliopsis binata, Pennisetum orientale*, species of *Cymbopogon, Apluda, Stipa, Heteropogon, Chrysopogon* etc. *Lantana camara, Parthenium hysterophorus, Zanthium strumarium* and *Ageratum conyzoides* are the obnoxious weeds covering patches of ground at many places.

2. Sub-tropical Pine Forests

These occur between 600 m and 1700 m elevation covering the districts of Sirmour, Solan, Shimla, Bilaspur, Hamirpur, Mandi, Kangra and part of Chamba. *Pinus roxburghii* is the predominant species associated at places with *Pyrus pashia, Quercus leucotrichophora, Rhododendron arboreum, Lyonia ovalifolia, Acacia catechu, Terminalia chebula, T. arjuna, Syzygium cumini, Emblica officinalis, Mallotus philippinensis, Albizia chinensis, Carissa opaca, Dodonaea viscosa, Woodfordia fruticosa, Indigofera* spp., *Rhamnus triquetrus, Casearia tomentosa, Berberis lycium, Prinsepia utilis, Murraya koenigii, Rubus ellipticus, Rosa moschata, Moghania fruticosa, Leptodermis lanceolata, Myrsine africana, Colebrookia oppositifolia, Meriandra sterobilifera, Lespedera stenocarpa* and several grassy species as undergrowth constituted chiefly of the species of *Chrysopogon* and *Heteropogon* supported by *Themeda* and *Arundinella*. Exposed sites support *Euphorbia royleana* and *Carissa carandus* followed by *Randia tetrasperma. Lantana camara*, as a chronic shrubby weed has colonised the ground cover in exposed sites for miles together. The shady and moist ravines support the luxuriant growth of evergreen species of *Persea dulthie, P. odoratissima, Litsea lanuginosa, Xylosma logifolia* and *Quereus glauca*. The climbers include the species of *Smilax*, Clematis, *Jasminum, Milletia, Sagerratia, Tinospora, Vitis, Cissus* and *Bauhinia vahlii*.

3. Sub-tropical Dry Evergreen Forests (including Sal Forests)

These forests occur in lower parts of Sirmour, Kangra, Una, Hamirpur, Mandi, Solan, Nalagarh and Kunihar divisions etc. and are dominated by *Shorea robusta, Anogeissus latifolia, Terminalia tomentosa, Buchnania lanzan, Bauhinia variegata, B. vahlii, Grewia optiva, Ougeinia oojeinensis, Cassia fistula, Bombax ceiba, Butea nonosperma, Mallotus philippinensis, Dendrocalamus strictus, Semecarpus anacardium, Aegle marmelos, Feronia limonia, Diospyros tomentosa, Helicteres isora, Kydia calycina, Solanum verbascifolium, Holarrhena antidysenterica, Olea ferruginea, Acacia modesta, Pistacia integerrima* associated with *Pyrus pashia, Coriaria nepalensis, Rhus parviflora, Cotinus coggygria, Murraya koenigii, Colebrookia oppositifolia, Meriandra strobilifera, Callicarpa macrophylla, Clerodendrum viscosum, Ardisia solanacea, Abutilon indicum, Indigofera pulchella, Desmodium* spp., *Prinsepia utilis, Zanthozylum armatum, Rubus ellipticus* etc. Grasses are dominated by the species of *Themeda, Eulaliopsis, Heteropogon, Cymbopogon, Saccharum* etc. *Vetiveria zizanioides, Typha elephantina*, species of *Cyperus* and *Scirpus* are commonly met with in swampy areas.

4. Himalayan Moist Temperate Forests (1500-3600 m)

These forests are predominantly found in major parts of the state especially in the districts of Chamba, Kangra, Mandi, Kullu, Shimla, some parts of Kinnaur and Sirmour. The vegetation is composed of *Cedrus deodara, Pinus wallichiana, Taxus wallichiana, Quercus floribunda, Q. semecarpifolia, Q. leucotrichophora, Rhododendron arboreum, Acer* spp., *Betula alnoides, Corylus colurna, Aesculus indica, Abies pindrow, Picea smithiana, Persea gamblei, Buxus wallichiana, Carpinus viminea, Euonymus tingens, Juglans regia, Cornus macrophylla, Prunus cornula, Ulmus wallichiana, Ilex dipyrena, Celtis caucasica, Litsea umbrosa, Pyrus pashia, Alnus nitida*, etc. The shrubby element constitutes of *Cotoneaster bacillaris, Berberis* spp., *Princepia utilis, Desmodium tiliaefolium, Indigofera* spp., *Sarcococca pruniformis, Daphne cannabina, Skimmia laureola*, species of *Deutzia, Vitis, Clematis, Arundinaria, Viburnum, Rosa, Rhamnus, Ribes, Lonicera, Salix, Spiraea*, etc. Herbaceous elements constitutes the species like *Salvia glutinosa, Fragaria vesca, Duchesnea indica;* species of *Geranium, Impatients, Senecio,*

Polygonum, Bistorta, Polygonatum, Persicaria, Iris, Anemone, Thalictrum, Ainslaea, Valeriana, Primula, Aconitum, Cynoglossum, Viola, Arisaema etc. *Cannabis sativa, Girardinia diversifolia* are common in waste grounds, nalahas and near habitations. *Angelica glauca, Dioscorea deltoidea, Podophyllum hexandrum, Fritillaria oxypetala, F. roylei, Heracleum candicans, Typhonium diversifolium, Roscoea* spp.; *Alliums* etc. are found sporadic to frequent in the region. *Adiantum, Pteris, Diplazium esculentum* (Lingru), *Pteridium aquilinum* (Bracken), *Morchella esculenta* (Guchhi), *Agaricus campestris* (Chhatri), etc. constitute the ferns and mushroom elements. Species of *Parmelia* and *Usnea*, found as crustaceous and hanging growth of fruticose Asco-lichens are of common scene on the rocks and tree trunks/branches especially of oaks. This region is quite rich in having diversified and lush green vegetation, majority of which constitutes the medicinal wealth.

5. Himalayan Dry Temperate Forests

These forests occur in the districts of Kinnaur, Lahaul-Spiti and Chamba (Pangi-Bharmour) region. The main tree species occurring in this type are: *Pinus gerardiana, Cedrus deodara, Pinus wallichiana, Picea smithiana, Abies spectabilis, Juniperus macropoda, Populus ciliata, Salix viminalis, Quercus baloot, Q. floribunda, Acer pentapomicum, Alnus nitida, Celtis caucasica* forming the top canopy; whereas the second story constitutes of *Fraxinus xanthoxyloides, Rhus succedanea, Parrotiopsis jaguemontiana, Olea ferruginea, Corylus colurna* etc. Shrubby and the scrub element mainly have the species of *Salix, Berberis, Rosa, Viburnum, Lonicera, Hippophae, Myricaria, Daphne, Rubus, Caragana, Sorbus, Rhododendron, Juniperus, Colutea* etc. *Rabdosia rugosus, Ephedra gerardiana* and *Artemisia brevifolia* form dense scrub covering vast tracts of the slopes in this region. *Hippophae salicifolia* forms dense patches along Baspa catchment, whereas *H. rhamnoides* covers vast tracts along water courses in Kinnaur, Lahaul and Spiti regions. *Capparis spinosa* is conspicuous for its spreading patches, distantly, distinctive in cold desert regions of Pooh and Spiti regions.

6. Sub-Alpine Forests

These forests form a borderline between the temperate and the alpine zones in Kullu, Kangra, Chamba, Kinnaur Churchandni peak of Sirmour and Shimla districts. This zone has the tree species of *Quercus semecarpifolia, Betula utilis, Pinus wallichiana, Picea smithiana, Taxus wallichiana, Abies spectabilis* and *A. pindrow* chiefly and are associated with the rambling shrubs of *Rhodo-dendron campanulatum, R. anthopogon, R. lepidotum, Juniperus communis, J. recurva, Salix* spp., *Cotoneaster micro-phylla, Sorbus acuparia, Rosa macrophylla, Myricaria germanica, Hippophae rhamnoides, Lonicera rupicola, Ribes* spp. etc. This area has luxuriant growth of herbaceous elements exhibiting a variety of floral colours, morphological modifications and propagation mechanisms. Most prevailing genera are: *Anemone, Corydalis, Draba, Thalictrum, Cynoglossum, Astragalus, Polygonum, Potentilla, Sedum, Saxifraga, Primula, Anaphalis, Geranium, Valeriana, Senecio, Saussurea, Androsace, Gentiana, Swertia, Viola, Pedicularis, Rheum, Lepidium, Delphinium, Aconitum, Allium, Trillium, Ranunculus, Juncus, Carex* and *Poa. Eremurus himalaicus* (Himalayan Foxtail Lily) is a majestic plant near glaciers in Lahaul and Spiti. Under the Desert Development Programmes in operation in Pooh and Spiti regions, plantations of the species of *Populus, Salix, Prunus* and *Malus* have been raised which have created green patches in the otherwise rugged, treeless terrain.

7. Moist Alpine Scrub (above 3600 m elevation)

Such forests occur in areas other than the districts of Kinnaur, Lahaul and Spiti and Pangi-Bharmour of Chamba district, where cold dry conditions prevail. In this zone, the *Salix* spp. (dwarf willows), *Junipers* and *Rhododendrons* predominate. Other shrubby element is formed of the species of *Berberis, Lonicera, Cotoneaster, Astragalus, Ribes, Rosa, Viburnum* etc. The ground flora is dominated by the species of *Artemisia, Potentilla, Geranium, Fritillaria, Corydalis, Anaphalis, Aquilegia, Thalictrum, Paraguilegia, Bergenia, Aconitum, Aster, Stellaria, Silene, Jurinea, Saussurea, Tragopogon, Sedum, Saxifraga, Pedicularis, Epilobium, Leontopodium, Waldhemia, Primula, Iris, Euphorbia, Cynoglossum, Mertensia, Swertia, Gentiana, Ranunculus, Delphinium, Lagotis, Phlomis, Stachys, Valeriana, Pleurospormum, Nardostachys* etc. *Nardostachys grandiflora* is sporadically found in Manjiban, Jakha top and Tangankhai areas falling in Dodra-Kawar region bordering Kinnaur on one side and Uttarakashi of U.P. on the other and has recently been collected from Great Himalayan National Park area in Kullu district.

8. Dry Alpine Scrub

This type of vegetation is found predominantly in the cold desert areas of Pooh, Lahaul-Spiti and Pangi areas of Kinnaur, Lahaul-Spiti and Chamba districts respectively above 3600 m elevation going upto 5500 m i.e. the last vegetational limit. The vegetation is mainly dominated by the species of *Juniperus, Ephedra, Myricaria, Rhododendron,* supported by *Rosa webbiana, Ribes orientale, R. alpestre, Lonicera spinosa, L. rupicola, Clematis venayii, Capparis spinosa, Cassiope fastigiata, Hippophae rhamnoides, Elaeagnus* spp., *Salix viminalis, S. hastata* and *S. flabellaris. Populus ciliata, P. euphratica, P. alba, P. balsamifera* and *Salix viminalis* are planted near village habitations. Herbacious element is dominated by *Saussurea jacea, Aquilegia fragrans, Bergenia stracheyii, Gernaium wallichianum, Corydalis govaniana*; the species of *Artemisia, Allium, Gentiana, Draba, Lepidium, Polygonum, Leontopodium, Sedum, Arenaria, Potentilla, Glechoma, Arnebia, Androsace, Thalictrum, Aconitum, Saxifraga, Taraxacum, Carex, Kobresia, Epilobium, Anaphalis, Pedicularis;* valuable drugs like *Mentha longifolia, Hyssopus officinalis, Hyoscyamus niger, Thymus serphyllum* and the species of *Rheum* and *Aconitum.* The grasses are represented by the species of *Agrostis, Agropyron, Bromus, Deyeuxia, Dactylis, Festuca, Phleum, Poa* etc.

Resources of Medicinal and Aromatic Plants in H.P.

Although no scientific and systematic inventory of available resources of medicinal and aromatic plants with regard to their quantity, qualitative parameters and their regeneration under natural conditions is available, yet a fair amount of information has since been generated by various workers during the past couple of decades. Notable studies in respect of distribution and enumeration of medicinal and aromatic plants in different parts of the state have been conducted by Ahluwalia (1952) in Kangra district; Gupta (1963), Shabnam (1964) in Chamba division; Sarin (1967) in Lahaul; Uniyal and Chauhan (1971, 1972 & 1973) in Kangra, Mandi, Kullu; Chauhan (1970-1998) in Chamba,

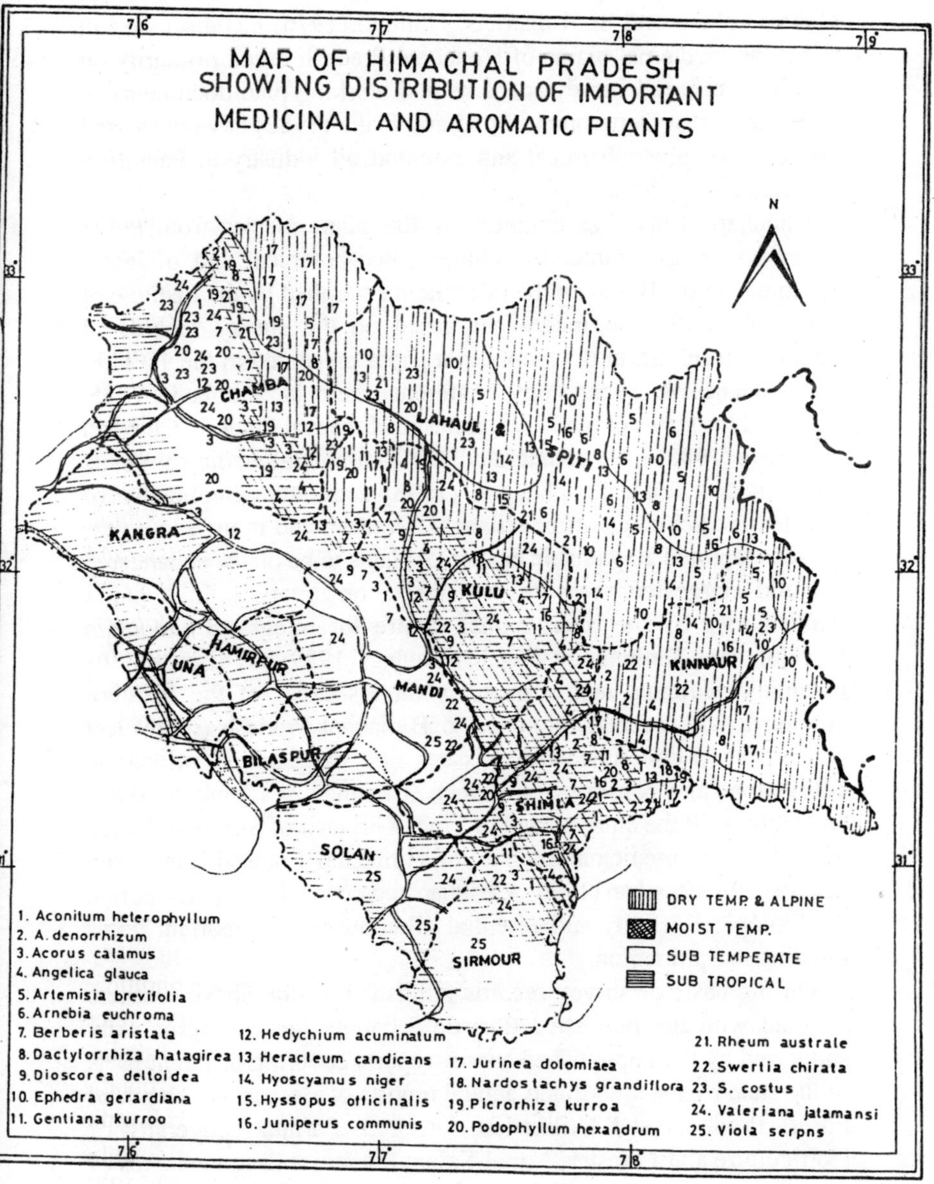
MAP OF HIMACHAL PRADESH
SHOWING DISTRIBUTION OF IMPORTANT
MEDICINAL AND AROMATIC PLANTS
N
CHAMBA
LAHAUL & SPITI
KANGRA
KULU
HAMIRPUR
UNA
MANDI
KINNAUR
BILASPUR
SHIMLA
SOLAN
SIRMOUR
DRY TEMP. & ALPINE
MOIST TEMP.
SUB TEMPERATE
SUB TROPICAL
1. Aconitum heterophyllum
2. A. denorrhizum
3. Acorus calamus
4. Angelica glauca
5. Artemisia brevifolia
6. Arnebia euchroma
7. Berberis aristata
8. Dactylorrhiza hatagirea
9. Dioscorea deltoidea
10. Ephedra gerardiana
11. Gentiana kurroo
12. Hedychium acuminatum
13. Heracleum candicans
14. Hyoscyamus niger
15. Hyssopus officinalis
16. Juniperus communis
17. Jurinea dolomiaea
18. Nardostachys grandiflora
19. Picrorhiza kurroa
20. Podophyllum hexandrum
21. Rheum australe
22. Swertia chirata
23. S. costus
24. Valeriana jatamansi
25. Viola serpns

Kangra, Una, Hamirpur, Shimla, Solan, Sirmour, Kinnaur, Kullu, Mandi, Lahaul and Spiti districts. Dhiman (1976) has discussed in Hindi the medicinal plants of Himachal Pradesh based primarily on the above referred survey accounts and working plan documents of the State Forest Department. Sood and Kalia (1982) have discussed the scope of phytochemical and essential oil industry in Palampur area (Kangra valley).

Chauhan (1984) has enumerated 400 plant species from Pabar valley covering commercial, ethnomedico-botanical and of lesser economic value. Based on the information supplied by the Institute of Research in ISM Jogindernagar based on the survey records of Chauhan *et al.*; Sarin & Chopra in a paper presented in the 1st H.P. State Science & Technology conference held at Shimla in 1985, discussed the status of medicinal plants in Himachal Pradesh. Chauhan & Khosla (1988) have enumerated 104 commercial medicinal plants of Himachal Pradesh. Chauhan (1988) reported ethnobotany of selected medicinal plants of Himachal Pradesh and in the same year has also reported 20 valuable medicinal plants to be on the endangered list and suggested measures for their conservation. In a paper requisitioned by the International Centre for Integrated Mountain Development, Kathmandu, Nepal, Chauhan (1989) has discussed the potential medicinal and aromatic plant resources of the Western Himalaya with special reference to Himachal Pradesh (reprospect and prospect) covering various issues like status scope, utilization cultivations and conservation aspects. Chauhan & Chauhan (1988) have discussed the ethnobotany of Trans-Giri area of Sirmour district; 12 species of medicinal orchids of Himachal Pradesh have been discussed by Chauhan (1990). Chauhan and co-workers have reported the biological activity and essential oil contents of important herbs growing in Spiti region.

On the basis of survey records published by the above workers coupled with the published floristic accounts for the region duly supported by the unpublished survey reports covering different parts of the state by the author and annual reports of Herbarium, Deptt. of Forest Products and Utilization, Dr. Y.S. Parmar University of Horticulture and Forestry, Nauni-Solan; the major drug resources of the state, which in fact, are the representatives for the entire Western Himalayan region have been enumerated and discussed under different categories in the chapters to follow.

Out of around 3500 known plant species, there are around 500 medicinal, 150 aromatic and quite a good number of potent alternative and substitute drug plant sources available for being harnessed at present. If the plants used in household remedies (ethnomedico-botanical importance) too are carefully enumerated, the number of useful medicinal and aromatic plants may easily number around 1500 species. Almost all these plants grow wild in valleys, forests, pastures or occur as weeds near fields and village habitations.

In the lower areas under the Shivalik ranges are found the drugs of Dashmool group, the three Myrobalans; *Helicteres isora, Celastrus paniculatus, Centella asiatica, Acacia catechu, A. nilotica, Cassia fistula, C. absus, Cissampelos pareira, Withania somnifera, Mucuna pruriens, Pueraria tuberosa, Justicia adhatoda, Holarrhena antidysenterica, Bacopa monnieri, Plumbago zeylanica, Calotropis procera, Evolvulus alsinoides, Citrullus colocynthis, Bryonopsis laciniosa* and *Mallotus philippinensis* (Hooker 1875-97, Anonymous 1948-70 and Chauhan, 1974).

In the temperate zone are found commercially exploitable quantities of herbs like: *Valeriana jatamansi, Selinum tenuifolium, S. vaginatum, Berberis aristata, Skimmia laureola,* species of *Viola, Morchella esculenta, Taxus wallichiana, Hedychium acuminatum, Symplocos paniculata, Bergenia ciliata, Thalictrum foliolosum, Angelica glauca, Dioscorea deltoidea, Heracleum candicans, Podophyllum hexandrum, Thymus serpyllum, Polygonatum verticillatum, P. cirrhifolium, Polygonum aviculare, P. recumbens* and the life rejuvenating drugs of Ashtavarga group (Collett 1921; Uniyal and Chauhan 1971; 1972; Chauhan, 1984, 1988, 1992, 1995, 1996).

The tree limit roughly corresponds to the perpetual snowline, i.e. 3700 m above mean sea level. Conifers like silver fir with rhododendrons, birch, kharshoo oak, junipers and dwarf salix form the demarcation line between the temperate and the alpine zones reaching the highest limit. The alpine grasslands, usually above 3800 m remain snow-covered for 4 to 8 months and are remarkable for the variety of beautifully coloured flowers emitting pleasing as well as nauseating odours and leading a short, but glorious life in the face of the exigencies of climate and form a rich store-house of the medicinal and aromatic plants. Valuable medicinal herbs like *Aconitum heterophyllum, A.violaceum, A. deinorrhizum, Ephedra gerardiana, Rhododendron anthopogon, R. lepidotum, Nardostachys grandiflora, Dactylorhiza hatagirea, Picrorrhiza kurrooa, Corydalis govaniana, Jurinea dolomiaea, Heracleum candicans, Aconogonum rumicifolium, Valeriana pyrolaefolia, Pleurospermum brunonis, Delphinium vestitum, Saussuria obvallata, S. gossypiphora, Tanacetum longifolium, Bistorta macrophylla, Achillea millefolium, Rheum australe, Senecio amplexicaule, Gentiana kurroo, Arnebia benthami, A. euchroma* and species of *Allium* are commonly found growing in this zone (Sarin 1967; Uniyal and Chauhan 1971, 1972; Chauhan, 1984, 1988, 1989, 1992, 1997).

Vast areas at higher altitudes in Kullu, Chamba, Lahaul-Spiti, Kangra, Kinnaur and Shimla districts form the 'Alpine Deserts' with little or no vegetation of any kind owing to the peculiar edapho-climatological conditions prevailing there.

As per list released by the Ayurvedic Pharmacopoeial Committee, Government of India for 350 plants frequently used in Ayurvedic preparations out of around 1100 single drugs as ingredients of indigenous medicines in India, there are just over 225 species growing in Himachal Pradesh which are available for commercial extraction and export to the outside markets at Delhi, Amritsar, Bombay, Hoshiarpur and even Dehradun for which proper record is not available. However, some selected species falling under this category including other potential resources are depicted in Appendix 1. Drugs like, Valerian, Chirata, Berberis salts etc. are exported to foreign markets too. It is estimated that raw drugs including Katha worth over 10 crores of rupees are exported annually out of the state, if all the items are accounted for at the prevailing market rates.

Himachal Pradesh is the largest supplier of seed potato, quality apple, chilgoza, chicory, kuth, dioscorea, dhoop, picrorrhiza, valerian and ephedra in the country. It is also becoming a pioneer state in tea production. Out of the large resources of medicinal and aromatic plants which are credited with medicinal properties, only 80-85 plants are collected and marketed annually (Appendix 2). The pressure on these traditionally known herbs is so great that many of them have entered the list of endangered species. Some such herbs are: dioscorea, patish, salampanja, chora, daruhaldi, karu, kutki, pushkarmool, bankakri, revandchini, chirayata and tejpatra (Appendix 3). On the other hand, hundreds of herbs which can form potent sources to meet the growing demands of the drug industry are not being utilized or are exploited on relatively lesser degree (Appendix 4). Rare species of medicinal and aromatic plants so far unexploited, but having promising value are enlisted in Appendix 5. Lack of marketing facilities and reluctance of the user industrial units to consume material within the state can be attributed as a cause for this.

Among the cultivated ones, mention can be made of *Saussuria costus* (Kuth), *Bunium persicum* (Kalajira), *Cichorium intybus* (Chicory), *Crocus sativus* (Kesar) and *Humulus lupulus* (Hops) in Lahaul and Spiti, Kinnaur and Pangi-Bharmour; whereas *Zingiber officinale* (Adarak) is commercially cultivated in Sirmour, Solan and Bilaspur. People cultivate *Allium cepa* (Piaz), *A. sativum* (Lahsun), *Ocimum sanctum* (Tulsi), *Amomum subulatum* (Bari illaichi), *Foeniculum vulgare* (Mithi Sonf), *Trigonella foenum-graecum* (Methi), *Curcuma longa* (Haldi), *Mentha spicata* (Pudina) etc. only to meet their domestic requirements in the lower valley areas. Cultivation technology for exotics like *Glaucium flavum, Solanum laciniatum, Rosmarinus officinalis, Glycyrrhiza glabra, Salvia sclarea, S. bisor, Matricaria chamomila* and indigenous drugs like *Withania somnifera, Mimosa pudica, Dioscorea deltoidea, Digitalis lanata* etc. has been developed by the Dr. Y.S. Parmar University of Horticulture and Forestry, Nauni-Solan.

The small land holding capacities of our farmers and fear of no return for their produce has also restrained them to go for cultivation of medicinal herbs. Even Kuth grown on a large scale in Lahaul had to suffer fluctuating fortunes and same is the case with Hops. Establishment of marketing infrastructure within the state is of prime importance to foster the economic gains through this valuable resource.

Regarding the industrial utilisation of medicinal and aromatic plants within the state, there are around 50 units/pharmacies in existence which manufacture the Ayurvedic medicines. Out of these, the two bigger units at Jogindernagar and Majra are run by the government department of Ayurveda. But, it is so disheartening that these units have to purchase raw material from markets situated outside the state at exorbitant rates. The genuineness and quality of the material also is subject to scrutiny. The requirement of raw herbs by the Govt.

Ayurvedic Pharmacies at optimum production capacity is depicted in Appendix 6. The potential sources of dhoop and incense; phytochemicals essential oils insecticides and pesticides as well as yielding tannins are enlisted in Appendix 7. The plants which can be integrated in the cropping patterns of our farmers in different agroclimatic zones are suggested in Appendix 7. The agrotechnology developed by the University for important exotic and indigenous medicinal plants in brief is given in Appendix 8. The plant species which cause nausea and vomiting by their aroma in the alpine passes/ pastures have been identified and enlisted in Appendix 14. The potent substitutes for indigenous and exotic drug material available in the state have been included in Appendix 15.

Descriptive Profile of Medicinal and Aromatic Plants

Abies spectabilis (D. Don.) Spach

Family	–	PINACEAE
Local/Common name(s)	–	Himalayan high-level Silver-fir, Kolroi, Tosh, Talispatra.
Distribution	–	Afganistan to Bhutan, 2800-4000 m spreading over large areas in the upper tree zone. Common in Himachal Pradesh between 3600-4200 m forming climatic climax and often goes upto 4500 m (Chauhan, N.S. 1550).
Flowering/Fruiting	–	April-May. Cones ripen during September-October.
Description	–	A lofty tree of great height. Leaves flattered, linear, 2.5-7.5 cm, spirally arranged around the branches; tip notched, upper surface dark green, shining; lower with a silvery white line on each side of the midrib with incurved margins. Mature female cones are cylindrical 10-20 cm × 4-7.5 cm erect, dark purple with compact bracts. Male catkins cylindrical, upto 5 cm long. Its allied species, *A. pindrow* Royle can be distinguished by the leaves borne on upper side of the branchlets and by two slender sharp, tapering points at the apex of leaves which are 4-7 cm long, shining dark green above and with two faint, silvery lines beneath.
Parts used	–	Leaves/needles.

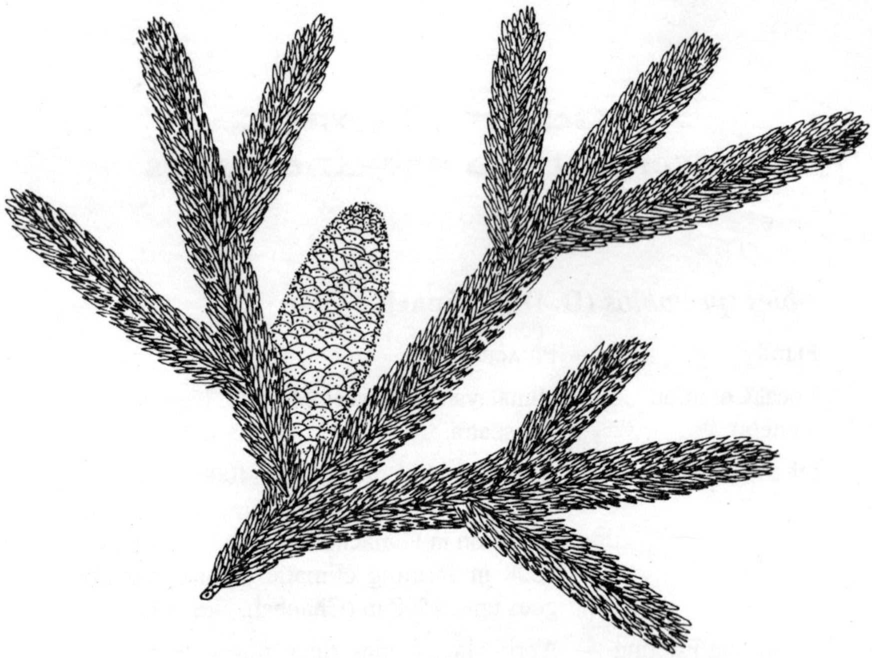

Abies spectabilis (D. Don.) Spach (Himalayan Silver-fir)

Current market rate – Rs. 6/- to Rs. 10/- per kg for dried needles in local markets.

Chemical contents – Contains essential oil and an alkaloid (0.16% dry wt.)

Properties and uses – The leaves in the form of decoction or infusion are used in chronic bronchitis, phthisis and other pulmonary affections. It is also reported as antiflatulent, antispasmodic, appetiser, astringent, carminative, digestive, diuretic, expectorant, febrifuge, stomachic and tonic. It is considered poisonous in large doses. Powdered leaves alongwith juice of *Justicea ahatoda* (vasaka) and honey are used in haemoptysis, phthisis and in catarrh of bladder. Infusion of leaves is given to infants suffering from fever and chest infection; also in hoarseness and during dentition. In Ayurveda, the plant is also mentioned as contraceptive. Leaf juice is antiperiodic. The name talispatra given to it is still controversial. In many parts of the country it is used; whereas in others *Taxus wallichiana* leaves are valued as talis, and still in some parts *Rhododendron anthopogon* leaves are used as (broad-leaved talispatra) in indigenous preparations.

Preparations – Talisadi churna, Talisadi-vati, Sudarshan-churna, Lavanbhaskar churna, Kanakasava.

Cultivation – Propagation is through seeds. The percentage of fertile seeds produced by firs is much smaller than that yielded by spruce and pines. Seeds cannot be safely stored for more than 1-2 years. Seeds are raised in carefully prepared nursery beds or polythene bags covered to a depth equal to the thickness of seeds in the soil. Seedlings begin to appear in 2-3 weeks. Seedlings are sensitive to heat of the sun and need protection by using bush screens. It can also be propagated by side grafting.

Abrus precatorius Linn.

Family – LEGUMINOSAE

Local/Common name(s) – Ratti, Gunja, Chuntli, Indian liquorice, Crabs-eye, Red bead vine, Prayer beads.

Distribution – It is found in South China, the Pacific Island, Tropical Africa, the West Indies and almost all tropical countries in moist places probably introduced (Bentley and Trimen, 1881). It has been collected from Nalagarh, Hamirpur,

Kangra, Una, Mandi and Paonta Sahib range and Nalagarh-Kunihar division below 1200 m. elevation (Chauhan, N.S.: 1259, 1301 & 1530).

Description – A much branched climber with a woody stem. Leaves are paripinnate, 5-7.5 cm long, leaflets 20-40, opposite, deciduous, oblong, ligulate, membranous, glabrous or thinly silky beneath. Flowers are reddish or white. Calyx, campanulate, thinly silky; corolla ovate, wing narrow; keel larger than wings; stamens 9 united in sheath. Ovary many ovuled; Style small. Pods oblong, 3 to 5 seeded; seeds ovoid or sub-globose, generally two third scarlet or white and the rest jet black giving appearance of a black eye.

Flowering/Fruiting – September to January.

Parts used – Root is often known as Indian liquorice. It is found in pieces of varying lengths, which are generally more or less twisted and branched, covered by a very thin pale brown or somewhat reddish bark. This surrounds a yellowish, woody portion, the fracture of which is short and somewhat fibrous. It has sweetish and mucilaginous taste resembling with ordinary liquorice roots.

Uses – *Seeds:* Purgative, emetic, tonic aphrodisiac, used in nervous disorders and cattle poisoning. The seeds which are of a brilliant scarlet colour with black mark on one side are much employed as beads. In the past, jewellers made use of these seeds as a unit of measurement for weighing gold as 'Ratti'. Young girls used to string these into necklaces and also for decoration of baskets, bracelets and rosaries.

Poultice of seeds: Used as suppository to bring abortion.

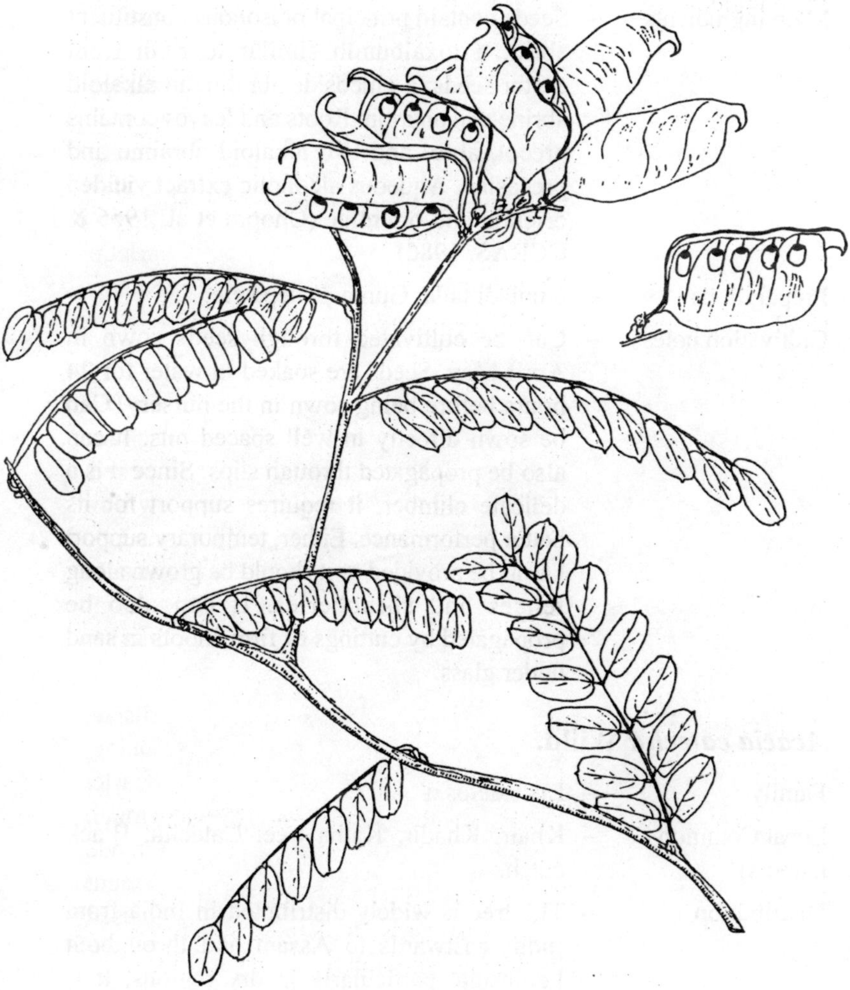

Abrus precatorius Linn. (Ratti, Gunja)

Root: Emetic, alexiteric and the alcoholic extract of the seed kernels has analgesic and central nervous system depressant activity. The drug has spasmogenic effect and was found to have relaxant effect on rat uterus (CCRAS 1986).

Main ingredients – Seeds contain principal poisonous constituent abrin, a toxalbumin similar to ricin from castor seeds; a glucoside abralin, an alkaloid abrine and abarnin. Roots and leaves contains precol, abrol and two alkaloid abraime and precasine. Aqueous alcoholic extract yielded orientin and orientine (Chopra et al. 1956 & CCRAS, 1986).

Preparations – Gunjadi taila, Gunja-pralepa etc.

Cultivation notes – Can be cultivated through seeds sown in April-May. Seeds are soaked in water for 24 hours before being sown in the nursery. Can be sown directly in well spaced pits. It can also be propagated through slips. Since it is a delicate climber, it requires support for its better performance. Either, temporary support has to be provided or it should be grown along fences and shrubberies. It can also be propagated by cuttings of firm shoots in sand under glass.

Acacia catechu Willd.

Family – LEGUMINOSAE

Local/Common name(s) – Khair, Khadir, Katha tree, Catechu, Black cutch.

Distribution – The tree is widely distributed in India from Indus eastwards to Assam and throughout Peninsula, particularly in dry regions. It is also common in plains and in northern India. In H.P., the tree is common in Una, Hamirpur,

Mandi, Bilaspur, Nalagarh and Solan regions below 1300 m elevation. (Chauhan, N.S.: 881, 1022, 1485, 1932).

Description – *A. catechu* is a moderate sized tree attaining a height of 10-13 m. In favourable localities, it attains a cylindrical stem upto 1.5 m in girth with 6 m bole. Ordinarily; it is a small tree with 60-90 cm girth and 2.4-3 m bole. Its greyish brown bark is nearly 1.25 cm in thickness, exfoliating in long, narrow strips; brown or red inside. Branches are unarmed. Leaves are bipinnate, armed with stipulary thorns, occasionally having numerous leaflets. Flowers are small, white or pale yellow in axillary spikes (1-4 together). Corolla 5 cleft, petals united, stamens distinct. Pods thin, flat, glabrous, 4-8 seeded.

Flowering/Fruiting – May-June to October.

Parts used – Heart wood, extract (Katha). The sapwood is thick red, turning brownish red to nearly black with age.

Current market rate – Khair bark Rs. 10/- per kg.
Katha Rs. 190/- to 370/- per kg.

Main ingredients – Contains catechin, catechutanic acid and tannin. Wood contains α (alpha) β (beta) and γ (gamma) catechion, and 1-epicatechin.

Uses – The cutch obtained from the heartwood of over 20-30 yrs old trees is used as important ingredient of 'Paan' chewed extensively in India. It is a useful raw material for plywood adhesives and is extensively used for dyeing canvas, fishing nets and ropes.

Various plant parts are used in sore-mouth, chest pain, cancer, colic pain, gravel, pthisis, bronchial asthma, consumption and strangulation of intestine, Bak juice alongwith

Acacia catechu Willd. (Khair, Katha)

asafoetida is used in haemoptysis. Important drug for leprosy in Ayurveda.

Bark is used as astringent, digestive and useful for cough and diarrhoea. It is useful in bronchial affections, cancer, piles, sore throat, ulceration, obstinate skin diseases including different forms of Leprosy. Gum from its bark is considered a good substitute for gum arabic. (Ahuja, 1965; Umrao Singh et al., 1983; Asolkar et al., 1992). The tree also serves as a host for lac insect. 'Kheersal', a product met with in some older trees in the form of white crystalline deposit of catechin is used for the treatment of cough and sore throat. Mixture of flower tops, cumins etc. is given in gonorrhoea.

Preparations – Khadiradi Vati, Khadirarishta, Khadir-Sar, Lavangadi-vati, Chandanadi-vati, Kutajarishta etc.

Cultivation Notes – Seeds can be collected during December-January and sown in February-March in Polythene bags or nursery beds. Before sowing, the seeds should be immersed in water for 24 hours. Seedlings are ready for transplanting after 6 months. Direct sowing of the seed is successful for afforestation as transplanting can not be relied upon, because of its fragile root system. Soaking of seeds in cold water for 24 hours gives good germination. The tree can also be propagated through root & shoot cuttings (Stumps) made from seedlings of about 15 months old. Planting of stumps should be done soon after the break of rains. Sowing khair seed in lines along with the agricultural crops acts as the cheapest and the best method. (Ghosh, 1977). The seeds and young plants demand plenty of light, moisture lack of weeds and should be planted at a spacing of 2.5 m × 2.5 m, to 3 × 3 m. Where conditions are favourable, they grow to a height of 1.5 to 1.8 m within a year. Two regular thinnings upto the age of about ten years are required and are repeated at intervals of five to six years.

Acacia nilotica Delile
spp. *indica* (Benth.) Brenan.
Syn. *A. arabica* Willd.

Family	– LEGUMINOSAE
Local/Common name(s)	– Kikar, Babul
Distribution	– Common in warmer areas in Nahan, Nalagarh, Una, Bilaspur, Hamirpur and Kangra divisions, where it grows to a full tree, sporadic in Solan, Kunihar, Mandi, Chamba and Rajgarh divisions, where it is found as shrub or a small tree (Chauhan, N.S., 1155, 1267, 1890, 1927, 1966).
Flowering/Fruiting	– July-September to April-June.
Description	– A moderate sized tree with a large spreading crown, bark dark-brown almost black, much fissured, spines 1.25-5.0 cm. long, straight white, sharply pointed. Leaves 2.5-5.0 cm long, rachis and pinnae downy pinnate 3-6 pairs, 1.25-3.0 cm long with several glands at their insertions; leaflets 10-20 pairs, 0.25-0.50 cm long, linear, glabrous. Flowers yellow, fragrant in axillary, globose heads which are about 1.25 cm diameter; peduncles in fascicles of 3-5, short slender, grey-downy, with a whorl of bracts above the middle. Calyx cup-shaped, 0.1 cm long. Corolla twice the length of calyx. Pod generally solitary,

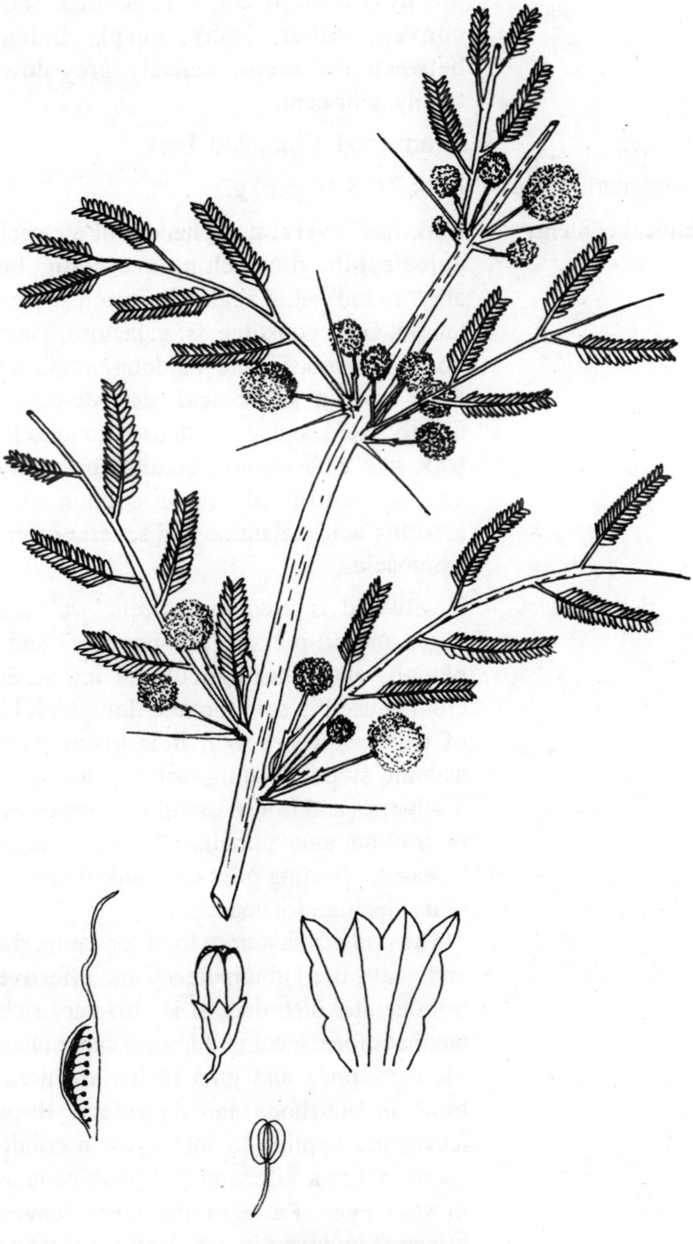

Acacia nilotica Delile

4-6 by 1.25-1.50 cm, 8-12 seeded, stalked convex, rather fleshy, deeply indented between the seeds, densely grey-downy, tardily dehiscent.

Parts used – Heart wood, Gum, Pod, Bark.

Current market rate – Bark Rs. 8-10 per kg.

Chemical contents – Bark has several polyphenols like catechin, epicatechin, dicatechin, quercetin; beta-amyrin and gallic acid. It is a rich source of tannin. Gum constituents: galactose, 1-arabinose, 1-rhamnose, and 4 aldobiouronic acids. Pods contain gallic acid and Me-ester, n-digallic acid and 2 condensed tannis. Root bark has octacosanol, betulin, beta-amyrin and beta-sitosterol. Seeds contain niacin, ascorbic acid, thiamine and several essential aminoacids.

Properties and uses – Heartwood is used for wheels, well-curbs, sugar and oil-presses, rice-pounders and for ploughs and other agricultural implements. Gum is used in sore throat. Along with latex of *Calotropis procera*, it is given to cure asthma; stops bleeding, urinary and vaginal discharges and is also useful in diabetes. Bark is cooling and anthelmintic. It cures skin diseases, bleeding piles etc. Bark is also used as a substitute for soap.

The pod is a favourite food for cattle, sheep and goats, used in impotency and effective in urinogenital disorders. It is also very rich in tannin before ripening. Flowers cause allergy. Flowers, pods and gum resins are used as tonic in diarrhoea and dysentery. Bruised leaves are applied to sore eyes in children, eaten in throat infection and poultice is used in sore eyes. Paste of the burnt leaves is effective ointment in itch. Various plant parts are used against hairfall, earache, syphilis, cholera, dysentery, leprosy and rinderpest. Ethyl alcohol (50%) extract of bark is antiprotozoal, hypotensive and spasmolytic.

Preparations – Babul-arista and Lavangadi-vati, Babool tooth-paste.

Cultivation notes – Can be raised through seeds. Seedlings can be planted as roadside avenue, windbreaks and shelterbelts as well as a biofence in field borders.

Achillea millefolium Linn.

Family – COMPOSITAE (ASTERACEAE)

Local/Common name(s) – Birnjasif, Millfoil, Yarrow.

Distribution – Pakistan to Uttar Pradesh. Temperate Eurasia 1800-3600 m in meadows, cultivated areas. In Himachal Pradesh, it has been found growing in Khabal-Soundari (2700 m), Gajjiyani, Thatch (3800 m) Winch-Camp Forest (2350 m), Chherna-Devi Bardhar (2300-2700 m). Swarhkothi slopes (2500 m); Pattan valley, Sangla valley, Chansal slopes, Rohtang and Churdhur slopes (2600 m-3600 m). (Chauhan, N.S., 1145, 1388, 1452, 2097, 2139, 2426).

Flowering/Fruiting – July-September.

Description – A small, perennial, tufted herb upto 50 cm tall, with a slender cropping rootstock throwing numerous roots and stolons with blunt succulent scale at each node. Leaves alternate, oblong-lanceolate, 3-pinnatisect, minutely divided into linear, dentate, mucronate more or less hairy segments. Flowers are numerous, small, white in corymbose, ovoid flat topped

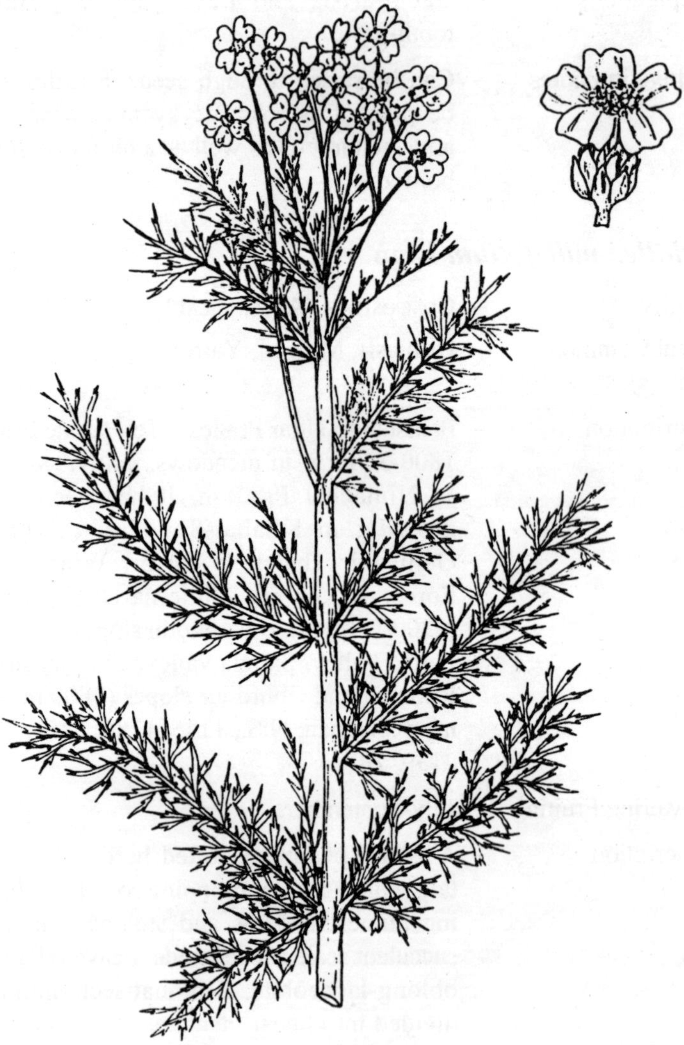

Achillea millefolium, Linn.

heads at the ends of stems and branches. Fruit shining.

Parts used – Leaves and flower heads/entire plant.

Chemical contents – The milfoil oil is obtained by distillation of the herb especially inflorescence which yields 0.049 per cent of oil on dry material basis. It contains volatile oil (0.25%), containing azulene (1-5 per cent), achilloic acid, an alkaloid-achilleine, a bitter principle, ivain and tannin. Also contains gum, various salts as malates, nitrates, phosphates, chlorides of potassium and calcium and ash 15-18 per cent (Martin-dale, 1977; Khory and Katrak, 1984).

Properties and uses – It has been used from early times for a great variety of medicinal purposes and has been given as infusion or liquid extract. It is stated to have diuretic, stimulant and haemostatic properties in the dose of 2-4 gm (Martindale, 1977). In many parts of Europe, a decoction of the leaves and flower heads is employed as a carminative, tonic and aromatic stimulant (Dey, 1980). Due to the presence of choline, it has hypotensive effect (Chopra et al., 1969). The leaves and flower heads are also used as a tonic and vapour bath for fever. The herb is employed as a substitute for hops in some parts of Sweden in the preparation of beer and is thought to render it more intoxicating. Its hot infusion is used as an emmenagogue in France (Watt 1972). Yarrow tea is good remedy for severe colds, fevers and obstructed perspiration. A decoction of the whole plant is employed for bleeding piles and is good for kidney diseases, also prevents baldness (Grieve, 1959). Medicated vapour bath containing yarrow, lavender flowers, wheat bran and milk is used in nasal catarrh and fever. It has a specific action on the pelvic organs and is used in amenorrhoea, menorrhagia and leucorrhoea. As a bitter tonic, it is given during convalescence from fevers and in atonic dyspepsia to promote appetite. Its chief use is in relaxation of the sphincter and with discharge of mucus or blood during defecation. The infusion is applied to relax throat and sore nipples with benefits (Khory and Katrak, 1984). Local inhabitants chew the leaves to relieve tooth-ache as it produces a pungent, but tingling and cooling sensation bringing immediate relief.

Preparations – Liv. 52, Livergen, Livomyn etc.

Cultivation – Propagated by divisions of roots and through seeds. Since the plant produces a number of suckers, these can be separated and transplanted in the field at a spacing of 30 cm × 30 cm during spring. It produces profuse quantity of seeds and can also be grown through seeds sown in nurseries and transplanted during pre-monsoons in the field having loamy, humus rich soils. It thrives best in full sun and can be grown in borders and smaller spaces around the fields. Seed sown plants bear flowers in the second year.

Achyranthes aspera Linn.

Family – AMARANTHACEAE

Local/Common name(s) – Puthkanda, Umblokando, Latjira, Chirchita, Apamarga.

Distribution – Throughout India upto 1300 m. As a weed, it is found in perennial crops, grasslands, waste places., shaded or unshaded areas all over Himachal Pradesh upto 1500 m elevation, (Chauhan, N.S., 1288, 1396, 1848).

Description – An erect herb about 0.5-2 m high with often a woody rootstock and much branched hairy stems. Leaves are opposite, elliptic, obovate or sub-orbicular, acute, obtuse or acuminate 4-10 cm × 2-5 cm, sparingly hairy with a short petiole. Flowers are green or pinkish, crowded in 2-4 cm terminal spikes which elongate in fruit and reach upto 50 cm in length with young and upper flowers ascending and older and lower flowers descending flat against the stem. Perianth glabrous and shining; stamens 5, alternating with truncate, staminodes. Fruit is an oblong utricle enclosed in the hardened perianth.

Flowering/Fruiting – April-September.

Parts used – Whole plant.

Main ingredients – From the seeds, two saponins; saponin A and saponin B have been isolated. Saponin α-methyl ester, β-methyl ester, have also been found. Alcoholic extract of the root yielded saponins. Saponins from the seed have cardiotonic activity. Further, it showed diuretic activity. Saponins possess anti-inflammatory activity besides potentiating the inhibitory response to the adrenaline (CCRAS, 1990).

Preparations – Agastya-haritaki, Kaphaketu-rasa, Apamarga kshara.

Uses – The seeds are given in hydrophobia, in cases of snake bite, in ophthalmia and cutaneous diseases. The flowering spikes rubbed with a little sugar are made into pills and given internally to patients bitten by mad dogs. The pulp of fresh leaves applied externally is a good remedy for scorpion stings. The ash of the plant mixed with conjee is a remedy in dropsy. Saponins obtained from the seeds have

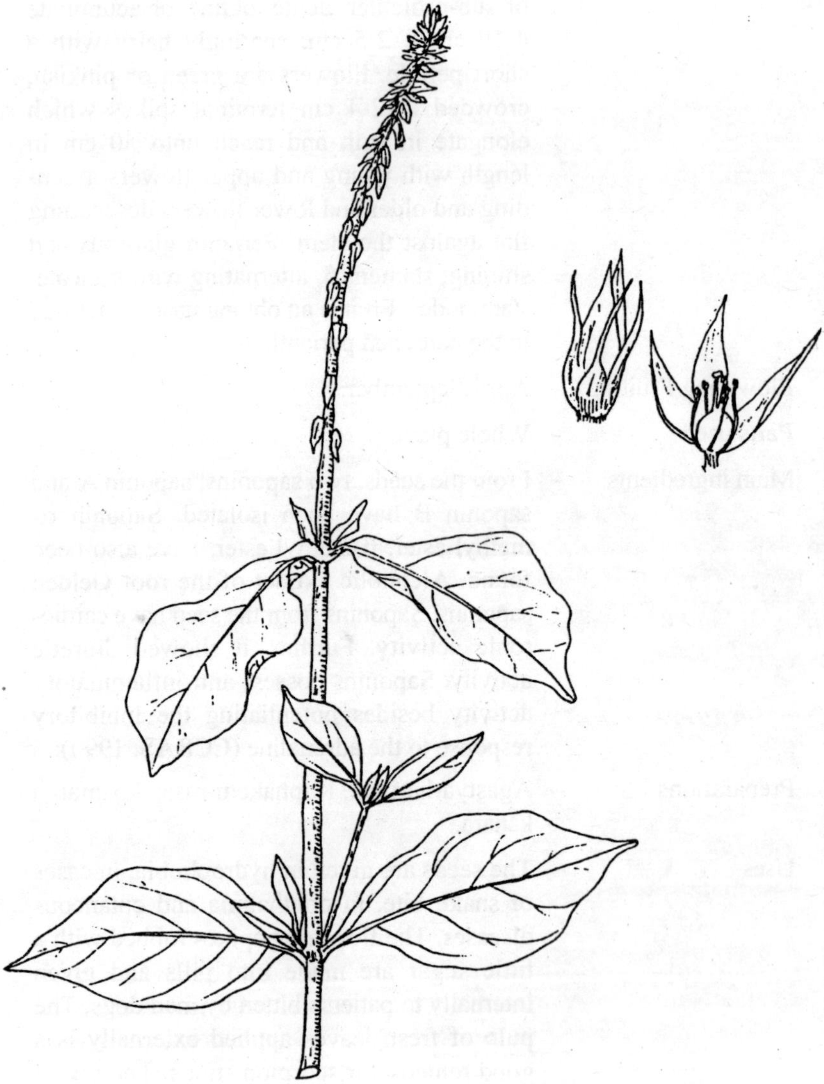

Achyranthes aspera Linn. (Puthkanda)

cardiotonic and diuretic activity. Saponins from the plant have been reported to possess anti-inflammatory activity besides potentiating the inhibitory response to the adrenaline. Solvent extract of the aerial parts showed contraceptive and hormonal properties in rat (cf. CCRAS, 1990:16). Root is used as an abortifacient in local folklores and the kheer prepared of its seeds allays appetite and one does not require food for a couple of days on having its kheer.

Cultivation notes – It is a very hardy plant and can grow in any type of conditions upto 1500 m elevation. Propagation is through seeds which can be broadcast in the field or sown in lines at a spacing of 45 × 45 cm during March-April. It's growth has been found quite luxuriant even without much of the inputs at Nauni (1250 m). Harvesting of aerial shoots is done at full fruiting state. Rootstock, being perennial throws out vigorous shoots in the growing season.

Allied species – *Achyranthes bidentata* Blume.

Aconitum deinorrhizum Stapf.

Family – RANUNCULACEAE

Local/Common name(s) – Mohro, Mouro, Mohra bish, Safed bish, Vatsnabh. Indian aconite.

Distribution – The plant is common in the sub-alpine and alpine zones of the Himalaya from Indus to Kumaon, occurring at altitudes between 2400 m - 4500 m. (Chowdhery & Wadhwa, 1984). In Himachal Pradesh, the plant is sporadic in Marui thatch (3200 m), Lamba thatch (3800 m); Odi-shagali thatch (3200-3500 m); Gorju thatch (3800 m) in Rorhru Forest Division;

Kinnaur and Shimla (3300 m - 3900 m) above mean sea level (Chauhan, N.S., 726, 1123, 1468). Reported endemic to Bushahr Himalaya (Nair, 1977).

Description – A tall, glabrous, biennial herb with paired tuberous roots. The daughter tubers are conical rather elongate upto 15 cm long and 3.5 cm in diameter at the crown with few root fibres. The surface is dark brown and coarsely wrinkled. Mother tubers are similar, but have longer filiform root fibres. The drug is very hard and horny and its starch is gelatinized during drying. Leaves are scattered, reniform or ovate reniform, 5-10 cm long, acute or obtuse, 5 pedatipartite almost to the base. Cauline leaves, sharply toothed, lowest long petioled; flowers more than 2.5 cm long, bright blue to greenish blue with purple veins in terminal racemes which go upto 45 cm. long. Fruit is a capsule bearing numerous seeds.

Flowering/Fruiting – July-September.

Parts used – Dried conical daughter tubers. The roots of *A. chasmanthum* and *A. balfouri* are also sold as Vatsanabh. Since *A. chasmanthum* contains pseudoaconitine in larger proportions, its use in medicines requires careful purification.

Current market rate – Rs. 70-110/- per kg.

Main ingredients – Air-dried roots contain upto 1.2 percent of alkaloids chiefly aconitine and pseudoaconitine. Pseudoaconitine is highly toxic and biologically 1.5 times as active as aconitine. A sample of roots from Bushahar was reported to contain 0.9 per cent alkaloid of which pseudo-aconitine was 0.4 per cent (Chopra et al., 1956 & 1965).

Aconitum deinorrhizum Stapf. (Vatsanbh, Mohra bish)

Uses – Together with *A. balfouri*, it is the principal constituent of *A. ferox* and is imported into England. It is a poison and affects both the heart and central nervous system. In case of extreme toxicity, the heart stops and respiration is depressed. With large doses, death is almost instantaneous, resulting in paralysis of heart or the respiratory centre. With moderate toxic doses, there is a tingling of the tongue, mouth, stomach, skin; followed by numbness, anaesthesia, nausea, vomiting, diarrhoea, excessive salivation, weak and irregular pulse, difficult respiration and muscular weakness may result. Externally, it is used with mustard oil as massage in neuralgia, paralysis and muscular rheumatism. Its application should be avoided in wounds or abrased skin as it may get absorbed and cause poisoning (Chauhan, 1984; 98-99). Locally much used for remo-ving body pains, but carefully. Root is smoked in toothache and body pains. Leaves are employed to improve the flavour of country liquor. Also used in leprosy, cholera, rheu-matism, as tonic in diarrhoea and cough (Mahandhar, 1980).

Preparations – Agnitundivati, Laghu vishgarbhataila, Swalp-kasturi-bhairavrasa, Sanjivanivati, Mrityun-jayarasa, Kaphketu-rasa etc.

Cultivation notes – At present entire aconite supply is obtained from wild sources more so from Rohru Forest Division. In order to obtain drugs of standard quality, a more careful collection and cultivation of this herb is required. It is endemic in distribution in erstwhile Bushahar State (parts of Rohru, Rampur and Kinnaur). Its collection, marketing and export has also been banned being an endangered species.

Propagation is effected easily by division of roots either in late autumn or early spring. It prefers rich humus and deep loamy soil with partial shade. Can be propagated through seed and splits of root with collar having vegetative buds, in the areas situated between 2500-3200 m ele-vation in forest margins. Seeds should be sown in spring in small seed beds and are transplanted when they are 5 cm tall. Harvesting/collection from the wild sources is done during autumn and calls for precaution and expertise to avoid poisoning incidents. Roots should be kept at a safer place away from the knowledge of children, women and ignorant persons till its disposal.

Aconitum heterophyllum Wall.

Family – RANUNCULACEAE

Local/Common name(s) – Atvisha, Atish, Patish.

Distribution – The plant is common in temperate to alpine zones of the Himalaya, from Indus to Kumaon occurring at altitudes between 3000-5000 m (Anonymous, 1948); 2500-4000 m (Rawat et al. 1985). In H.P., earlier common, but now getting rarer and rarer in Chamba, Kangra, Kinnaur, Kullu, Lahaul and Spiti, Shimla and Sirmour (Chowdhery & Wadhwa, 1984). Chansil (3500-4000 m), Sheuthru-Shagali thatch (2800-3800 m); Bajouni thatch, Kullu (3900 m); Dainasar and Thamsar in Chhota and Bara Bhangal areas of Kangra (3000-4500 m); Panju-Talra in Chhajpur (2800-3000 m); Mani Mahesh, Chamba (3500 m). In Khashadhar range of Rohru forest division, it is found growing in the field borders of villages at or around 2400 m elevation,

(Chauhan, N.S., 765, 1257, 1464, 1663, 2024, 2240, 2305, 2323).

Description – An erect, leafy herb having biennial, paired, tuberous roots. Stems upto 1 m tall with broad, ovate or orbicular, cordate, lobed and toothed, shortly stalked or sessile, amplexicaule leaves. Lower leaves are long petioled, while the cauline leaves are sessile, amplexicaule. Flowers bright-blue to yellow greenish blue with purple veins in long, many flowered peduncles. Fruit is a 5-celled capsule. Seeds are dark brown.

Flowering/Fruiting – July-September, seeds can be collected during September-October.

Parts used – The daughter tubers are usually preferred. Shrivelled mother tubers are often sold as inferior grade drug. The dried roots are generally ovoid-conical in form and tapering down-wards; 2.5-4 cm long and about 0.5 cm in diameter at the upper end. Externally they have a greyish or light ash colour and internally they are milky-white. The surface is slightly wrinkled longitudinally and marked here and there with whitish scars of small, detached roots and their upper extremity bears the scaly remains of leaves. Besides the official root, several other drugs are also sold in bazaars of India under the name of 'Atis' like the roots of *Asparagus sarmentosus* and *Aconitum napellus*. The roots of *Chaerophyllum villosum* have also been found as adulterants in the material sold as Atish.

Current market rate – Rs. 1300-2200/- per kg.

Main ingredients – The root contain non-toxic, amorphous alkaloids, atisine (0.4%), dehydroatisine, heteratisine and hetisine. It also contains aconitic acid, tannic acid, pectin and starch.

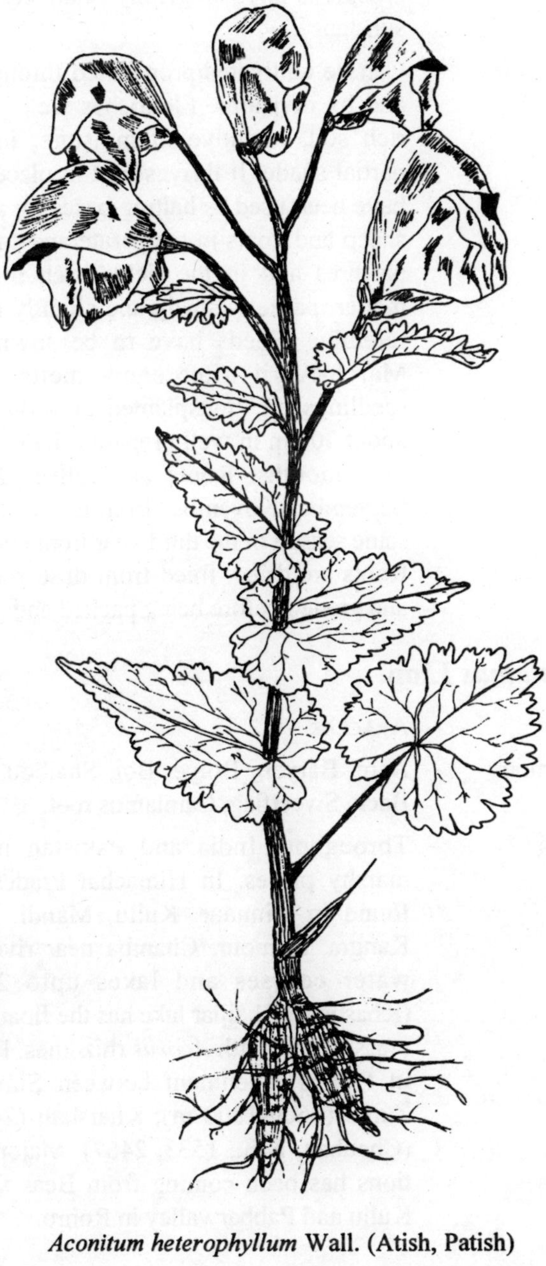

Aconitum heterophyllum Wall. (Atish, Patish)

Atisine is physiologically relatively inactive substance.

Cultivation notes – Can be cultivated/propagated through seeds and by divisions. Flourishes well in humus rich soil, retentive of moisture, in sun or partial shade. It thrives best in places which have been used as halting bases for flocks of sheep and goats in temperate and sub-alpine pastures and in the fields, rich in organic matter, partially shady and slightly moisture retentive. Seeds have to be sown during March-April after snow melts and the seedlings are transplanted at a distance of about 30 cm in well prepared fields after 1½ to 2 months. Roots are collected during September-November from the wild and the same season in the third year from cultivation. Roots are dried, freed from dust, root fibres and graded before being packed and stored.

Acorus calamus Linn.

Family – ARACEAE

Local/Common name(s) – Bare, Baryan, Boiye, Boi, Shalbou, Vacha, Bach, Sweetflag, Camlamus root.

Distribution – Throughout India and Pakistan in damp marshy places. In Himachal Pradesh, it is found in Kinnaur, Kullu, Mandi, Shimla, Kangra, Sirmour, Chamba near river beds, water courses and lakes upto 2600 m (Khashali). Khajjiar lake has the floating land mass strewn with *Acorus* rhizomes. Frequent in Pabbar catchment between Simoli and Zundi (1600-1800 m); Kharshali (2400 m), (Chauhan, N.S., 1533, 2467). Major collections has been coming from Beas valley is Kullu and Pabbar valley in Rohru.

Description – A semi-aquatic or marshy, perennial herb with indefinitely branched rhizomes, creeping in mud and have stout joints with large leaf scars. Rhizomes are cylindrical or somewhat compressed, about 1.5 cm in diameter, smooth, pinkish or pale green; white and spongy within. Leaves are few, distichously alternate, forming erect tufts at the extremities of rhizomes. Scapes arise from the outer leaves. Rhizomes are aromatic.

Flowering/Fruiting – April-August.

Parts used – Cut and dried pieces of the aromatic rhizomes, which are generally flattened, somewhat curved pieces, usually some inches in length and from about 1.2 cm to 2.5 cm in diameter. Externally, it has yellowish brown or fawn colour and is more or less shrunken and wrinkled; marked on its upper surface with the scars and fibrous vestiges of the leaves and on its under surface with numerous little elevated pale coloured circular dots with a dark centre (indicating the points from where the roots arose).

The common adulterant of *Acorus calamus* is orris roots (*Iris* species) which are readily distinguished by their darker colour, different structure and aromatic odour and taste. Rhizomes are collected during autumn and early spring.

Current market rate – Rhizomes Rs. 20-22/- per kg.

Main ingredients – The dry rhizomes contain 1.5-3.5 per cent of yellow bitter aromatic volatile oil. It has a mellow odour resembling that of patchouli. The Indian oil has much higher asarone content than the commercial oil. In addition to asarone, it contains α-pinene, d camphene, calamene, calamenol, calamenone and

Acorus calamus Linn. (Bach, Bare)

alcohols. The roots also contain a glucoside, 'Acorin', calamene, tannin, mucilage, starch, vitamin C, fatty acids, sugar and calcium oxalate.

Uses – Powdered roots are used as vermifuge. Rhizomes are emetic in large doses, stomachic and carminative in smaller doses; useful in bronchitis and remittent fever. It is a simple and useful remedy for flatulence; colic or dyspepsia and a pleasant adjunct to tonic or purgative medicines. The rootstock and leaves are used in preparation of perfumes and hair powders and for flavouring liquors, beer, vinegar; snuff etc. The leaves and roots are also used as an efficacious insecticide for the control of household pests and of insects attacking field crops, stored grains, and woollens. The fresh rootstock is used in confectionery and as a substitute for ginger. Fresh rhizome is inhaled in common colds as anti-allergic and small pieces if chewed empty stomach help in curing asthma (Manandhar, 1980, Chauhan, 1984).

Preparations – Bacha-churna, Sarasvata-churna, Ashwagandha-rishta, Yogaraja-guggulu, Sanjivani-vati, Chandraprabhavati, Unmadgajakesari-rasa, Mahashankhavati.

Cultivation notes – Clayey loams and light alluvial soils are suitable for its cultivation. The field is irrigated and ploughed with green manure before planting. The growing ends or tops of the previous year's crop are planted at 30 cm apart, leaving the leafy portion well above the ground. The crop is ready for harvesting in about a year. The plants are dugout and tops of the stocks are kept for next plantation. The rhizomes are dried in sun, before they are marketed. An acre of crop yields about 15 quintals of dry marketable rhizomes. (Anonymous, 1948). In subsequent years the crop regenerates itself as same pieces of rhizome which remain in the soil sprout up in the growing season.

Adiantum venustum G. Don.

Family – ADIANTACEAE

Local/Common name(s) – Hansraj, Hanspadi, Maiden hair fern, Damtuli

Distribution – In shady and moist rocky slopes, edges of ravines, on stone walls and comparable situations in most parts of the state upto 2800 m elevation (Chauhan, N.S., 1149, 1487, 3094 and 3504).

Description – A dark brown fern with fronds 3-4 times pinnate. Rachis slender, polished, naked, segments rigid, prominently veined and toothed, upper edge rounded, lower cuneate into the petiole. Sori 1-3, large, roundish, placed in a distinct hollow on the upper edge.

Parts used – The whole plant.

Current market rate – Rs. 15-25/- kg.

Chemical contents – Contains tannis and mucilage.

Properties and uses – It possesses astringent and aromatic properties, is emetic in large doses; tonic, febrifuge and expectorant. In Yunani medicines, it is used in ophthalmia, hydrophobia, tumours, colds, headache and is also considered aphrodisiac. In Chamba (HP), it is pounded and applied to bruises, boils and vitlago. The plant is useful as a mild tonic especially during convalescence from fevers. It is also used for the prevention of hair from falling.

In Yunani system of medicines, the oil obtained from the plant is applied to piles and tubercular glands and wounds and also to bring out a thorn which has penetrated into the body. A few other species growing in the area are also used in similar manner e.g. *A. capillus-veneris* L., *A. pedatum* L. and *A. lunulatum*.

Preparations – The herb is employed as an ingredient in several Ayurvedic and Unani systems of medicines especially in dermatological pharmaceutical preparations.

Cultivation – The plant grows naturally in shady and moist slopes, especially along walls, rock-boulders with sufficient soil deposition on it; along water channels, nalahs, natural springs and village bawlies (source of drinking water). Similar conditions are required for its plantation under domestication. Can also be grown as indoor pot plant.

Aegle marmelos (L.) Corr.

Family – RUTACEAE

Local/Common name(s) – Bel, Bael fruit, Mahaphala, Bilpatri.

Distribution – Sub-montane regions and plains, almost throughout India. Almost very much planted around temples and in gardens. Cultivated fruits are larger than the fruits from wild plant (Day 19). In Himachal Pradesh, it has been collected from Una, Kangra (Chauhan, N.S., 1914; Uniyal and Chauhan, 356, 420). Also found in Nalagarh, Bilaspur, Mandi, Kunihar, Nahan & Paonta Sahib forest ranges.

Description – A medium-sized, deciduous tree, having strong axillary thorns. Leaves trifoliate, stalked, alternate; leaflets oblanceolate, crenate, gland dotted; middle leaf is stalked and larger while the later one's are sessile. Flowers greenish, fragrant. Fruit is globose with woody shells.

Flowering/Fruiting – Flowers: May-June.
Fruits: May-June of the following year.

Parts used – The ripe and unripe fruits. Fruits are globose and the pulp is mucilaginous, aromatic, sweet and astringent. Dried slices are having on the outerside a smooth, greenish-brown shell enclosing a hard orange brown gummy pulp, in which cells and seeds may be seen. Dried pulp has a mucilaginous, acidic and slightly astringent taste and have agreeable aroma.

Uses – The Bel fruit is valuable chiefly for its mucilage and pectin; It is very useful in chronic dysentery, diarrhoea, particularly for patients having diarrhoea alternating with spells of constipation. Sweet drinks (Sharbats) prepared from the pulp of the fruits are useful as soothing agents for intestines of patients who have just recovered from bacillary dysentery. The unripe or half ripe fruits improve appetite and digestion. The antibiotic activity of the leaf, fruit and root of this plant has been confirmed in some recent experiments.

Main ingredients – Fruit contains marmalosin, young bark contains coumarin, and umbelliferone; old bark has umbelliferone and coumarin. Leaves are having essential oil, consisting of α and β-phellandrene. Matured bark contains fragrine, umbelliferone marmesin, sterols and triterpenoids. The leaves contain 0.6% essential oil mostly compound of d-limonene (CCRAS 1990).

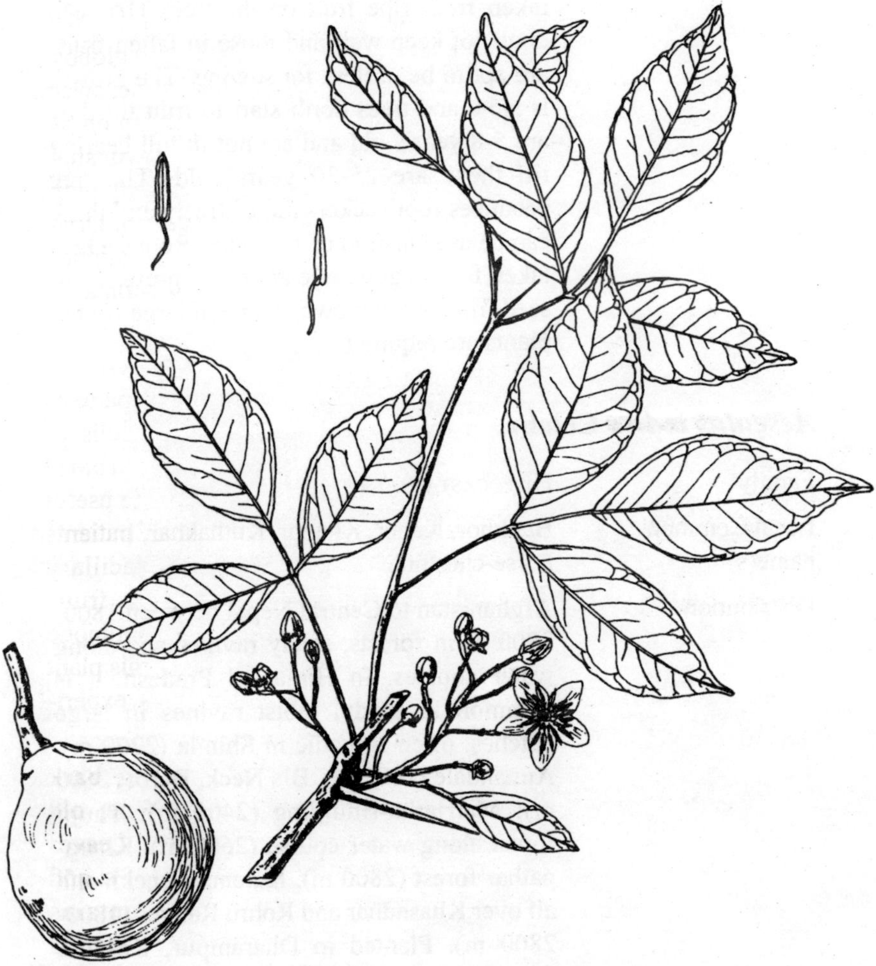

Aegle marmelos Corr. (Bilpatri)

Preparations – Dasmula-taila, Dasmularista, Bilvachurna, Gangadhara-churna, Amritarista, Chyavana-prasa, Mahanarayana taila, Dhanya Panchaka Kvatha, Pusyanuga churna, Vatsakadi Kvatha.

Cultivation notes – Propagation is through seed which should be taken from ripe fruit on the tree. The seed does not keep well and those in fallen fruits are apt to be useless for sowing. The growth is slow and trees don't start to fruit till they are 5-8 year's old and are not in full bearing till these are 25-30 years' old. The tree produces root suckers rather freely and these can be used instead of seedlings. Root suckers taken from a good tree are much preferred to seed from an unknown source, if large fruited plants are required.

Aesculus indica Colebr.

Family – HIPPOCASTANACEAE

Local/Common name(s) – Bankhor, Kanor, Khanor, Kuthakhar, Indian horse-chestnut.

Distribution – Afghanistan to Central Nepal between 1800-3000 m in forests, shady ravines and along water courses. In Himachal Pradesh, it is common in shady, moist ravines in large patches, often sporadic in Shimla (2200 m), Annandale (2000 m), B's Neck, Barot (2600 m); Marriarha-Bhujjling (2400 m); Thujji forest along water course (2600 m); Kharapathar forest (2800 m), Kulenu-munekh and all over Khashdhar and Rohru Ranges (1800-2800 m). Planted in Dharampur, Kasauli, Dharamsala and various other places. (Chauhan, N.S., 1384, 1981, 2073, 2251).

Aesculus indica Colebr. (Khanor)

Flowering/Fruiting – April-June
Fr.: July-October.

Parts used – Seeds.

Current market rate – Not available.

Main ingredients – The seeds contain a mixture of saponins, one of which is described as aescine, which easily crystallises. Also contains flavonoid glycosides, aesculine, albumin and fatty oils. (Anonymous, 1948).

Uses – Aesculine is used medically and extracts from the seeds are used industrially. As medicine, it is especially used for the complaints of the veins, such as phlebitis, haemorrhoids, varicose veins; in ulcers; to prevent thrombosis; in some cases of migraine, effusions of blood; for limb complaints and frost bite. Fruits are given to horses in colic. Oil from the seeds is externally applied in rheumatism. The hydroxycoumarin glycoside aesculin from the bark of the branch absorbs ultra-violet rays and is an ingredient for suntan oil (Chopra et al. 1956; Manandhar, 1980).

Preparations – Suntan oil and veterinary formulations. Presently, the seeds are being used in the preparation of shampoo.

Cultivation notes – Can be easily raised through seeds sown in nursery and transplanted. Seeds are collected during October-December and dried in shade after removing the fruit coat. The seeds should not be stored as they are attacked by moulds and insects and should be immediately sown in well prepared nurseries at a distance of 20 × 10 cm (30 cm apart with seed spaced at 25 cm apart). The seed can best be sown in drills in sunken beds during October-November before snowfall. 80-90% germination is reported and the seedlings are ready for transplantation after 1 year. 3 weedings are required in the nursery (Ghosh, 1977, Sharma & Sharma, 1988).

Ajuga bracteosa Wall. ex Benth.

Family – LAMIACEAE (Labiatae)

Local/Common name (s) – Neelkanthi, Leelkounthe, Ratpacho, Lilkounthe

Distribution – Found as a weed in field borders, along roadsides, open slopes and even on rock crevice upto 1500 m above mean sea level. It's allied species *A. parviflora* is also found sporadically (Chauhan, N.S., 721, 1510, 2195, 2781 & 3081).

Description – A perennial, erect or ascending hairy herb, often prostrate, with oblanceolate or sub-spathulate, obtuse, sinuate toothed leaves; purplish violet tinged from lower surface. Flowers are purplish-violet, tinged from lower surface in distant, axillary whorls in a spike.

Flowering/Fruiting – April-October.

Parts used – The herb (chiefly leaves)

Current market rate – Rs. 15-25/- per kg.

Chemical contents – Contains glycoside and tannin (Karnick and Pathak, 1982). Also yields ceryl alcohol β-sitosterol, Y-sitosterol, cerotic and palmitic acids alongwith a glucosidic constituent (Chopra *et al.* 1969).

Uses – The aromatic leaves are regarded as stimulant, diuretic, tonic and apparient. It is said to be a bitter astringent given in the treatment of fevers and also used as febrifuge. As an aromatic tonic, it is useful in ague. Also used to kill lice, regarded as depurative and useful

Ajuga bracteosa Wall. ex Benth.

in cases of malarial fever as a substitute for cinchona, (Kirtikar and Basu, 2027; Watt, 153; Wealth of India I: 42; Karnick and Pathak, 150-162). The herb is also used in gout, rheumatism, palsy and amenorrhoea. Ethyl alcohol extract has been found to have anticancer activity.

Preparations – Mostly consumed in traditional medicines locally and also in Ayurvedic formulations.

Cultivation – Though it occurs as a seasonal weed frequently, yet, if required, can be propagated through seeds and by transplanting naturally raising seedlings in appropriate spacings and adopting required cultural operations like application of fertilizers, irrigation, hoeing and weeding.

Albizia lebbek Benth.

Family – LEGUMINOSEAE (MIMOSACEAE)

Local/Common name(s) – Shirish, Kala sirish, Siris, Sirinh, Indian Walnut.

Distribution – The tree is common throughout India, much planted as avenue tree and also in tea gardens for shade (Dey, 1980). It occurs scattered in mixed deciduous forests and ascends to 1200 m in the Himalaya. (Chowdhery & Wadhwa, 1984). In Himachal Pradesh, it is found growing in Nauni (1200 m), Nalagarh-Ramshahar, (upto 1200 m), Nahan Division, Kangra, Chamba, Una, Hamirpur, Bilaspur, Mandi, below 1100 m. Also found planted at many places (Chauhan, N.S., 1571, 1574, 1884).

Description – A large, deciduous tree, bark brownish grey, rough with numerous, small cracks; leaves bipinnate; leaf axis 10-30 cm long with

2 glands; pinnae 1-4 pairs; leaflets 4-9 pairs, unequal sided, dark green, slightly pubescent beneath. Flowers are white, fragrant, in globular, umbellate heads on short pedicels, 1-4 together; calyx long, tubular; petals 5, united to beyond the calyx, stamens very long, greenish, 10-30, long, thin, dehiscent, monadelphous, straw coloured. Pods flat and thin, 8-10 seeded.

Flowering/Fruiting – April-May; pods ripen during Sept.-October and remain on the tree for a considerable time.

Parts used – Bark, leaves, flower and seeds.

Current market rate: – Not available.

Main ingredients – Gum, Saponin and tannin (7-11 per cent)

Uses – Siris is considered as alterative, aphrodisiac, astringent, expectorant, restorative and tonic. It is effective in asthma, reduces enlargement of cervical gland; in cough and cold; strengthens gums, thickens and retains seminal fluid. The seed are used in the treatment of piles and as an astringent in diarrhoea. The flowers are employed to cure boils, eruptions, swellings and act as antidote to poisons. Leaves are useful in ophthalmia. The powdered bark is useful in ulcers and in snake-bite wounds. Oil obtained from the seeds is useful in leucoderma. Powder of shoot bark strengthens gums. The wood is durable, of light colour and fairly strong. It is excellent for high class furniture, interior decoration and is also much used for house building, agriculture implements, rollers, the oil and sugar cane mills, well curbs, boats, carts, carriages and toys manufacturing. (Anonymous, 1948). The bark is used for tannin and dyeing and as detergent.

Preparations – Mahasirisagad, Sirisarishta, Dashanglepa.

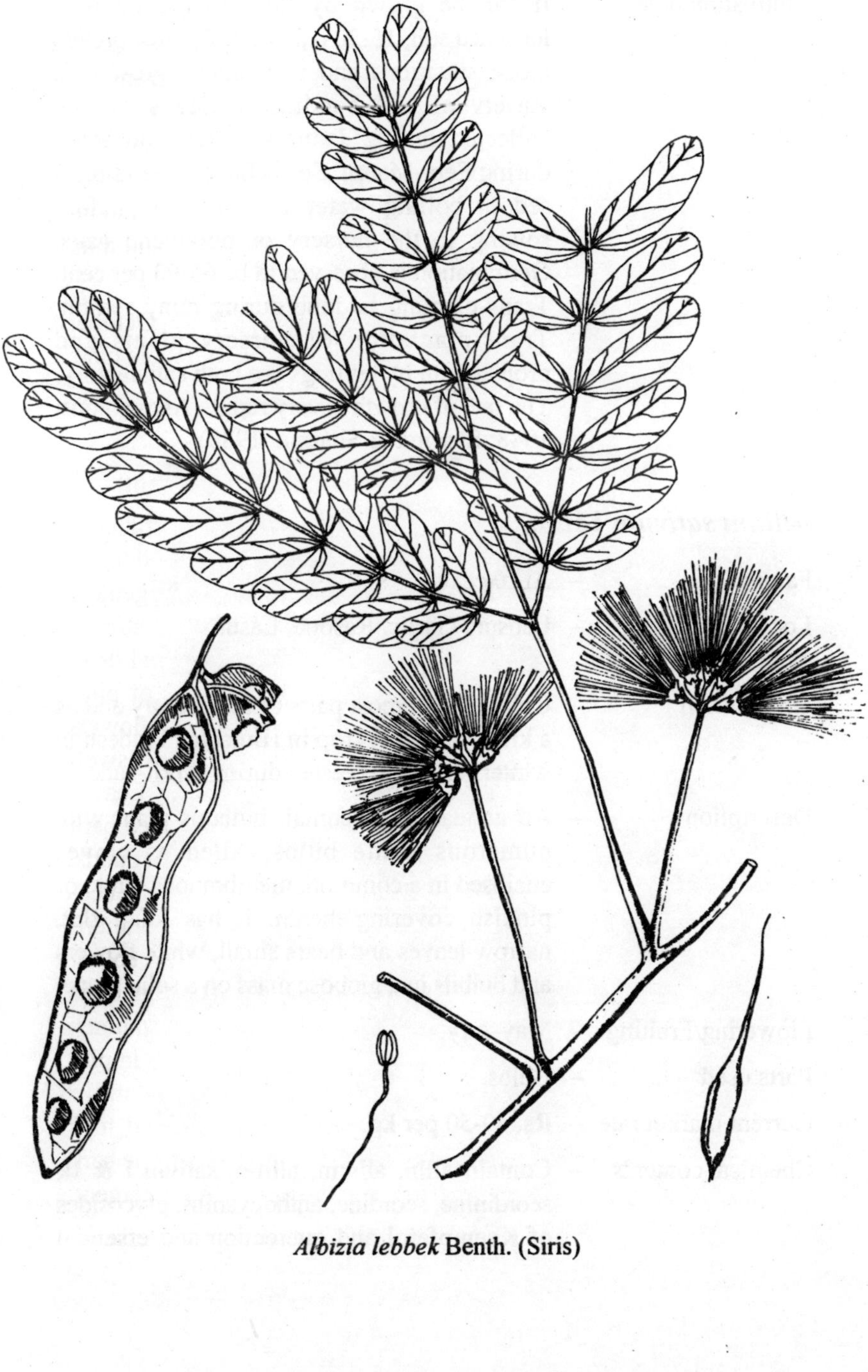

Albizia lebbek Benth. (Siris)

Cultivation notes – It can be grown by line sowing on well loosened soil, 15-20 cm apart. The tree prefers moist situations and is found to grow in a variety of soil conditions. Seeds can be collected during January-March and sown during Feb.-March. Seeds have to be immersed in boiling water for 24 hours before sowing in the nursery or polythene bags. Germination is observed to be 60-90 per cent. Trans-planting is done during rainy season. Trans-plantation of nursery plants and propagation by cuttings are both satisfactory. The rate of growth is very rapid. (Anonymous, 1948; Sharma & Sharma, 1988).

Allium sativum **Linn.**

Family – ALLIACEAE

Local/Common name(s) – Lahsun, Garlic, Rasona, Lasun

Distribution – Cultivated in most parts of the country and as a kitchen garden crop in Himachal Pradesh in winters and is harvested during April-June.

Description – An annual or perennial, bulbous plant with numerous white bulbs called as cloves enclosed in a common, membranous white or pinkish covering/sheath. It has long, flat, narrow leaves and bears small, white flowers and bulbils in a globose mass on a solid scape.

Flowering/Fruiting – May-July.

Parts used – Bulbs.

Current market rate – Rs. 40-50 per kg.

Chemical contents – Contain allin, allicin, allitin, sativin I & II, seordinine, scordine, anthocyanins, glycosides of Kaempferol and querection and essential oil. Vitamins like thiamene, riboflavin and niacin have also been found.

Properties and uses – As a medicine, garlic was held in high esteem by the ancient physicians of India. It is considered hot and stimulant and is administered in fevers, coughs and other debilitating conditions. It has also a reputation as febrifuge in intermittent fevers. Externally, the juice is used as rubefacient in skin diseases and as ear drops in earache and deafness. It is also used in fairly large quantities in western medicines. Various proprietary preparations of garlic have been given in pulmonary phthisis, bronchiactasis, gangrene of the lung and whooping cough. Laryngeal tuberculosis and duodenal ulcers are treated by garlic juices. Inhalation of the fresh juice is useful in pulmonary tuberculosis. It has been found useful in atonic dyspepsia, flatulence and colic. The juice diluted with water can be used as a lotion for washing wounds and foul ulcers. Recently, the garlic pills (pearls) are extensively and effectively used in high blood pressure. It has been found to lower the blood cholesterol. Garlic juice is larvicidal against coulex mosquito. The oil of garlic is appreciated as flavouring agent for meat preparations, soups, canned goods and table sauces (Martingale; 600; Watt, 173; Chopra et al., 1958; 273-74; Guenther, 6: 69). Garlic has proved to be an antibiotic, antiseptic, antidiabetic, antiarthritic, anticholesteremic, anticancer, antibacterial, antifungal, insecticidal and pesticidal.

Preparations – Rason-vati, Rason-pind, Rasonashtka, Lahsunadyaghrita, Garlic capsules and Garlic pearls.

Cultivation – Frequently cultivated for its use as condiment an spice all over the country. Cloves from the bulb are planted in well manured fields during September-October and the crop is harvested during April-July.

Aloe barbadensis Mill.

Family – LILIACEAE

Local/Common name(s) – Kumari, Ghrita-Kumari, Ghee-Kuwar, Kwar patha, Musabbar, Kware, Aloe.

Distribution – Kerala, Tamil Nadu and hotter provinces of India, found planted in Indian gardens. Many of the forms of this species are naturalized in India and are found in semi-arid states in all parts from the dry westward valleys of the Himalaya upto Cape Comorin. It is commonly found in field borders along paths, slopes in various parts of Himachal Pradesh below 1500 m elevation.

Description – A perennial fleshy plant. Leaves sessile, crowded, lanceolate, erect, spreading, spiny toothed at the margin, fleshy. Flowers: on a scape, scaly branched and longer than leaves, yellow in colour. Fruit: Capsule.

Parts used – Fleshy leaves.

Current market rate – Re. 0.70-1.00 per kg.

Flowering/Fruiting – July-September/October-November.

Uses – The plant is bitter, cooling, purgative, abortifacient, cathartic, stomachic, blood purifier, emmenagogue, alterative aphrodisiac, anthelmintic, useful in eye diseases, and uterine complaints. Fresh juice of leaves used as cooling and in fevers. Dried juice is useful in constipation. Pulp is edible as confection, vegetable and pickle.

Main ingredients – Pulp has been reported to contain a glucoside, barbaloin, also β-barbaloin and aloe-emodin (CCRAS, 1990).

Preparations – Kumaryasava, Bhringaamaladi-taila, Kumaryadi-vati, Rajapravartain-vati, Loha-rasayana, Aloes compound.

Cultivation notes – A healthy mother plant may have as many as 4-10 suckers. These seedlings are removed from mother plants when they are two months old. They are planted along slopes or on a level ground. The spacing between two plants is 60 cm-60 cm. Weeding is called for once in two months. Watering is necessary for a few days after planting. A good drainage is necessary as the waterlogging can lead to the decay of the plants.

The yield can be taken in about one to one and a half years. Roughly, one plant of one and half years old weighs around 5-8 kgs.

Even without moisture, the growth of the plant is not that poor. The plant grows luxuriantly in xeric conditions as well.

Angelica glauca Edgew.

Family – UMBELLIFERAE (APIACEAE)

Local/Common name(s) – Chora, Choraka, Chamchora, Gandrayan, Chura.

Distribution – Western Himalaya from Kashmir to Shimla, 2700-3000 m (Chopra et al., 1956). In H.P., it is reported from Dhauladhar range, Hattu, Narkanda, Chhota Bhangal and Bara Bhangal, Chamba, Kalatop, Luwayin-Lohardi forest (2600 m), Mariarha forest (2700 m), Panju onway to Talra Thatch, Polri Thatch (2800 m), Jaldayan Thatch (3200 m), Munekh forest (2600 m). Dhola-Bagi (Rohru) in the shade of broadleaved mixed spruce forest (2400 m). (Chauhan, N.S., 679, 722, 1999, 2060, 2236, 2312, 2346, 2450 and 2621).

Description – A, tall, robust, erect, aromatic, perennial herb 1 to 1.5 m tall with thick, stout, perennial rootstock. Roots are brownish, long, fleshy branching, thick, annulate, fusiform, resinous and pungently aromatic. Leaves are pinnately divided, the basal ones are ternate pinnate. Leaflets oval to irregularly mucronate, serrate; upper surface dark green, lower glaucous. Flowers white or purple in long stalked compound umbels. Fruit: glabrous, flattened, oblong, 12 × 7 mm. The roots and rootstock become thick and increase with age.

Flowering/Fruiting – July-September.

Parts used – Dried roots and rootstocks.

Current market rate – Not available.

Main ingredients – Dry roots yield 1.3% essential oil, resin and valerianic acid. The oil contains lactones and sesquiterpenes besides small amount of pinene, (phellandrane etc.). Roots and fruits contain several furocoumarins, such as angelicin, bergapten, xanthotoxin, umbelliferene and some phenols. (Anonymous, 1948), Chopra et al. 1956: 19 and 1969: 6). Choudhry et al. (1960) also reported a lactone in the ethereal extracts of the fruits of this plant.

Uses – The root is pungent, aromatic, stomachic, tonic, stimulant, carminative, diphoretic and diuretic. It increases appetite and combined with tonics, it is given in typhoid conditions, bronchitis, flatulence, colic and pain in the stomach. Roots are used as spice and condiment. Also used in obstinate constipation and

Angelica glauca Edgew. (Chora)

biliousness. The aromatic root is a flavouring agent for food etc. Roots are also used in the preparation of gin and the liqueurs known as bitters (Khory & Katrak, 1984, Mehra et al. 1969; Maheshwari & Singh, 1965; Chopra et al. 1974; Chauhan, 1984).

Preparations – Beside indigenous preparations, it is used in dry gin blending.

Cultivation notes – Can easily be cultivated through seeds. It prefers deep soil, rich in organic matter and partial shade. It can also be raised by making splits/divisions of the collar with vegetative buds alongwith root.

Arctium lappa **Linn.**

Family – ASTERACEAE/COMPOSITAE

Local/Common name(s) – Jangli Kuth, Burdock.

Distribution – Occurs widely in Western Himalaya, 1800-3800 m., Western Asia, and Europe. In Himachal Pradesh, the plant is commonly met near villages in high hill areas, often forming patches in sheep and goat camping sites in temperate forests in Chamba, Kullu, Shimla, Kinnaur and Lahaul & Spiti (Chauhan, N.S. 1144, 1198, 2652, 2801, 5575).

Description – An erect, branched, biennial to perennial; upto 1 m. or more tall. Leaves ovate, cordate, margins undulate, usually cottony beneath, stalked, 75-30 cm across. Flower heads globose, 1.8-3.7 cm across, in terminal clusters. Involucral bracts with rigid awl-like spreading barbed tips. Corolla and stamens purple-pink; style white, Achenes oblong, ribbed, angled.

Flowering/Fruiting – June-September.

Parts used – Root

Current Market rate – Not available.

Chemical contents – Main constituents are acetic acid, butyric acid, caffeic acid, chlorogenic acid, lauric acid, linoleic acid, oleic acid, palmitic acid, propionic acid, stearic acid and tiglic acid; aldehydes, (Acetaldehyde, benzaldehyde, butyraldehyde, isovaleraldehyde, propionaldehyde and valeraldehyde); carbohydrates, inulin (upto 45-50%), mucilage and pectin sugars. Other constituents are fats (0.4-0.8%) fixed and volatile oils (0.07-0.18%) bitters (Lappatin), resin tannins etc. leaves yield germacranolids, bitter sesquiterpenes, lactone, arctiopicrin, fukinone, arcitol and taraxasterol. Seeds yield sesquilignans and fruit yield arctin, arctigenin, mateiresinol and lappaol.

Properties and uses – It has diuretic and arexigenic properties and it has been used for cutaneous eruptions, rheumatism, cytitis, gout and specifically for eczema and psoriasis. The plant extract has been found to cause sharp, long lasting reduction of blood sugar within increase in carbohydrate tolerance and less toxicity. 50% ethyl alcohol extract was found as cardiac, stimulant, diuretic and spasmolytic.

Preparations – It is an official drug of British Pharmacopoeia. Used in tumor growth inhibitory mixtures, liquid compositions for treating psoriasis and seborrthoic eczemas.

Cultivation – It can easily be cultivated through seeds, which have very high germination capacity. Other requirements are same as for *Saussurea costus* with the added advantage that the former is not successful in monsoon prone areas, whereas the later flourishes well in temperate regions with good rainfall.

Arnebia benthami (Wall. ex G. Don.) Johnston
Syn. Macrotomia benthamii DC.

Family	– BORAGINACEAE
Local/Common name(s)	– Ratanjot, Laljari, Masreen
Distribution	– Pakistan to West-Nepal between 3300-4300 m in open slopes and alpine pastures. In HP, the plant has been collected from Thamsarjot of Bara Bhangal in Kangra and Jakha Kanda of Dodrakawar in Shimla district between 3500-4500 m.
Description	– An erect, robust, herbaceous perennial, 30-90 cm tall with stout, woody, deep-red roots. Leaves narrow, lanceolate, hairy. Radical leaves are large, 28 cm × 1.3 cm; cauline leaves, 7.5 cm × 1.3 cm. Flowers pink purple in dense terminal spikes, bending downwards like a tail.
Flowering/Fruiting	– July-August.
Parts used	– The roots.
Current market rate	– Rs. 55.00 per kg
Chemical contents	– A red dye, essential oil.
Properties and uses	– The plant is considered expectorant and used for cardiac disorders. Aqueous extracts, syrup, (Sharbat) and jam prepared from the flowering shoots are considered useful in diseases of the tongue, throat and is also useful in fever. Used as a colouring matter in hair oils, cookeries and for dyeing of silk. Root is frequently used as an antiseptic and antibiotic in healing of the wounds when applied as a poultice and also for subsiding swellings of the organs. Its paste made in water is applied to fireburns and is a quick healer.

Arnebia benthami **Johnston**

Preparations – Used as a colouring matter for hair oils, foodstuffs and silk. Also in many ayurvedic formulations.

Cultivation – As per habit of the plant, it can be propagated through seeds as well as divisions of roots with stem portion in tact. In-situ conservation is best method for its regeneration.

Arnebia euchroma (Royle) Johnston, Contr.

Family – BORAGINACEAE

Local/Common name(s) – Ratanjot, Khamet, Dimo.

Distribution – Afghanistan to Central Nepal, Western and Central Asia between 3300-4500 m in drier areas. Frequently found in Lahaul & Spiti at Kibber, Attagru, Kee Gompa, Rongtong slopes, Burr peak; Kinnaur in Pooh division (Chauhan, N.S., 2026, 2051, 2157, 2507, 2848).

Description – An erect, caespitose, perennial, hairy herb, with thick purple roots. Stems many, arising from the axil of the leaves, which formed a sterile basal cluster the preceding year; fistular, striate, covered with white trichomes. Basal leaves lanceolate, acute; lower and middle cauline leaves sessile, lanceolate; upper ones usually shorter and broader. Flowers pink or purplish-White, in terminal, subglobular inflorescence. Bracts leaf-like, not exceeding the calyx. Calyx divided to the base. Corolla funnel-shaped. Nutlets irregularly and coarsely tuberculate.

Flowering/Fruiting – July-September.

Parts used – Roots.

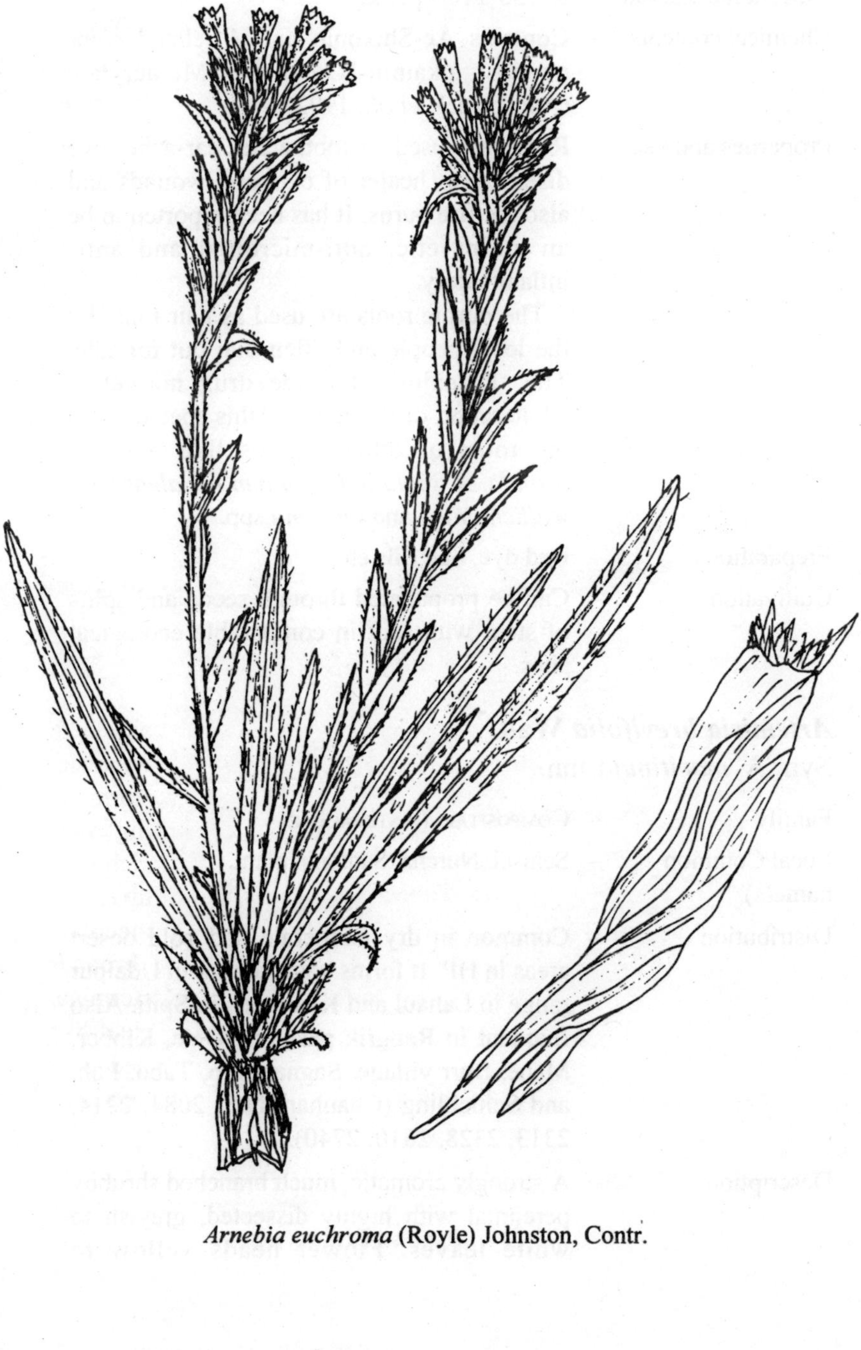

Arnebia euchroma (Royle) Johnston, Contr.

Current market rate – Rs. 55-110/- per kg

Chemical contents – Contains Ac-Shikonin and 4 related compounds; alkamin-B and β-di-Me-acrylate (Asolkar LV *et al.*, 1992).

Properties and uses – Roots are used in toothache, ear-ache, eye diseases, as healer of cuts and wounds and also the fire burns. It has been reported to be an antipyretic, anti-microbial and anti-inflammatory.

The purple roots are used as hair tonic by the local people and often dug out for sale. The 'Ratanjot' of crude drug market is adulteration of the roots of this species with the roots of other species like *Arnebia hispida, A. guttata, Geranium nepalense, G. wallichianum* and *Onosma* spp.

Preparations – Red dye, hair oils etc.

Cultivation – Can be propagated through seeds and splits of stem with root in comparable ecological sites.

Artemisia brevifolia Wall.

Syn: A. *maritima* Linn.

Family – Compositae (Asteraceae)

Local/Common name(s) – Seinski, Nurcha, Santonin plant.

Distribution – Common in dry temperate and cold desert areas in HP. It forms vast patches in Udaipur range in Lahaul and Kaa loops in Spiti. Also frequent in Rangrik pastures, Gete, Kibber, Mane, Burr village, Sagnam top, Tabo, Poh, and Schichling (Chauhan, N.S. 2084, 2214, 2313, 2328, 2510, 2740).

Description – A strongly aromatic, much branched shrubby perennial with highly dissected, greyish to white leaves. Flower heads yellow to

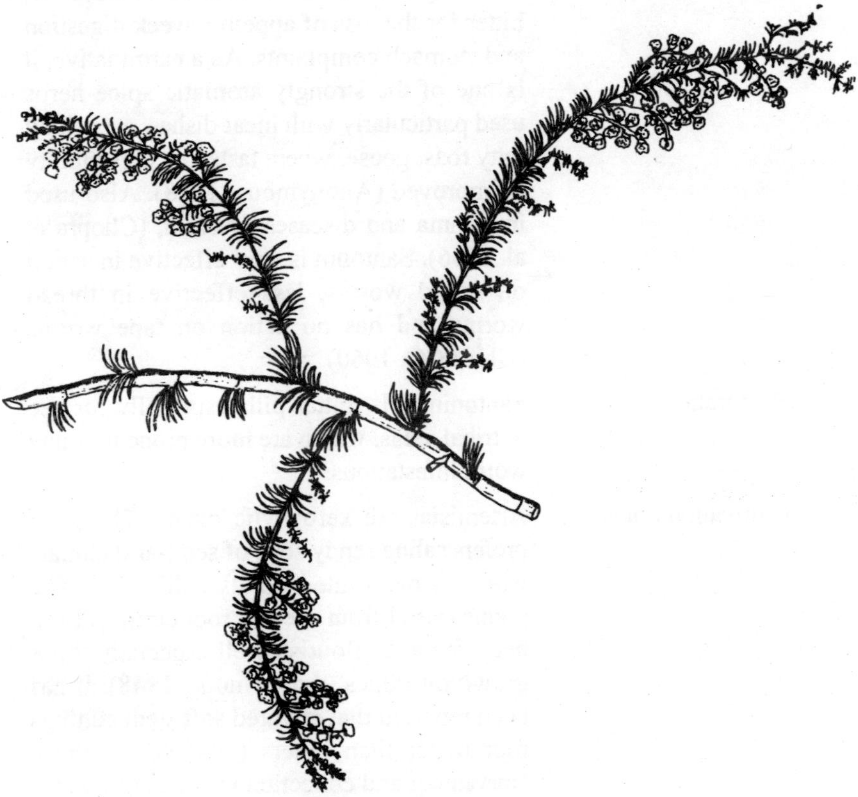

Artemisia brevifolia Wall. (Seinski)

		yellowish-red arranged in clusters of axillary spikes.
Flowering/Fruiting	–	July-September.
Parts used	–	Flowering heads (6-7 cm, terminal shoots) and young branches which are full of buds.
Current market rate	–	Not available.
Main ingredients	–	It contains 1-1.6 percent santonin, a bitter principle. The essential, oil contains about 9.3 per cent cineol as the main constituent.

Uses – The drug is now rarely used as an aromatic, bitter for the loss of appetite, week digestion and stomach complaints. As a carminative, it is one of the strongly aromatic spice herbs used particularly with meat dishes, especially fatty roast goose, where taste and digestibility is improved (Anonymous, 1948). Also used in asthma and diseases of brain, (Chopra et al. 1956). Santonin is very effective in action on round worms, less effective in thread worms and has no action on tape worms (UNESCO, 1960).

Preparations – Santonin pills, bitter pills especially for use in tribal areas, which are more prone to round worm infestations.

Cultivation notes – Artemisias are xerophytic plants. The plant prefers saline sandy soils of semi-arid climate which is best-suited for its cultivation. The plants raised from seed or root cuttings have been found to flourish well especially those grown on ridges (Anonymous, 1948). It has been reported that matured soft stem cuttings root better than others (UNESCO, 1960). Harvesting and collection is done during mid summer to early autumn when the buds have fully developed, just before opening, for maximum santonin content, which decreases thereafter. 60-70 cm of flowering herbage is cut off, bundled and dried. The drug has pleasant and spicy smell and an aromatic, slightly bitter taste.

Asparagus abscendens Roxb.

Family – LILIACEAE

Local/Common name(s) – Shatawari, Shatamuli, Shatavar, Sahasrapal, Sainsarbuti.

Distribution	–	Distributed throughout tropical and sub-tropical parts of India ascending upto 1300 m in the Himalayas. In Himachal Pradesh it is found growing sporadic in the districts of Una, Chamba, Bilaspur, Hamirpur, Kangra, Mandi and Solan. (Chauhan, N.S. 1765). Its allied species—*A. racemosus* Willd has been introduced in the Nauni Herbal Garden and is performing well.
Description	–	A struggling or scandent, much branched spinous shrub with woody terete stems. Cladodes—0.6-1.2 cm long, linear; spines, stout, straight, densely crowded; flowers small, white mildly fragrant; 3-4 mm across, in solitary or fascicled, simple or branched, copious racemes; fruits 0.8 cm in diameter, globose, often obscurely 3 lobed berry, with only one seed.
Flowering/Fruiting	–	August-November, Fr. December-January. It often flowers 2-3 times in a year.
Parts used	–	Tuberous roots.
Current market rate	–	Asparagus roots Rs. 55/- to Rs. 70/- per kg depending on the quality.
Main ingredients	–	Root contains asparagine, saponin (steroidal). Fruits contain β-sitosterol, sarsasapogenin and diosgenin.
Uses	–	Root is used as appetizing, diuretic, aphrodisiac, laxative, astringent useful in dysentery, diarrhoea throat complaints and leprosy. (Manandhar, 1980). Used as demulcent in veterinary medicines (Umrao Singh et al. 1983). Root bask is taken with milk for vitality and strength. It is an ingredient of GERIFORTE used against fatigue and senile pruritus. *Domestic use:* Delicious dishes are pre-

pared out of its tender shoots. It is regarded as a sacred plant and considered auspicious on marriage, ceremonies at other oracle rites.

Preparations – Shatavarighrita, Narayana taila, Vishnu taila, Satamulyadiloha, Shatavaripanaka.

Cultivation notes – Asparagus thrives best in well drained, deep, loose soils. It requires liberal use of fertilizers. It is propagated both by seeds and suckers. The seeds are soaked in warm water for a day before sowing and seedlings are ready for transplantation in 2 months. Vigorous seedlings are transplanted in pits 0.5 m × 0.5 m × 0.5 m at a spacing of 0.5 m with soil, sand and FYM in the ratio of 2:1:1. Once established, asparagus will continue to yield for many years. Proper support has to be given to the plant. It can best be grown along the fence. It has minimum water requirement and when the plant suffers from acute shortage of water, the cladodes begin to turn yellow. A plant of one year produces 6-10 suckers. They are removed by splitting and planted in polythene bags and kept in shade during April. Watering is required for its survival. Transplanting is done during monsoon in pits. Yield can be taken after 2 years.

A. racemosus and *A. filicinus* are the other useful species available in H.P.

Atropa acuminata Royle

Family – SOLANACEAE

Local/Common name(s) – Jharka, Indian belladona, Jalgi, Deadly night shade.

Distribution – Earlier, it used to grow sporadically in forest shade between 2500-3000 m in Kullu, Chamba, Kangra and Shimla districts. But now, it is difficult to locate. Some considerable populations exists in Panju forest, Chhajpur block of Jubbal division at 2800 m. (Chauhan, N.S. 5727).

Description – A tall, straight plant about 1 m high. Leaves are stalked, elliptic, lanceolate, acuminate 7-16 cm long and 5-10 cm broad. The aerial parts die every year in autumn and the new ones appear in the following year. The plant has a large tap root with many lateral rootlets. Flowers are yellowish coloured, solitary, shortly stalked. The fruit on ripening is a purple black berry of the size of a cherry.

Flowering/Fruiting – June-September.

Parts used – Leaves and roots.

Current market rate – Not available.
Seeds Rs. 200/- per kg.

Chemical contents – Atropine, hyocyamine, hyoscine (Scopolamine) are the most important alkaloids present in leaves and roots. Concentrations of tropane alkaloids in leaves decrease with maturity. Roots also contain cuscohygrine, but no flavour, which are present in leaves and flowers. Seeds contain fatty oil (25%) and also essential oil.

Properties and uses – Indian belladona is used in India for the manufacture of tinctures, plasters etc. and is used in the similar manner as the European *A. belladona* in the treatment of intestinal and biliary colic and are often prescribed with vegetable laxatives to prevent gripping. Also used in asthma, whooping-cough, bladder and ureteral spasms. It is considered as sedative and narcotic. Ethyl alcohol extract (50%) of leaves is antiprotozoal, antiviral and hypoglycaemic.

Atropa belladonna Linn.

Preparations – Atropine salts, tincture belladona, belladona plaster etc.

Cultivation – Can be cultivated through seeds and root cuttings. Same operations are useful as applicable to *Atropa belladona* for which agro-technology is available in India with CIMAP, Lucknow and various other institutions in the country.

Azadirachta indica A. Juss.

Family – MELIACEAE

Local/Common name(s) – Neem, Nim, Nimba, Margosa tree.

Distribution – Found throughout India. As a common road-side plant, grown over the greater parts of India. In Himachal Pradesh found mainly in Nalagarh and Nahan forest divisions.

Description – Neem is a very well known tree of India. The tree has pinnate leaves, i.e. its leaves are divided into numerous smaller segments called leaflets, each looking like an ordinary leaf. Flowers are small, white, in short axillary bunches. Fruits 1.2-1.8 cm long green or yellow; seed one in each fruit.

Flowering/Fruiting – March-April/July-August.

Parts used – Whole plant, bark, root bark, young fruits, nut or seed, flowers, leaves, gum and toddy.

Uses – Used in leprosy, intestinal worms, piles and urinary diseases. Root bark and young fruits are astringent, tonic and antiperiodic. Bark is bitter tonic, astringent, antiperiodic and also vermifuge. Fruit is purgative, emollient and anthelmintic. Leaves are discutient, leaf juice anthelmintic. Oil from nuts and leaves acts as

Azadirachta indica Juss. (Neem)

a local stimulant, an insecticide and antiseptic. Flowers are stimulant, tonic and stomachic. The antibiotic activity of leaves and roots of the tree and their utility in skin diseases have been confirmed experimentally. Neem oil is used as a commercial insecticide and the oil cake is prized as a fertilizer.

Main ingredients – Nimbidin, nimbin, nimbinine, nimbosterol, and more unidentified steroids have been isolated. Tetranortriterpenoids like nimbin, salanin, nimbolide, azadirone, are found to be pre-sent alongwith some polyphenolic compounds (Pandey et al. 1990).

Preparations – Sankhapuspi-taila, Jvarasamhara-rasa, Mahamarichyadi taila, Arsoghna vati, Pathyadikvatha, Maha Manjisthadyarista, Punarnavadi-kvatha, Raktasodhaka-vati etc. Widely utilized in preparation of soaps and toothpastes.

Cultivation notes – Direct sowing of the seed is done either in patches or in lines. The soil is worked to a depth of about 15 cm and seeds are sown at a depth of 1.5 cm. Patches are spaced 3 × 3 m and spacing between lines may be about 3 m. A seed rate of about 3-4 kg/ha is considered sufficient. The seeds don't require any pre-sowing treatment. Since the seeds have a short viability, these have to sown immediately after collection, preferably within a fortnight.

Artificial propagation can be carried out successfully through direct sowing, planting out of entire plants or stump planting. The first one is preferred over the other two. Air layered branches treated with IBA or NAA in linolin paste at 0.1 per cent root satisfactorily. (Singh, 1982).

Baliospermum montanum **Muell-Arg.**

Family	– EUPHORBIACEAE
Local/Common name(s)	– Danti
Distribution	– Bihar, North Bengal, Nagpur outer Himalayas from Kashmir to Bhutan. In Himachal Pradesh it is found in moderate to sporadic quantity in all areas below 1300 m elevation. (Chauhan, N.S. 7149).
Description	– A leafy, stout usually monoecious under shrub 0.9 to 1.8 m high, with herbaceous branches arising from the root. Leaves are sinuate toothed; the upper small, lanceolate, penninerved, and the lower large, reaching 15-30 cm long, and often as broad, ovate. Flowers numerous in axillary racemes or contracted panicles (all male or with a few females below). Fruits capsule, of three, 2 valved crustaceous cocci. Seeds are smooth and mottled.
Flowering/Fruiting	– Almost all the year around.
Parts used	– Dried roots.
Current market rate	– Rs. 6/- per kg (Dried material).
Main ingredients	– Contains β-sitosterol. Seeds are rich in oil, resin and starch.
Uses	– The root is pungent, heating, purgative, anthelmintic, diuretic alexiteric, useful in pains, diseases of the skin and of the abdomen, in piles, enlarged spleen, itching, anaemia, leucoderma and jaundice. This eradicates the phlegm of the nasal orifices. It is digestive and appetiser and is best for ascites, hernia, oedema and other diseases due to vitiation of Vata and Kapha.

Preparations – Abhayarishta, Mahakalyanaghrita, Kalyana-gritham, Punarnava mandoor, Kaishore-guggulu, Chandrapabha-vati etc.

Cultivation notes – For propagation, stem cuttings should have a length of 14-15 cms. The cuttings should be taken some distance behind from the apical region and very hard, basal portion should be avoided. If the upper portion of the shoot is taken, the inflorescence should be removed and only terminal bud is to be retained for a satisfactory result.

Bauhinia variegata Linn.

Family – LEGUMINOSAE

Local/Common name(s) – Kachnar, Kovidara, Kariala, Karale, Kanchnar.

Distribution – Sub-Himalayan tract from the Indus eastwards, also dry forests of Eastern, Central and South India. Common in Una, Hamirpur, Mandi, Chamba, Bilaspur, Solan and Sirmour as an agroforestry species.

Description – A fairly medium sized to a large tree. Leaves deeply cordate, 11-15 nerved, sub-coriaceous, cleft through about a quarter, lobes obtuse. Flowers are large, fragrant, sessile or shortly pedicelled in few flowered lateral corymbs. Calyx lube is slender, the limb is spathaceous; petals obovate, or oblong, generally four white and one with red or purple lateral veins radiating from near the base. Fertile stamens 5, rarely 3 or 4. Ovary long stalked, style long stigma minute. Pods are hard, flat, glabrous, slightly falcate, dehiscent. Seeds 10-15 per pod.

Flowering/Fruiting – Flowering: April when leafless.
Fruiting: Rainy season.

Bauhinia variegata Linn. (Kachnar)

Parts used	– Root, bark, flowering buds and flowers.
Uses	– Bark is anti-inflammatory, useful in skin diseases, ulcers and scrofula. Dried buds are used in dysentery, piles and worms. Roots are used as antidote to snake poisoning. Decoction of roots is used in dyspepsia. Flower buds are pickled and also used as vegetable.
Main ingredients	– From the flowers, hentriacontene octacosanol reducing sugars, stigmasterol, β-sitosterol β-p glucophyranoside and one unidentified flavonol have been isolated (CCRAS, 1990).
Preparations	– Kanchnara-Guggulu, Chandanasava, Vidangasava etc.
Cultivation notes	– Natural reproduction takes place from the seed. The seeds are dispersed on dehiscence of the pods on the trees and germinate readily when provided with favourable conditions. Whereas the artificial propagation is carried out by direct sowing of the seeds and through stump planting. Both these methods have been found to be equally good and better than planting out one season old entire plants. Branch cuttings normally root with difficulty, but these root well in August, November and February with the application of auxins. Direct sowing is done in lines, spaced about 3 m apart. Germination starts in about a week after the onset of monsoon rains ensuring good soaking of the soil. Sown area needs protection against grazing. The entire plants have to be planted out with the ball of earth. Stumps, or container grown plants can be preferred for transplanting. To raise entire plants for planting out in July-August, sowing is done in March-April and previous years seed is used.

Berberis aristata DC.

Family	– BERBERIDACEAE
Local/Common name(s)	– Kashmal, Kushmol, Chothar, Daru-haldi, Daru-haridra; Rasanjan, Rasont.
Distribution	– The plant is distributed throughout the country between 1650-3400 m altitude. In Himachal Pradesh, it is present in Chamba, Kinnaur, Sirmour, Kullu, Mandi and Shimla. Found in rare and sporadic conditions in shady ravines, field borders and depressions between 1800-2800 m. The population of this species has decreased alarmingly. Its population patches can be found in Pekha-Diudi, Khabal, Dodra Kawar & Tikkri; Bagi, Palthari, Chiuni in Rohru (2000-2800 m); Chherna, Devi Bardhar in Chhota Bhangal area; Barot in Mandi, Parvati valley in Kullu, Mcleodganj, Dharamsala (1800-2600 m) (Chauhan, N.S., 676, 907, 1085, 1478, 1512, 1584, 2102, 2178, 2212, 2290, 2479, 2726; Uniyal & Chauhan, 208, 1045 and 1057).
Description	– A large, thorny, deciduous shrub nearly 2-5 m tall with pale rough, furrowed bark; leaves stalked, obovate or elliptic, fascicled in the axils of simple or branched spines, leathery, often with sharply toothed margin; flowers yellow, in compound racemes with stout pedicels; racemes are longer than the leaves; fruits 1.8 cm × 0.6 cm, ovoid, grey-white when young, bluish when ripe. Wood and roots are yellow from inside.
Flowering/Fruiting	– May-July/August-September.
Parts used	– The wood, root bark and extract known as Rasount. Rasount is a dark brown extract of the consistency of opium, having a bitter,

astringent taste, readily soluble in water forming a rich yellowish brown colour.

Current market rate – Extract (Rasount) Rs. 120-125/- per kg.
Wood pieces Rs. 10/- per kg.
Root bark Rs. 13-18/- per kg.

Main ingredients – Berberine is the principal alkaloid. (Chopra et al. 1956). Also contains oxycanthine and umbellatine, fat, resin and tannin. The fruits contains malic acid, citric acid and tannins.

Uses – The wood, rootbark and extract have been used in Hindu medicines from a very remote period. Its properties are said to be analogous to turmeric. Rasount and the plant are used as alterative, and deobstruent and are used in skin diseases, menorrhagia, diarrhoea, jaundice and above all in affections of the eye. In painful micturition from bilious or acrid urine, a decoction of the Indian barberry and amla is given with honey. A decoction of the root bark is used as a wash for unhealthy ulcers and is said to improve their approach and promotes cicatrization. Rasount, mixed with honey is an useful application to aphthous sores. It is regarded efficacious in chronic uterine troubles, piles and specific for cholera. Ripe fruits are edible and are also given as a mild laxative to children. Berberine in the form of its salts as Berberine hydrochloride and Berbein sulphate is used in dyspepsia, diarrhoea, malarial affections, amenorrhoea, enlargement of the spleen, anorexia, vomiting of pregnancy and intestinal catarrh. Infusion of berberis has bactericidal power (Kirtikar and Basu: 103; Wealth of India, 1:178; Dey, 139-40; Khory and Katrak, 34: I.P.C.; 41).

Preparations – Darbyadi-kwath, Darbyadi-taila, Rasanjanadi-Churna, Sudershan-Churna, Punarnava-

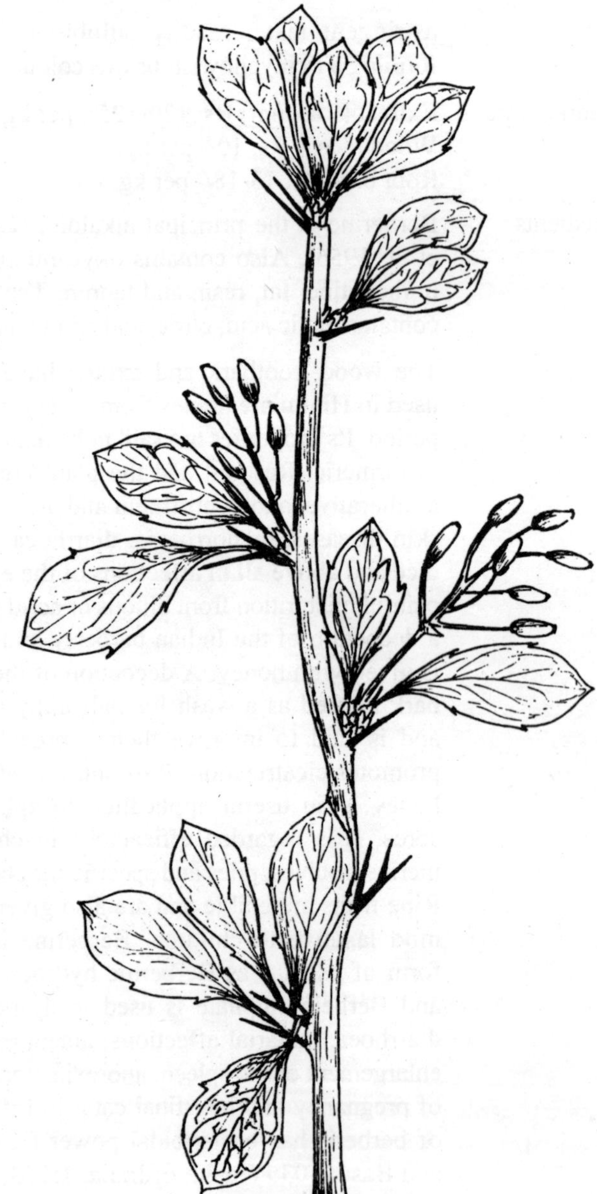

Berberis aristata DC.

Mandoor, Khadirarishta, Laghu-Vishgarbha taila, Dashanglepa etc.

Cultivation notes – Due to the increased activities in agriculture and horticulture in the State, the habitat of *B. aristata* have ruthlessly been cleared out and the plant has managed to survive in inaccessible slopes, banks of streams and near village *gharats* (rural floor mills). It is mainly propagated through seeds. Propagation through cuttings has also been reported. Work on techniques to standardise vegetative propagation is in progress at many places.

A closely allied species, *B. asiatica* is distributed in Shimla, Sirmour, Mandi, Kullu and Kangra districts (Chauhan, N.S. 1336). It is used in similar manner as *B. aristata*.

Berberis lycium Royle

Family – BERBERIDACEAE

Local/Common name(s) – Kashmal, Kashwale, Daruhaldi.

Distribution – It occurs from Kashmir to Garhwal on the outer Northern-Western Himalayas on clearances and along roadside between 800-2500 m (Anonymous 1948). In Himachal Pradesh, it is present in Chamba, Kangra, Kinnaur, Kullu, Shimla, Sirmour. It is common throughout the state from valley areas extending upto 2600 m in ghasnies/ plains especially in depressions and field borderlands (Chauhan, N.S., 682, 824, 891, 1422, 1696).

Description – It is an evergreen shrub of average height. Twigs are pale yellowish, bark greyish, rough, rather deeply furrowed; leaves are lanceolate or nearly obovate, oblong, entire. Flowers,

Berberis lycium Royle (Daruhaldi)

yellow in a simple raceme. Racemes are many flowered, corymbose, shorter than the leaves; pedicels are elongated and single flowered; berries are purplish-blue on ripening.

Flowering/Fruiting – March-June Fruit: June-October.

Parts used – Root, bark, stems and extract known Rasount.

Current market rate – As in case of *B. aristata*.

Main ingredients – Berberine and Umbellitine are the main alkaloids. The Rasount contain moisture, (25 per cent) and berberine (9.4 per cent). The fruits contain malic, tartaric, citric acids and tannin.

Uses – The root in Yunani system of medicines is considered bitter with an unpleasant taste; used in splenic troubles; tonic, febrifuge; intestinal astringent; good for cough, chest and throat troubles, eye sores and itching of the eye; piles, menorrhagia; useful in chronic diarrhoea; allays thirst. As a gargle it strengthens the gums and is a good application to boils. The extract of either stem or root 'Rasount' or 'Rasanjana' has since old times been used for skin diseases, sores and leprosy; as an antidote for poisoning, useful in eye and ear troubles; as remedy for oral ulcers and in piles. It is also used for disorders of the urinary tract and to treat enlargement of the spleen. A simple decoction of it, with honey is given in jaundice. Rasount is used as a purgative for children, as a blood purifier and as an external medicament in conjunctivitis in combination with opium. Umbellitine has a specific inhibitory action on the growth of causal organism of oriental sore. It has a depressant action on the cardiovascular system.

Preparations	–	Same as for *B. aristata* DC.
Cultivation notes	–	Same as for *B. aristata.* Collection of roots is done during autumn and winter. Thousands of truck loads of roots are exported annually out of the state. To avoid exploitation of collectors by contractors and to generate local employment potentials, it is necessary to establish small processing units for Rasount and berberine salts, in the vicinity of the plant populations itself.
Preparations	–	Used in indigenous preparations in the similar fashion as *B. aristata.*
Cultivation notes	–	Through seeds. It has been propagated through cuttings. Work is in progress in Dr. Y.S. Parmar University of Horticulture & Forestry, Nauni-Solan, H.P.

Bergenia ciliata **(How.) Sternb.**

Syn.: B. *Ligulata*. (Wall.) Engl. L.

Family	–	SAXIFRAGACEAE
Local/Common name(s)	–	Pashanbheda, Pakhanbed, Dhoklambu, Patharchat.
Distribution	–	It is found throughout temperate Himalayas from Kashmir to Bhutan 2700-3000 m and in Khasi hills at 1300 m (Anonymous, 1948). In Himachal Pradesh, it is abundant in Sirmour, Shimla, Mandi, Kullu, Kangra, Chamba & Kinnaur districts between 1800-2800 m elevation on moist, rocky slopes and more frequently on rock boulders in ravines and nalahs. (Chauhan, N.S. 1476, 2737).
Description	–	A rhizomatic, perennial herb on rocky and shady places. Rootstock is thick; dark-brown from outside, pinkish from within; covered with dark sheaths of withered leaves. Stem

Bergenia ciliata (How.) Stternb. (Pashanbheda)

short, thick and fleshy; leaves stalked and sheathing at the base; ovate or orbicular, cordate; margin fringed with short and stiff hairs. Flowers are white, pink or purplish in terminal corymbs. Fruits are cone-like capsules.

Flowering/Fruiting – March-April Fruit: June-July.

Parts used – Rhizomes. These are brown in colour, wrinkled and scaly, bearing scars of rootlets. Internal portion are reddish in colour.

Current market rate – Dry rhizomes Rs. 11-13/- per kg.

Main ingredients – Rhizome contains gallic acid, tannic acid, mucilage, wax, glucoside, albumens, starch, calcium oxalate and mineral salts. It is a good source of tannins.

Uses – Rhizomes and root taste bitter, act as astringent, diuretic, tonic, also used in fever and applied to boils and ophthalmia. (Anonymous, 1948). It is an important drug for dissolving kidney and bladder stones.

Domestic uses: The flowers are pickled by Gorkhas and rhizomes are used to cure fever and swollen joints. Leaves are kept on head under the cap for relieving fever. Roots are also consumed in preparing local liqueur.

Preparations – Calcury tablets, Pashanvedadi Kwath, Haritakyadi kwatha etc.

Cultivation notes – It is easy to propagate the plant through rhizome cuttings. Prefers rocky/stony, but moist shady sites with a layer of humus rich soil to facilitate horizontal spreading of the rhizomes. Rhizomes are harvested in autumn, cut into small pieces, dried in sum, graded and stored in sacks.

Bergenia stracheyi **(Hook f. & Thoms) Engl.**

Family	–	SAXIFRAGACEAE
Local/Common name(s)	–	Pashanbheda, Pakhanbed, Kandaro-dhokeru, Gatikpa.
Distribution	–	The plant is reported from Afganistan to Uttar Pradesh between 3300-4500 m in alpine slopes. This is a common alpine species of Western Himalaya. In Himachal Pradesh, it forms vast patches in sub-alpine and alpine areas in Chhota & Bara Bhangal area of Kangra, Pangi-Bharmour of Chamba, Lahaul-Spiti, Kinnaur, Dodra-Kwar, Khashadhar and Rohru Ranges of Shimla districts.
Description	–	A rhizometic herb on rocky and shady places. Rootstock is thick, dark-brown with short stem, thick and fleshy. Leaves are stalked and sheathing at the base; ovate or orbicular, cordate, margin fringed with short and stiff hairs. Flowers are white, pink or purplish in terminal corymbose cymes. Fruit is a conical capsule having minute seeds.
Flowering/Fruiting	–	June-August.
Part used	–	Rhizomes
Current market rate	–	As in case of *B. ligulata*.
Main ingredients	–	Rhizomes contains gallic acid, tannic acid, glucoside, mucilage, wax, starch, calcium oxalate and mineral salts.
Uses	–	*Rhizomes and root:* bitter, astringent, diuretic aphrodisiac, tonic, also used in fever and applied to boils and ophthalmia (Manandhar, 1980). Bark is antiseptic, carminative, and is used to heal up the wounds (Chopra, et al. 1956).
Preparations	–	Pashanvedadi kwath, Haritakyadi kwath etc.

Bergenia stracheyi (Hook f. & Thoms) Engl.

Cultivation notes – Propagation can easily be done through rhizome cuttings.

Betula utilis D. Don.

Family – BETULACEAE

Local/Common name(s) – Bhojpatra, Bhurjapatra, Bhuj, Bhooj, Bhurja-granthi (fungal growth on the trees), Birch.

Distribution – The tree is found in alpine regions of the country between 2900-3900 m altitudes. It forms forests at upper limit of tree growth in Himachal Pradesh in Chansal, Gajjyain Thatch (3900 m); on way to Chandranahan (3800 m); Rupin catchment, Khabal kanda (3200-4000 m); Chhota and Bara Bhangal, Pangi-Bharmour, Parwati valley in Kullu, Churdhar in Sirmour, Chitkul and Sangla Kanda of Kinnaur. In Spiti, a dilapidated patch occurs near Poh bridge (Saran) on way to Mane Lake, composed of rambling/ crooked/deformed trees (Chauhan, N.S. 728).

Description – A moderate sized, deciduous tree upto 20 m tall or often a mere shrub forming the upper most limit of the vegetation. Bark is smooth, reddish brown, with whitish, linear, horizontal, lenticels. Leaves are stalked 5-10 cm long, ovate, unequally serrate, acute, slightly hairy, along the mid ribs, pubescent when young and have a rounded base. Male catkins are few, borne at the tips of long shoots; while the female catkins are short, solitary or in pairs; erect, terminating dwarf shoots. Perianth is 4 parted in male flowers and missing in female flowers.

Flowering/Fruiting – May-July, Fruits: September-October.

Parts used – Papery bark (Bhojpatra) and fungal growth known as Bhurjagranthi, (Bark is thin, papery,

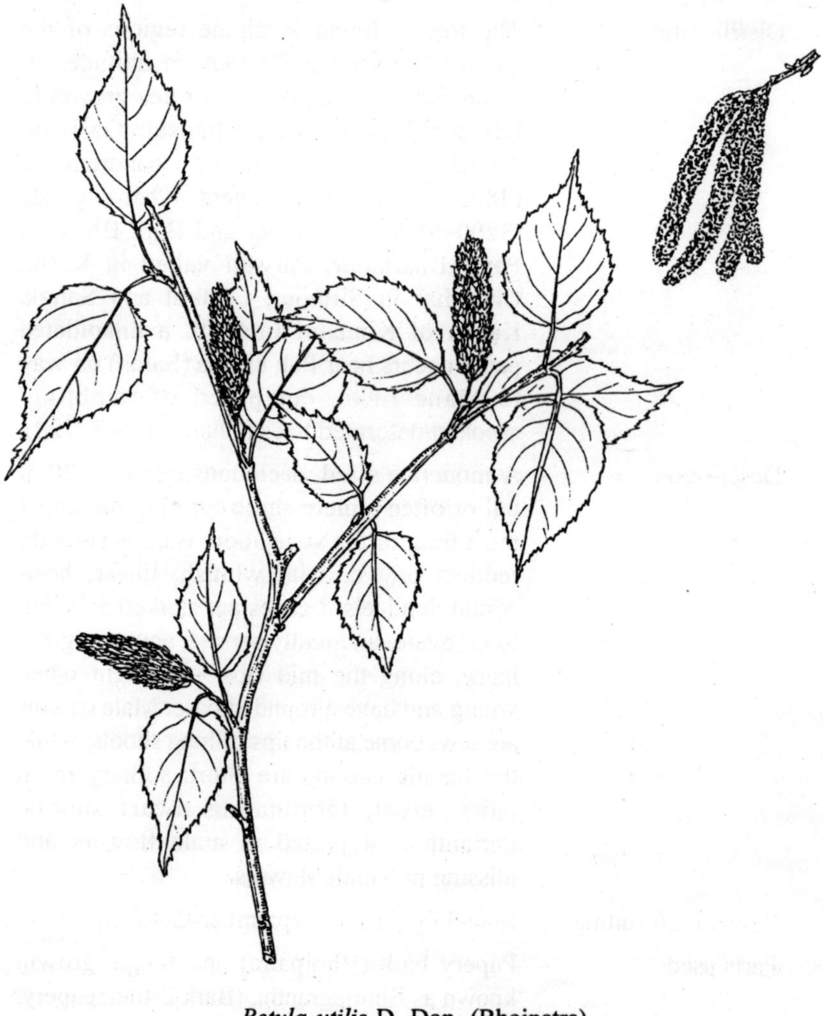

Betula utilis **D. Don. (Bhojpatra)**

shining, reddish white or white. Bhujragranthi is a fungal growth in the form of a black amorphous lump weighing 1.5-1 kg.

Current market rate – Rs. 25-26/- per kg depending upon the quality. For religious purposes, the retail shopkeepers sell it at a fairly high price even exceeding Rs. 100/- per kg.

Main ingredients – The plant contains betulin, lupeol, olenolic acid and acetyle olenolic acid in addition to leucocyanadin in the outer bark and polymeric anthocyanidins in the inner bark. (Malhotra et al. 1990). No analysis seems to have been done on bhurjagranthi—a fungal out-growth found on the tree and used by the natives.

Uses – In Ayurveda, the bark is acrid, pungent, heating, tonic, alexiteric; useful in convulsions, bronchitis, diseases of the blood and the ear, leprosy etc. In Yunani system, the bark is good for ear-ache. The decoction of the bark is used as a wash in otorrhoea and poisoned wounds. It is also antiseptic carminative given in anaemia, cough, obesity, urinogenital diseases, hysteria, toxemia and various infections. Fumes of the birch bark burnt on charcoal are passed into the vagina in order to expel the flacenta. Its paste is also applied on the vaginal wall for the same purpose. Infusion of the bark also has aromatic and antiseptic properties. The thin, papery bark peeling off in horizontal rolls was once used as a writing paper. Its use for literacy purpose is attested by Kalidasa, Sushruta and Varahamihira etc. It also formed the garment of the attendants of Lord Shiva. Even today, bhojpatra is used by Hindu Tantriks for preparation of Talishman and amulets. The fungal outgrowth, "Bhurja-granthi", is locally used against muscular pains and swellings. Its decoction is used as a vaginal douche to avoid conception. Internally, it is given for hidden injuries and is also a good veterinary medicine. Paper bark is also valued for packing, covering umbrellas, thatching, roofing and hukka tubes.

Preparations – Not available.

Cultivation notes – Betula is often cultivated as an ornamental plant in temperate region and thrives best in moist, sandy soils. Propagation is by seeds sown as soon as they mature or sown after stratification in sandy soil which is kept moist and shady. Seedlings should be transplanted when about a year old. It can also be propagated by layers, green wood cuttings under glass and grafting or budding on seedling stock (Bailey and Bailey, 1977).

Bhojpatra of best quality should be collected during March-April, when it comes out in large flakes, dried, packed and stored.

***Bombax ceiba* L.**

Syn: *Bombax malabarica* DC

Family – BOMBACACEAE

Local/Common name(s) – Simul, Semul, Salmali, Semulmusli, Silk-cotton tree or Indian Kapok tree.

Distribution – Throughout India, Burma and Ceylon, ascending upto 1000 m or more. In Himachal Pradesh, it is sporadic to scattered in all parts of the warmer region extending upto 1400 m (Chauhan, N.S. 872).

Description – A large deciduous tree. Stem straight, more or less buttressed at the base when old, covered with large conical prickles when young; branches whorled, horizontally spreading. Leaves digitate, leaflets 5-7, lan-ceolate, oblanceolate or obovatc, acuminate, more or less coriaceous, glabrous. Stipules are small, caducous. Flowers are fleshy, Calyx very fleshy, cup-shaped, bright silky hairy inside and dark-brown outside. Petals-crimson or orange, oblong, white tomentose outside. Stamens usually 5 adelphous (aggregated in 5 different bundles); filaments numerous, the innermost forked. Ovary 5-celled; style 5-lobed at the apex Capsule 12.5-17 cm, oblong ovoid; seeds are obovate, smooth, surrounded by white silky hairs.

Flowering/Fruiting – Flowering: January to March.
Fruiting: April-May.

Parts used – Root, bark, flowers and fruits.
Gum (MOCHRAS). Mochras is not a normal juice, but it is the product of a diseased action resulting in proliferation of the parenchymatus cells of the bark which increase in size and exude whitish mass turning red and finally dark brown.

Current market rate – Not available.

Main ingredients – From petroleum ether extract of the bark three compounds (two of naphthaquinone type and a lactone) were isolated. In addition to the two napthaquinones already isolated, 3 methyl 1-5 isopropyl, 1-6 hydroxy 7-methoxy naphthalene and 1-8 lactone have also been isolated. Petroleum extract yielded lupeol and naphthalactone. Gum contains catechu-tannic acid. Seeds contain crude fat having stearin. Roots of young plants (semul-musli) contains proteins, arabinose, glactose, pectose matters, starch and mucilage.

Uses – Various parts of the plant are used in small pox, bleeding gums, toothache and carries,

Bombax ceiba Linn. (Semul)

sores in mouth, pain in leg, fever enlarged spleen, atrophy, emaciation, rheumatism, spermatorrhoea, cholera, pneumonia, pleurisy, intercostal neuralgia, leprosy and rinderpest disease. Semul-musli (root of young sapling) is stimulant, tonic and forms the chief ingredient in the musla-semul or semul-musli, a medicine which is aphrodisiac and given in impotence. Root and bark are emetic, Gum known as Mochras is astringent, tonic haemostatic, aphrodisiac and useful in diarrhoea and dysentery. Silk-cotton is a reputed filling material for life jackets and sleeping bags. Young flower buds are pickled and also cooked as vegetables. Flowers and fruits are used in snake bite. Paste of flowers and also of leaves is applied in cutaneous troubles. Seeds are used in gonorrhoea, chronic cystitis and catarrhal affections. Young tap root is used against dysentery.

Preparations – Salmali-ghrita, Laghugangadhar-churna and several aphrodisiac, restorative and expectorant preparations.

Cultivation notes – Natural reproduction is abundant in some regions. Artificial regeneration of the tree is by direct sowing or by transplanting one years old nursery raised seedling and by stump planting. The tree starts fruiting at the age of 8 years. The fruits are collected in March-April before they dehisce and the cotton is removed. Soaking the seeds in water for about 12 hours is reported to hasten germination. The growth of the seedling is fairly fast. The sapling have to be protected from cattles, weeds, insect pests and diseases. The roots of young saplings are harvested and used as semul-musli in medicines.

Bunium persicum **(Boiss.) Fedt.**

Syn. *Carum bulbocastanum* Koch.

Family	–	UMBELLIFERAE/APIACEAE
Local/Common name(s)	–	Kala-jirah, Umbu, Siahzira, Krishnajeerak.
Distribution	–	Native of Baluchistan, Afghanistan, Kashmir and Himachal Pradesh. In Himachal Pradesh, it is found in Chamba, Kinnaur, Lahaul and Spiti districts. It is mostly found between 2000-3500 m under domestication and Sangla in Kinnaur has a reputed farm for saffron and jira (Chauhan, N.S. 3273).
Description	–	A glabrous herb with globular tuberous roots; profusely branched, attaining a height of 50-60 cm and a spread of 40-50 cm. Leaves are dissected, having narrow segments. Flowers are small, white, in compound umbels. Fruit are darkish-brown, ribbed with pointed ends and have deep aroma.
Flowering/Fruiting	–	May-July.
Parts used	–	Fruits (commonly called as seeds). Kalazira is a low volume, high value, non-perishable commodity. Its fruits commonly called seeds are used as prized spice in cooking for flavouring of dishes. In the indigenous systems of medicines, the fruit are regarded as stimulants, carminative and are useful in curing diarrhoea, dyspepsia, fever, flatulence, stomach-ache, haemorrhoids and obstinate hiccup.
Current market rate	–	Rs. 300-400 per kg.
Chemical contents	–	Contains 2.5% essential oil having 45-65% carvone, a mixture of ketone, carvone, terpene and traces of carvacrol and cuminaldehyde (40.6%).

Preparations – Hingwashtak churna, Lavanbhaskar churna, Yograj-gaggulu, Vidangadi-loha etc.

Cultivation – Kalazira is propagated through seeds initially, but its subsequent growth and production is maintained by tubers/bulbs, which are formed 10-15 cm deep in the soil by the germinating seedlings. It requires four years to produce new seed crop, while raising it through seeds. The seeds require pre-chilling for germination and thus, the seeds sown during October-November germinate after the melting of snow in March. Plants raised through 7-8 years old tuber produced more branches, an average height of 50-60 cm with 40-50 cm spread. It prefers sandy-loam soils rich in organic matter, high hill, temperate, dry regions between 1850-3100 m elevation experiencing 1-5 m of snow during winters and very low rainfall in summer. Seeds and tubers both require long chilling period for better germination, sprouting and initiation of desirable floral primordia. Low rainfall during summer at flowering and seed setting stages contributes towards high yield, better flavour and quality of seeds. 1-1.5 kg of seeds are required at first sowing and re-seeding in subsequent years require 200 gm seed per hectare to maintain optimum plant population.

Its related species, *Carum carvi* Linn. (caraway) also commonly known as kalazira is of inferior grade to *B. persicum*.

***Butea monosperma* (Lamk.) Taub.**

Family – LEGUMINOSAE (FABACEAE/PAPILIONACEAE)

Local/Common name(s) – Dhak, Palas, Palash, Tesu, Palah, Flame of the forest.

Distribution	– Common throughout the greater part of India upto 1300-1400 m except in very arid parts. In Himachal Pradesh, the tree is found sporadic/scattered or rare in Hamirpur, Kangra, Chamba, Mandi, Bilaspur, Solan, Sirmour districts. More frequently met in Nalagarh (Palah-da-jungle) (Chauhan, N.S. 631, 1213).
Description	– A moderate sized, deciduous tree; bark fibrous, bluish grey or light brown outside, exuding a red juice when cut. Leaves are pinnately 3-foliolate, rachis 12-22.5 cm long, leaflets hard, rigidly coriaceous, glabrescent above and silky tomentose beneath. Flowers are in terminal, thick racemes appearing before the leaves on naked branches, giving an appearance of fire flame from a distance. Calyx is dark brown, hairy outside and fleshy, grey-silky inside. Petals are bright orange-red; stamens, 2-adelphous; ovary shortly stalked. Pod 10-15 cm, bearing single seed at the end of the pod. Seeds resemble the old copper coin, are flat, 3.8 cm long and 2.5 cm broad and 1 mm thick. Testa (seed coat) is thin, smooth veined and dark reddish brown.
Flowering/Fruiting	– April-May/June-July.
Parts used	– Seeds, gum, leaves, flower, bark. (It's gum is known as 'Gum Kino').
Current market rate	– Not available.
Main ingredients	– Flower of *B. monosperma* have seven flavonoid glucosides, butrin, isobutrin; three glucosides—coreopsin, isocoreopsin and sulphurein; two new compounds—monospermoside and isomonospermoside. Gum and bark have been reported to contain tetramers of leucocyanidin (CCRAS, 1990: 158-159). Root bark contains β-sitosterol, leucoanthocyanidin, amyrin, betulinic acid, stigma-sterol

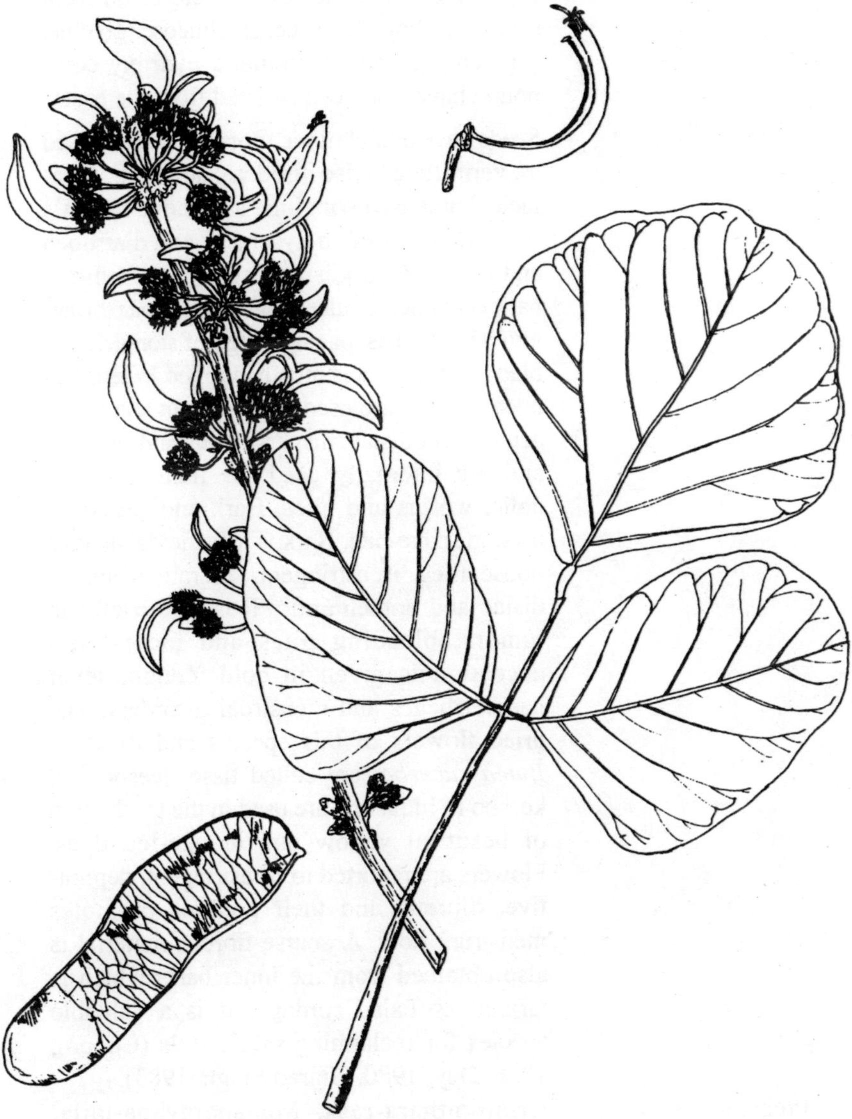

Butea monosperma (Lamk.) Taub. (Dhak, Palash)

and an active principal palasonin. Leaves contain glucosides and a yellow tasteless oil is obtained from the seeds (yield of oil from seeds are about 18 percent). Glucose, glycine, a glycoside and an aromatic hydroxy compound have also been isolated from the roots.

Uses – Seeds have anthelmintic property and are used as vermifuge. Also used as aperient, rubefacient and used with lime to remove Dhobi's itch. Gum is used in dysentery and diarrhoea and also as aphrodisiac. Fresh gum from the bark is applied to ulcers and septic soar throat; given in phthisis, haemorrhage of stomach and bladder. Infusion is locally applied in leucorrhoea. Leaves are used as astringent and tonic; act as aphrodisiac, disperse boils and pimples and are internally used for flatulence and colic, worms and piles. Bark and seeds are used in snake-bite. Bark is insecticidal against house flies, is astringent, alterative, aphrodisiac and anthelmintic. It is also useful in tumors, bleeding piles and ulcers. It's decoction is given in cold, cough, fever haemorrhages and menstrual disorders. The dried flowers of this species and those of *Butea superba*, are called tisso, teesoo and keesoo in India and are used in the production of beautiful yellow and orange-red dyes. Flowers are reported to be astringent, depurative, diuretic and their poultice promotes menstrual flow. A coarse fibrous material is also obtained from the inner bark, which is termed as Palas cordage. It is a valuable species for reclaiming saline soils (Chopra, 1956, Dey, 1980, Umrao Singh, 1983).

Preparations – Krimikuthara-rasa, Mahanarayana-taila, Janma-ghutti, Palasabijadi-Churna, Palasa-ksharaghrita.

Cultivation notes – It is frost hardy and drought resistant and prefers saline soil. Propagation is by seeds or through root suckers (Bailey and Bailey, 1977).

Calotropis gigantea (Linn.) Ait. f.

Family – ASCLEPIADACEAE

Local/Common name(s) – Aak, Madar, Akanda, Arka, Calotropis.

Distribution – Common throughout India in warm and dry places. In association with *C. procera* (Ait) R.Br., it is available in Nalagarh, Paonta valley, Bilaspur, Hamirpur, Una, Kangra, Mandi, Chamba in sandy ravines and waste places upto 1000 m elevation (Chauhan, N.S. 1828, 1859, 1967).

Description – A small to large shrub, bark ash coloured, wrinkled; leaves are thick, ovate-oblong or broadly oblong, acute covered with white tomentum. Flowers are white in *C. gigantea* and pink with purple spots in *C. procera.* The root bark occurs in short, corky, deeply vertically cleft pieces. The outer surface is yellowish grey, soft and corky, fissured longitudinally and can be easily separated from the middle cortical layer, which is white, friable and transversed by narrow, brown liber rays. The plant yields white milky latex *C. procera,* is smaller in size and more commonly found in the state as a weed of waste places.

Flowering/Fruiting – Throughout the year, but chiefly in the cold season, i.e. November-January.

Parts used – Roots, leaves, latex and powdered flowers.

Current market rate – Not available.

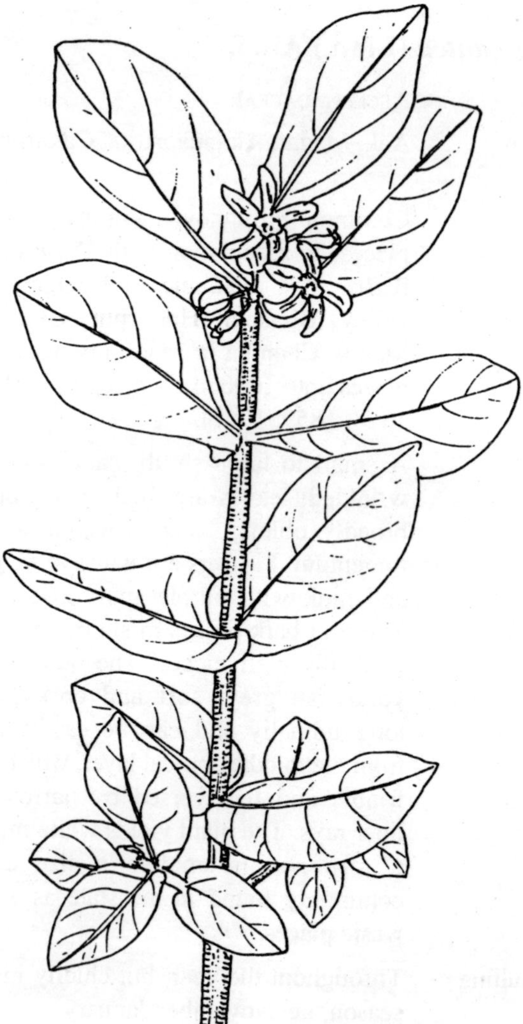

Calotropis gigantea (Linn.) Ait. f.

Main ingredients – Leaves and stalks contain calotropin and calotropagenin. Latex contains uscharin (0.45%); calotoxin (0.15%) and calactin (0.15%). Lethal dose of the first two are 0.5 microgram and 0.7 microgram per g of frog respectively. (Chopra et al. 1956). From the stem bark, root bark, flowers and latex, β-sterol, taraxasterol, α & β amyrins; triterpenoids and a mixture of n-hydrocarbons have been isolated (CCRAS, 1990:29).

Uses – Root bark is useful in dysentery and is used as a substitute for Ipecacue; acts as diphoretic and emetic. Tincture of the leaves is used in intermittent fevers. Latex is irritant and in combination with *Euphorbia nerifolia* is used as purgative. Powdered flowers are useful in colds, cough, asthma and indigestion.

Extract of roots and leaves is anti-cancer, used in lupus, tuberculous leprosy, syphilis, rheumatism, scorpion sting and night blindness. Plant is also useful in smallpox, leg and chestpains, ailments of spleen, dropsy, epilepsy, tetanus, delirium, scabies, ringworms, hydrocele, cholera, neuralgia, pneumonia, dysentery etc.

Preparations – Arka-lavana, Arka-taila, and Arkakshara.

Cultivation notes – Propagation can be done by cuttings raised in sand. Seeds freed of silky-white floss can be raised in nursery in warm weather conditions.

***Cannabis sativa* Linn.**

Family – CANNABINACEAE

Local/Common name(s) – Bhang, Charas, Ganja, Hashish; Vijaya, Hemp. Bhangolu (seeds).

Distribution – It is widely distributed in the temperate world. It is also cultivated in temperate and tropical regions particularly in India for narcotic drugs, cordage fibres and seeds used for culinary purposes. Frequent in all parts of the state upto an elevation of 2500 m. (Chauhan, N.S. 112, 179, 589, 828, 1014, 1254, 1439, 1634, 1739, 2020 and 2627).

Description – An annual, tall herb going upto 5 m under cultivation. Stem is slender, grooved, finely tomentose; leaves are alternate, palmately divided; 5-20 cm long, toothed, rough above, pale downy beneath; leaflets 7-11; flowers dioecious, i.e. borne on different plants; are yellow-green. Male flowers are clustered in short, drooping panicles. Perianth is of 5 parts. Female flowers are sessile, erect. Perianth is a single, entire, leaf enclosing the ovary. Achene small, rounded, dark grey in colour.

Flowering/Fruiting – June-October.

Parts used – Charas, bhang, ganja etc. obtained from the flowering and fruiting parts of the plant. Charas, the narcotic resinous substance appearing on the leaves and inflorescence is collected from female plants especially the cultivated ones. Bhang consists of dried flowering/fruiting tops of female plant with resin. Ganja is the dried flowering tops of female plants without leaves.

Current market rate – Resin (Extract and flowering tops); in the form of powder or dry leaves have a lucrative, but under world market and fetch a handsome price under the names of Bhang, Charas, Ganja, Hashish and marijuana.

Main ingredients – Resin and volatile oil, cannabinol, pseudo-cannabinol, cannabinin, canin, cannabine, cannabinon, gum, sugar and potassium nitrate.

Uses – On the basis of the use, the Indian Hemp plant is popularly known as the, 'leaf of

Cannabis sativa Linn. (Vijaya, Bhang)

illusion', increaser of pleasure and cementer of friendship. The latest convention presents the definition of cannabis as 'Flowering or fruiting tops of the plant from which resin has not been extracted by whatever name they may be designated' (Coza, 1972). Ganja is the dried resinous flowering tops of the female plant free from the leaves. It is smoked with or without tobacco. Charas is the narcotic, resinous aerial exudation from all the parts of the plant and is collected chiefly from the female plants. Bhang consists of dried leaves and flowering tops. Bhang is least narcotic, whereas charas is the chief narcotic drug. Hashish and Marijuana are the other names of narcotic products obtained from the plant.

In action, the drug is anodyne, anti-spasmodic, anaesthetic, aphrodisiac, heating, appetizer, astringent, digestive, exciting, exhilarant, hypnotic, induces sleep, sedative, narcotic and toxic (Dastur, 1962; Dey, 1980). Seeds are fed to the birds and milk cattle as a lactagogue. The semi-drying oil is used as luminent, in paints, varnishes and soap. The plant yields a very useful fibre. Paste of fresh leaves is used to resolve tumors. Leaf juice removes lice and dandruff. Leaf powder is a useful dressing for wounds or sores. Seeds are roasted and eaten as culinary by the hill people. Bhang leaves mixed with other ingredients make sweet drink which is refreshing in summer season.

Preparations – Madnanandmodaka, Jatiphaladi-churna, Vijaya-vatika, Vedantak-rasa, Kameshwar-modak.

Cultivation notes – The plant is grown on rich, loamy soil in mild humid climate in well prepared fields liberally manured. The seeds are sown either broadcast or by drills. The seed rate varies from 20-30 kg per acre. Thick sowing gives a crop with short stalks and fine fibres. For seed purpose, the plants are thinned out after germination. The crop is harvested as soon as lower leaves are shed and the tops of stalks and flowers turn yellow. Harvesting at the time, the plants are about to flower, gives the optimum yield of high quality fibre. The fibre is extracted from stalks either by water wetting or dew wetting. An average crop yields 2-3 tons of dry stem per acre. (The Wealth of India, 1950). Bhang, Ganja and Charas are collected at full flowering stage, just before pollination for optimum quality yield. In the erstwhile Bushaher State, the bhang (Charas) collected on 20th of Bhadron i.e. 2nd week of September is considered most qualitative, potent and useful drug for home remedies.

Capsicum annuum **Linn.**

Family – SOLANACEAE

Local/Common name(s) – Lalmirch, Lanka mirch, Red Chillies.

Distribution – Cultivated as kitchen garden crop in Himachal Pradesh.

Description – An annual, much branched herb with ovate, elliptic, pointed leaves. Flowers are white and the fruit, a berry is of variable sizes and shape, red on ripening and many seeded. Most of the colouring matter is present in the outer layer, which also contains some fatty oil.

Flowering/Fruiting – May-November.

Parts used – The ripe and dried fruits.

Current market rate – Rs. 70-80/- per kg.

Chemical contents – The chillies contain capsicin, capsaicin and solanine. The pungent principle is capsicin and its maximum concentration is found in the inner walls of the fruit. Also contains ascorbic acid.

Properties and uses – In action, the fruit is carminative, digestive, hypnotic (to some extent), stimulant and stomachic. Useful in cholera, diarrhoea, dysentery, filaria, hoarseness, rheumatism, sore-throat, tonsillitis, pharyngitis, neuralgic affections and yellow fever. A pill made of capsicum, ginger and rhubarb is carminative and may be advantageously employed in atonic dyspepsia (Chopra, 1958: 667). As an antidote to dog poison, it is supposed to be the best drug, as it subsides the pain and stops puss formation. It is one of the most important food dyes and a condiment of daily use.

Preparations – Visham-jwaraghnivati, various liniments and embrocations.

Cultivation – The crop is cultivated through seeds. Seedlings are raised in nursery beds during March-April and transplanted in the field during May-June in Himachal Pradesh.

Carum carvi Linn.

Family – UMBELLIFERAE (Apiaceae)

Local/Common name(s) – Shia-Zira, Carway, Shingujira, Mawo, Dru.

Distribution – A herb, native to Europe and West Asia and is now cultivated in Bihar, Orissa, Punjab, Bengal, Andhra Pradesh and in the hills of Kumaon, Garhwal, Kashmir and Chamba.

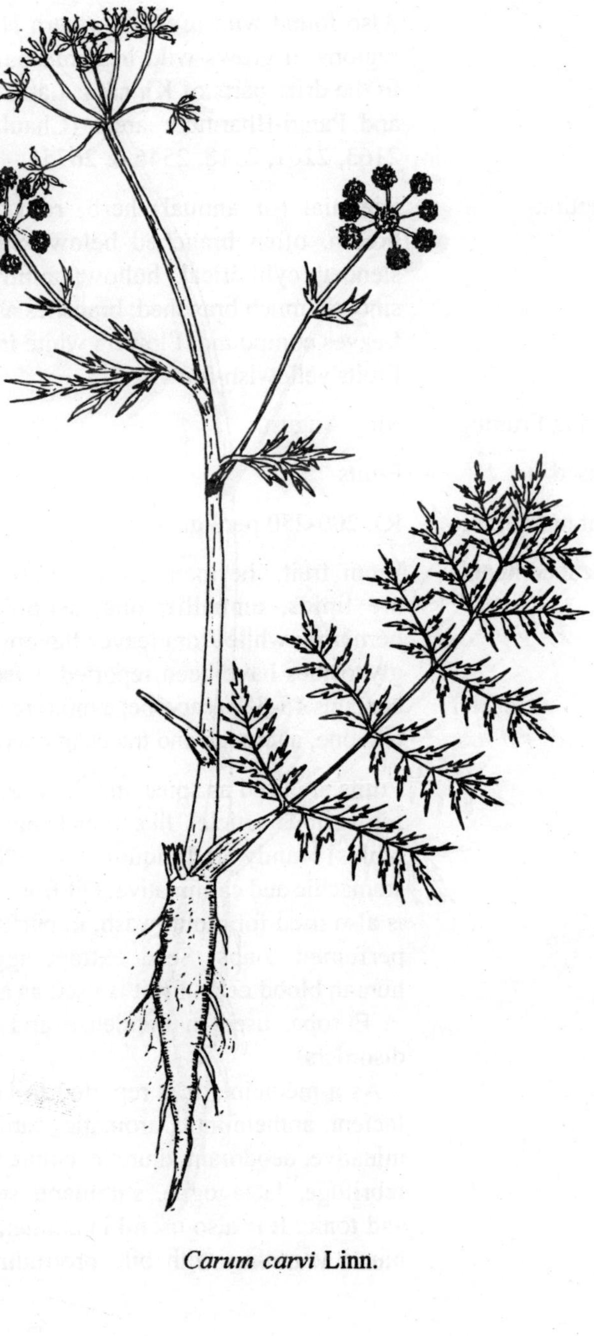

Carum carvi Linn.

		Also found wild in the Northern Himalayan regions. It grows wild in Himachal Pradesh in the drier parts of Kinnaur, Lahaul & Spiti and Pangi-Bharmour areas (Chauhan, N.S. 2163, 2201, 2518, 2546 & 2675).
Description	–	Biennial (or annual) herb, root tapering, brown, often branched below. Stem erect slender, cylindrical, hollow; faintly striate, smooth, much branched; branches ascending. Leaves compound. Flowers white in umbels, Fruits yellowish-brown.
Flowering/Fruiting	–	May-August.
Parts used	–	Fruits
Current market rate	–	Rs. 200-250 per kg.
Chemical contents	–	From fruit, the main constituents extracted are lipids, umbelliferone, scopoletin and herniarin; while from leaves flavonoides and glycosides have been reported. Essential oil contains 45-65% carvone; a mixture of ketone carvone, a terpene and traces of carvacrol.
Uses	–	Fruits are used as spice and flavouring agent for various articles like bread, meat, vegetables brandy and liquors etc. These are stomachic and carminative. Oil from the fruits is also used for mouthwash, in perfumes and perfumed soaps. Seed extract agglutinate human blood cells. Fruit is used as antiseptic in Europe; used in flatulence and stomach disorders. As a medicine, it is reported to be abortifacient, anthelmintic, aromatic, cardiac, carminative, deodorant, diuretic, emmenogogue, febrifuge, lactagogue, stimulant, stomachic and tonic. It is also useful in ammenorrhoea, blood vomiting with bile protruding piles,

		rheumatism, intermittent fevers and worm infestation.
Preparations	–	As a flavouring agent, it is used in various preparations: Hingwastak-churna, Lavan-bhaskar-churna, Yograj-guggulu, Vidangadi-louha etc.
Cultivation	–	It is cultivated through seeds.

Cassia absus Linn.

Family	–	LEGUMINOSAE (Caesalpiniaceae)
Local/Common name(s)	–	Chaksu
Distribution	–	An erect, branched, viscidly glandular, biennial herb, 15-45 cm tall. Leaves compound, leaflets 4, ovate, 2.5-5 cm; obtuse, a small, awl-shaped gland between each pair. Flowers few, reddish-yellow, 3-4 cm in diameter appearing in short racemes. Stamens 5, equal, perfect. Pod hairy, flat, seeds 6.
Flowers/Fruiting	–	August-September.
Current market rate	–	Rs. 55-65 per kg.
Chemical contents	–	Roots contain anthraquinones, chrysophanol, aloe-emodin; alkaloids-Chaksine and isochaksine; leaves contain alkaloids quercetin and rutin. Seeds have sitosterol glycosides.
Properties and uses	–	Powder of leaves is used in ulcer dressings. It is one of the constituents of Unani drug for leprosy. Seeds are also astringent; used in ringworm and other skin affections, in conjuctivities and ophthalmia. Mucilage from leaves is an active ingredient of medicaments useful for soothing digestive system and mucous tissues.

Preparations – Used in various formulations in Indian systems of medicines. In Unani, it enters into formulations for leprosy; whereas in Ayurveda, it is an ingredient of formulation concerning eye diseases.

Cultivation – It grows as a weed and bears profuse seeds. Easy to cultivate through seeds.

Cassia fistula Linn.

Family – LEGUMINOSAE (Caesalpiniaceae)

Local/Common name(s) – Amaltas, Aragwadha, Indian Laburmum, Purging cassia, Cassia pod., Aali-ra-tat.

Distribution – Found throughout India upto an altitude of about 1300 m and is more common in most of the evergreen forests. The tree is much favoured for planting in road side avenues and in gardens. In Himachal Pradesh, it is found in the forests in Nahan, Kunihar, Nalagarh, Solan, Mandi, Bilaspur, Una, Hamirpur, Kangra and Chamba divisions and can grow upto 1200 m elevation (Chauhan, N.S. 1885).

Description – A medium sized tree, nearly 15 m tall. Leaves pinnately compound; leaflets are 4 to 8 pairs, large, ovate-lanceolate, leathery, entire, acute. Flowers are bright yellow in long and drooping racemes appearing like Japanese Lantern. Fruit is a 20-40 cm × 3 cm cylindrical pod, dark-brown with a strong woody stalk having a blunt tip and the seeds are embedded in black, sweetish pulp.

Flowering/Fruiting – April-July. Fruits: August-November.

Parts used – Dried fruits, root bark, leaves.

Current market rate – Cassia pods Rs. 8/- per kg.

Main ingredients – Pulp, mucilage, proteins, volatile oil. Leaves contain anthraquinone derivatives and very

Cassia fistula Linn. (Amaltas)

little tannin. Root bark contains tannin, phlobaphenes and oxyanthraquinone substance. Pulp contains rhein, the major anthraquinone derivative, small amount of volatile oil and three waxy substances.

Uses – Used as emetic, febrifuge, laxative, reliever of thoracic obstructions and purgative. It is useful in boils, pustules, leprosy, ringworm, colic, gripping, flatulence, dyspepsia, constipation, indigestion, fever and heart diseases.

Preparations – Maha-Marichayadi-taila, Mahamanjisthyadyarista, Rasnadi-kwatha-Yoga, Aragva-dhadi-kwatha, arishta, leha and taila (CCRAS, 1990, Dey, 1980).

Cultivation notes – The plant is cultivated in many parts of India. It prefers light, porous soil and is immune from browsing by cattle and goats. The cheapest method of cultivation is by sowing seed in furrows made 5-12.5 cm apart and 10-15 cm deep. The furrow is closed by another furrow ploughed alongside. Broad cast sowing by close ploughing in one direction costs about twice, but gives 2.7 times as many seedlings in comparison to the first method in a given area. Thinning, weeding, is done in the first season of cultivation. The tree has good coppicing ability. The coppiced buds send out large number of shoots and a new harvest can be taken after a year. (Anonymous, 1950).

***Cassia occidentalis* Linn.**

Family – LEGUMINOSAE (Caesalpiniaceae)

Local/Common name(s) – Chakunda, Chakramard, Bara elwan.

Distribution – A common weed scattered from Himalayas to West Bengal, South India, Burma and Ceylon. Common in Himachal Pradesh in all the warmer parts upto 1350 m. (Chauhan, N.S. 858).

Description – A diffuse undershrub with loosely spreading branches, often growing 1-2 m tall. Leaves pinnate, elliptic-lanceolate, acuminate. Flowers yellow with reddish veins. Pod compressed, glabrous; 7.5-10 cm long and 5-6 cm broad.

Flowering/Fruiting – July-September.

Parts used – Leaves, Seeds and Roots.

Current market rate – Rs. 35-50 per kg. (Seeds)

Chemical contents – Sennosides have been reported from pods. Seeds contain N-Me-morphidine; anthraquinone glucoside; anthraquinones, polysaccharides; galactomannan. Leaves contain dianthronic heteroside; pericarp-apigenin, and roots have emodol, phycion and phytosterol. Flowers yield phycion, and its glucocides, emodin, and beta-sitosterol. Xanthone, casiolline have also been reported from the plant.

Properties and uses – It is one of the ingredients of 'Liv-52' used in hepatitis. Senna toxicity has been reported in cattle. Roots are used in skin diseases. Leaf and fruits are anthelmintic and the seeds are used in diabetes. Aerial parts and roots have exhibited antiviral and spasmolytic activity.

Preparations – Used in Ayurvedic formulations, Liv. 52 etc.

Cultivation – Can be cultivated with ease through seeds on any marginal and wastelands.

Cassia tora **Linn.**

Family – CAESALPINIACEAE (Leguminosae)

Local/Common name(s) – Panwarh, Chakunda, Chakramard, Reli, Elwan, Ringworm plant.

Distribution – Distributed throughout India, ascending to 1400 m. In Himachal Pradesh, it is found as a weed of waste places, roadsides, field borders etc. in all the warmer parts extending upto 1300 m elevation (Chauhan, N.S. 1330, 1848 and 3456).

Description – An annual, erect, shrub like herb, 15 cm to 1.5 m tall. Leaves pinnate; leaflets 6, hairy, ovate 2.5-5 cm, acute an awl shaped, yellow gland between the lowest pair. Flowers yellow, in pairs on short, axillary stalks. Stamens 10; lower 7, nearly equal, perfect; 3 upper minute, abortive. Pod linear, 15-20 cm; 4-angled, seeds numerous.

Flowering/Fruiting – August-October.

Parts used – Whole plant, Seeds.

Current market rate – Rs. 35-70 per kg.

Chemical contents – Leaves and stems contain d-manitol, myricyl alcohol and beta-sitosterol. Roots have anthraquinones and beta-sitosterol. Seeds yielded beta-sitosterol, emodin, subrofusarin, 1, 8-dihydroxy anthraquinone, chrysophanic acid, rein like aglycones, cassiaside, rubrofusarin, torosachrysone, quercetin and its analog, cassialactone etc. Pods contain sennosides.

Properties and uses – Leaves are pounded and applied on cuts and act like tincture of iodine. Seeds with turmeric and mustard oil are applied against eczema. Tender leaves are taken internally to prevent skin diseases, while the infusion acts as vermicidal. It has a curative effect in ringworm in cattle. Ethanolic extract (50%) of plant is having antiviral, spasmolytic and diuretic properties. Also used against epilepsy, scabies & sores. Seeds-ethyl alcohol extract; spasmolytic. Boiled with tea, it is taken for colds. Powder is given in abnormal delivery. Useful in fevers during teething trouble, itchy eruptions, leprosy, psoriasis, tumors, ulcers and hastening of suppuration.

Preparations – Solid concentrates of sennoside A & B. from pods. Various formulations in indigenous medicines for leucoderma, ringworm etc. i.e. Dadrughan-vati.

Cultivation – It is easy to cultivate the plant through seeds, which have high germination rate.

Cedrus deodara (Roxb.) Loud.

Family – PINACEAE

Local/Common name(s) – Devdaru, Cedarwood, Devdar, Kewli, Kelon.

Distribution – As one of the dominant conifer, it is distributed in most parts of the state between 2200-2800 m. Pure forests occur in Shimla and Chail catchment areas.

Description – A large, handsome, evergreen coniferous tree with spreading branches and dark green foliage. Needles arise in tufts at the ends of short branchlets. Wood is aromatic, light, yellowish brown in colour; taste oily and resinous. Male cones cylindrical 5-12 cm erect; female cones ovoid or ovoid-cylindric, 10-12 cm × 7.5 × 10.0 cm, dark brown when ripe (October-November) while the young cones are bluish purple. The tree reaches a gigantic diameter and height in the vicinity of religious places i.e. near temples etc. especially in Rantari (2200 m) and Kharshali (2600 m approx.) and such trees are conserved from religious point of view.

Flowering/Fruiting – September-November.

Parts used – Wood and the oil.

Current market rate – Cedar wood oil Rs. 125-150/- kg.

& Wood Rs. 14 per kg.

Chemical contents – Wood oil contains oleo-resin and essential oil while the needles contain ascorbic acid.

Properties and uses – The wood is carminative, diaphoretic and diuretic; given in fever, flatulence, heart palpitations, paralysis, pulmonary troubles, prolapse anni, dropsy and in urinary diseases like gravel. In ascites, it is given in combination with *Achyranthes aspera* etc. In gonorrhoea, syphilis, gout and rheumatism, the decoction is given as a powerful alterative with turmeric and guggulu. It's paste is applied to indolent swellings. The tar is used as a favourite alterative and given in chronic skin diseases and in large doses, it is used in leprosy and also applied externally to ulcers.

Preparations – Devdarvadi-kwath, Rasnadi-kwath; Sudarshan churna, Punarnava-mandoor, Laghu-Vishgarbha-taila, Khadirarishta Chandraprabha-vati etc.

Cultivation – Plants are raised from seeds in polythene bags and later transplanted in the field in pits at proper spacings.

Centella asiatica (Linn.) Urban

Syn: *Hydrocotyle asiatica* L.

Family – UMBELLIFERAE

Local/Common name(s) – Brahmi, Brahmamanduki, Budhbrahmani, Mandukparni.

Distribution – This plant grows in moist places throughout India. It is commonly seen in marshy banks of rivers, streams, ponds, in irrigated lawns and fields etc. below 1300 m elevation in most parts in Himachal Pradesh. (Chauhan, N.S., 251 RI Nursery, JNR).

Description – The plant trails on ground and its creeping stems bear roots on their nodes. Leaves are long stalked, 2-4 cm in diameter, sometimes more, rounded to broad or kidney-shaped, cordately based, deeply toothed margins.

Flowers are minute, pinkish and red, 3-6 in a cluster. Fruit is small like a grain of barley with 7-9 ridges. (Jain, 1968, Dey, 1980).

Flowering/Fruiting – July-September.

Parts used – Herbage, mostly leaves (Fresh and dried leaves). Stem, roots and seeds are also used. Collection is mostly done during February-March.

Current market rate – Rs. 18-25/- per kg.

Main ingredients – Contains bitter substance, a glucoside-asiaticoside, essential and fatty oil, sitosterol, tannin and resinous substance besides an alkaloid hydrocotylin. A bitter principle vallarine, pectic acid and resin are present in the leaves and roots of the plant. The plant also contains ascorbic acid in a concentration of 13.8 mg per 100 g. (Chopra et al. 1956; Dymock et al. 1976 (Rep), Jain, 1968).

Uses – The leaves or entire plant parts are boiled in water and this decoction is given in the treatment of leprosy. The glucoside asiaticoside is considered responsible for its use in leprosy. Experiments on animals have confirmed its property of inducing fast growing of skin, nails and hairs (Jain, 1968).

The plant is also considered useful in epilepsy and is a tonic for brain. (Chopra et al. 1956). It is a reputed tonic for tiresomeness after mental work and is specially used for loss of memory. It has also been reputed to improve the power of speech and poetic imagination (Lakshmipathi, 1973). It is a useful remedy for mentally retarded children.

Preparations – Brahmighrita, Brahmi-rasayana, Sarswatarishta, Brahmi-panak and in addition Brahmi-amla-taila etc.

Cultivation notes – *C. asiatica* is reported to be a cover crop in tea and rubber plantations. (Anonymous, 1950). It is cultivated/planted by means of suckers. Spreading branches bear roots at modes and new plantlets can be separated and planted closely in a well moistened soil. It carpets the ground very soon in optimum moisture.

Chenopodium ambrosioides Linn.

Family – CHENOPODIACEAE

Local/Common name(s) – Indian chenopodium, Sugandhvastuk, Banbathu, American wormseed.

Distribution – Native of West Indies and South America, the plant is widely distributed in South India, Bengal, Maharashtra, Kashmir etc. Frequent in waste places, roadsides and field borders in Himachal Pradesh below 1600 m elevation (Chauhan, N.S. 193).

Description – An erect, aromatic, much branched herb, upto 100 cm tall. The flowers are minute and clustered in pale green leafy spikes. The fruits are somewhat globular, slightly compressed with a thin pericap surrounding the seed. The entire plant possesses a comphoraceous odour. A volatile oil of medicinal value is found in the glandular hairs, specially of the pericap of the fruit.

Flowering/Fruiting – July-August/October-November.

Parts used – Fresh or half dried aerial parts (flowering and fruiting parts) and essential oil.

Current market rate – Not available.

Chemical contents – Leaves contain kaempferol-7 shamnoside and ambroide. Contents of calcium, phosphorous, oxalates and essential oil have been reported from the plant.

Properties and uses – The plant is an anthelmintic and its oil is used in medicines. The oil is effective against many forms of intestinal parasites. In hook-worms, 4 CC. of oil with tetrachloro-ethylene in the ratio of 1:3 has given good results. Ascaris infection has been effectively treated by a combination of chenopodium oil (1 CC) and santonin (5 grains). In combination with carbon tetrachloride, it has been recommended for the treatment of mixed intestinal infections. The oil has been found useful in the treatment of amoebic dysentery. The oil should however, be used with extreme care, otherwise it leads to serious toxic affects. It should not be given to children, mal-nourished persons and patients with renal, cardiac and hepatic diseases with ulceration of the gastro-intestinal tract (Wealth of India, II: 128). Chenopodium oil is used against intestinal worms in veterinary medicines also. In Mexico, the powdered seeds are given against intestinal worms. Shoot and root extract have shown nematocidal properties while the essential oil has exhibited antimicrobial and antifungal activity against human pathogenic fungi.

Preparations – Wormicidal formulations.

Cultivation – Since the plant grows as a common weed of the waste places; produces profuse seeds, it regenerates naturally. Propagation is through seeds.

Cichorium intybus Linn.

Family – COMPOSITAE (ASTERACEAE)

Local/Common name(s) – Kasani, Kasni, Kashini, Chicory, Wild Endive.

Distribution – It is a native of the temperate parts of the old world and is found wild in the Punjab, N.W.F.P., and Hyderabad. It has naturalized in Sonmarg area of Kashmir. The plant is successfully cultivated in Kinnaur of Himachal Pradesh and meets the major demand of the country. Also planted elsewhere, where it is tending towards naturalisation (Chauhan, N.S., 1103, 1203).

Description – A tall herb, 30-120 cm high, having grooved stem and wide spreading branches with long fleshy tap root. Leaves are oblong-lanceolate, basal leaves palmately lobed, cauline leaves entire, bract like, clasping. Flower heads bright blue, solitary, terminal, often clustered and the axillary flowers are arranged along stout, erect, green, nearly leafless, spreading stem. Flowers are 2.5-4 cm across with spreading ray florets. Involucral bracts are green, outer lanceolate, spreading and the inner one longer, erect.

Flowering/Fruiting – July-September.

Parts used – Roots.

Current market rate – Not available.

Main ingredients – The ash of root is rich in potash. The bitter principle is the glucoside of fructose and pyrocatechuic acid. The juice of roots contains stearin, mannite, and tartaric acid.

Uses – Chicory is said to be useful in fever, vomiting, diarrhoea, enlargement of spleen. The root is good stomachic and diuretic. The wild form is considered tonic and alexiteric.

Preparations – Not available.

Cultivation notes – The plant appears to grow on any type of soil. Sandy loam is, however, considered to

Cichorium intybus L. (Chicory)

be the best for its growth. The seeds are sown broadcase (7-12 lb/acre) on well prepared field. The crop lasts for 5-10 years.

To obtain root crop the practice in New-Zealand is to sow the seeds in mid October on ridges and 2-2.5 lb of seeds are required for one acre of land. The roots are lifted as soon as they are ripe with a special type of plough and left in the field for three days. The plant is not usually allowed to seed. The roots are either sun dried or kiln dried. The dried roots after roasting and powdering is used for admixture with coffee.

Cinnamomum tamala **Nces & Ebesm.**

Family – LAURACEAE

Local/Common name(s) – Tajpat (Leaf), Gurpatraj, Tejpatta, Tamalpatra, Patraj, Mithpatra.

Distribution – This is found in sub-tropical Himalaya, Khasi and Jaintia hills. It is also cultivated in certain parts of India. In Himachal Pradesh, it is present sporadic in Hamirpur, Tons valley in Shimla, Kangra, Chamba, Mandi, Solan, Nahan regions, bordering U.P. and is common in Jogindernagar and Palampur area in specific localities. (Chauhan, N.S., 691 and 1438).

Description – A small to medium sized evergreen tree, 2-12 m tall. Leaves are stalked, opposite, or sub-opposite, 10-20 cm × 5-7.5 cm; elliptic-oblong, 3-nerved from the base, shining, leathery, entire, long pointed. New leaves are slightly pinkish tinged. Flowers are small, 7.5 mm, yellowish. Fruit ovoid, black coloured when ripe in 5-15 cm long panicles.

Flowering/Fruiting – March-May, Fruits: June-October.

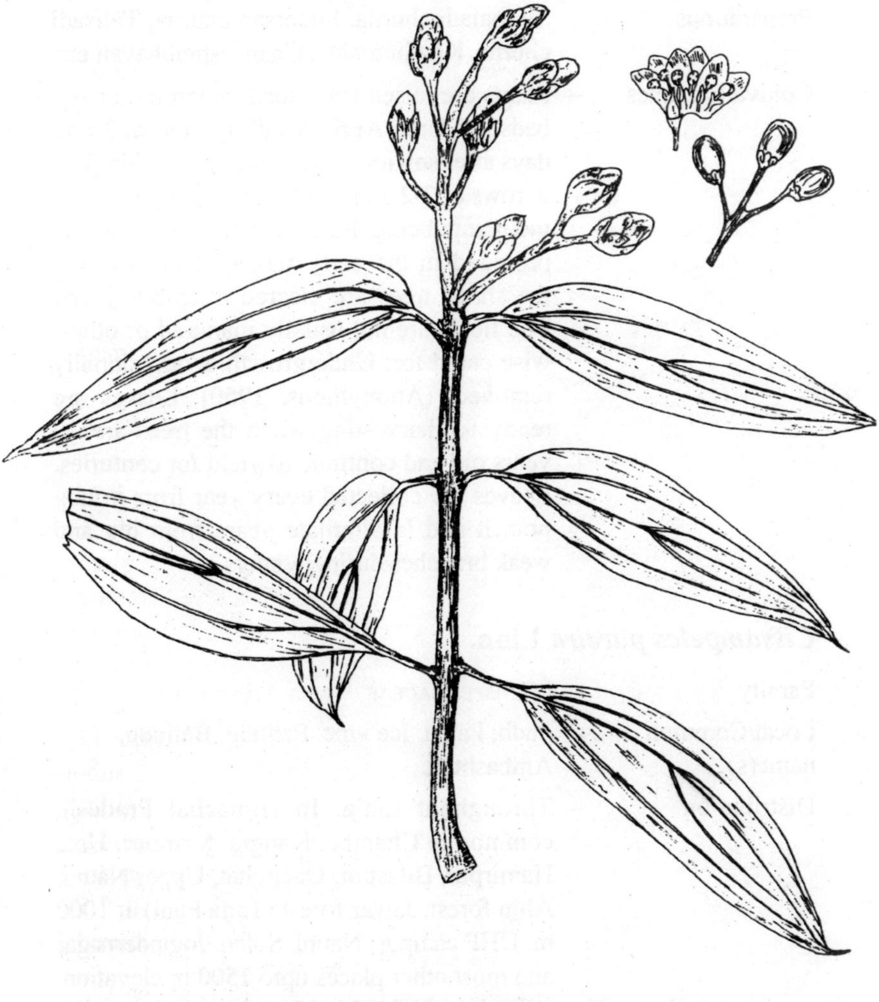

Cinnamomum tamala Nces & Ebesm. (Tejpatta, Tejpat)

Parts used – Bark, leaves.

Current market rate – Dry leaves Rs. 10-15/- per kg.

Main ingredients – Essential oil.

Uses – Therapeutic use: Bark is used in gonorrhoea. Leaf: Stimulant, Carminative useful in Colic, diarrhoea, rheumatism.

Preparations – Sitopaladi churna, Sudarsan churna, Talisadi churna, Khadirarishta, Chandraprabhavati etc.

Cultivation notes – Plants are raised from seeds sown in nursery beds in March-April. Seedlings appear 30-45 days after sowing and are transplanted in field in rows of 2-2.5 m apart. The spacing between the plants being 3-3.5 m. Sufficient shade is provided in the early stages of growth and the shade trees are cleared after 8-9 years. The fields are not usually manured or otherwise cared for. Undergrowth is occasionally removed. (Anonymous, 1950). Leaves are ready for harvesting when the trees are 10 years old and continue to yield for centuries. Leaves are collected every year from young branch and in alternate years from old and weak branches during winters.

Cissampelos pareira **Linn.**

Family – Menispermaceae

Local/Common name(s) – Padh, Patha, Ice vine. Patindu, Batindu, Ambashtha.

Distribution – Throughout India. In Himachal Pradesh, common in Chamba, Kangra, Sirmour, Una, Hamirpur, Bilaspur, Oachghat, Upper Nauni, Ahju forest, Jawar forest (Tatta Pani) at 1000 m, UHF campus, Nauni, Solan, Jogindernagar and most other places upto 1500 m elevation. (Chauhan, N.S. 834, 876, 1224, 1370, 1923).

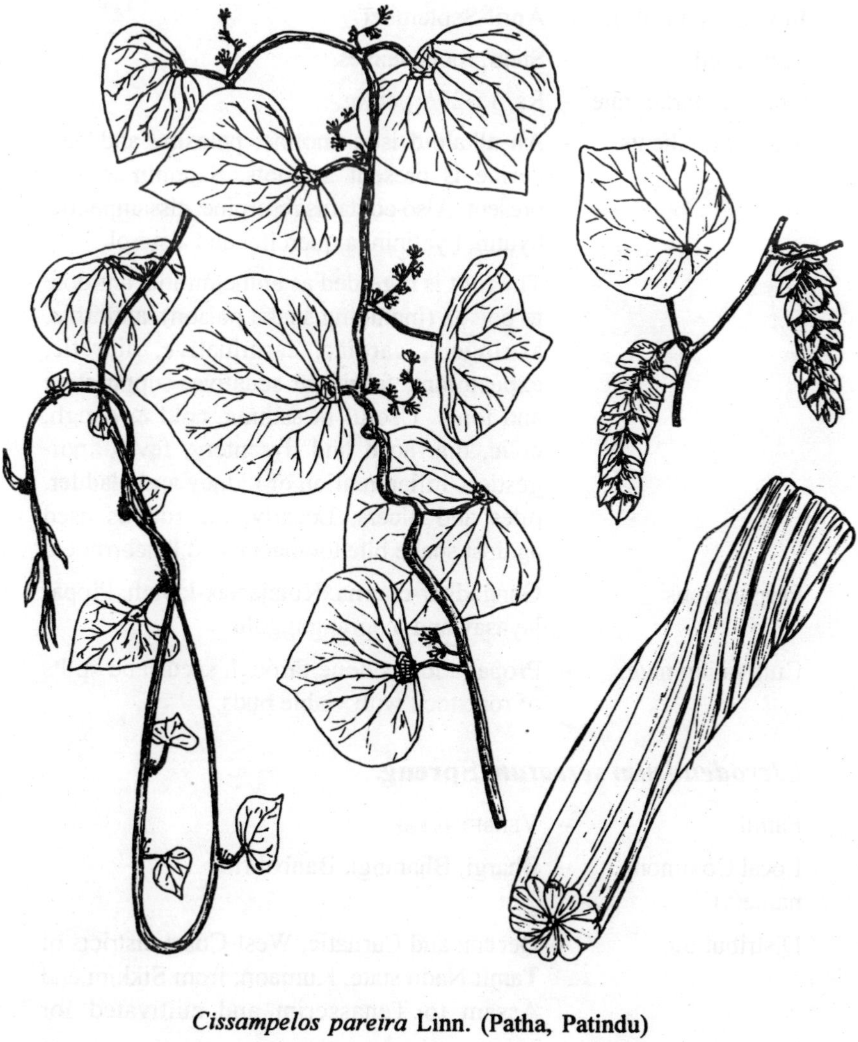

Cissampelos pareira Linn. (Patha, Patindu)

Description	–	A twining, perennial herb. Leaves are alternate, globular, cordate, kidney shaped, soft, hairy. Flowers are greenish, dioecious i.e. male and female flowers are borne on different plants. Male inflorescence in a raceme. Drupe ovoid, sub-globose, red in colour.
Flowering/Fruiting	–	April-September.
Parts used	–	Stem, roots, leaves.
Current market rate	–	Rs. 15-20/- per kg.
Main ingredients	–	An alkaloid isoquinoline, pelosine and berberine is present in roots. Saponin is also present. Also contains reserpine, cissampeline, hyatin, hyatinin, a quercitel and a sterol.
Uses	–	The root is regarded as anthelmintic, antidote to poison (including snake poison), antilithic, astringent, cardiac, carminative, diuretic, expectorant, febrifuge sedative, suppurative and tonic. Useful in asthma, cold & cough, colic, diarrhoea and dysentery, fever, indigestion, inflammation of kidney and bladder, piles and ulcers. Locally, the root is used against snake bite toothache and leucorrhoea.
Preparations	–	Gangadhar-churna, Kutajastak-kwath, Pippalayasav and Yograj guggulu.
Cultivation notes	–	Propagation is done through seeds and splits of rootstock with viable buds.

Clerodendrum serratum Spreng.

Family	–	VERBENACEAE
Local/Common name(s)	–	Bhargi, Bharangi, Banbakri.
Distribution	–	Deccan and Carnatic, West-Coast districts of Tamil Nadu state, Kumaon; from Sikkim and Assam to Tenasserim and cultivated for

ornamental purpose. In H.P., it grows in warmer regions of Una, Hamirpur, Kangra, Nahan-Paonta, Nalagarh, Deothal etc. (Chauhan, N.S., 3130).

Description – A perennial shrub, 1-2 m high, with bluntly four angular stems. Leaves in whorls, sessile, narrow lanceolate, sub-entire, glabrous, rather hard. Flowers are bluish-purple often white, in pyramid shaped terminal panicles.

Flowering/Fruiting – July-August. Fruiting: Cold season.

Parts used – Roots and leaves.

Current market rate – Not available.

Main ingredients – Alkaloid. Anthelmintic property is due to a bitter principle present in the leaves (Chopra et al. 1956). Roots contain sapogenin and triterpene, starch and resin.

Uses – *Root:* Useful in bronchial asthma, cough and scrofulous affections; loss of appetite, dropsy, and fevers.

Resin: Employed in syphilis and rheumatism.

Juice of leaves: Used with ghee as an application to herpetic eruptions.

Leaves: Vermifuge and bitter tonic (Chopra et al. 1956).

Preparations – Bharangi-gud, and Bharangi-kwath, Kanakasav, Sudarsan churna, Sringayadi churna, Yograj-guggulu (Dey, 1980).

Cultivation notes – Propagated by seed or cuttings of new year shoots.

Coriandrum sativum Linn.

Family – UMBELLIFERAE (APIACEAE)

Local/Common name(s) – Dhaniya, Coriander

Distribution – Cultivated all over the state for its seeds and leaves as a culinary herb and a condiment.

Description – An annual, well branched herb, upto 90 cm tall. The lower leaves are broader with crenately lobed margins, while the upper ones are narrow, finely cut with linear lobes. Flowers are small, white, often pinkish-purple tinged in terminal compound umbels. The fruits are aromatic, globular, ribbed, yellowish-brown in colour and when pressed, separate into 2 halves (pericarps), each containing a seed.

Flowering/Fruiting – February-April.

Parts used – The fruits. Green leaves are used as flavouring agent in salads.

Current market rate – Rs. 25-60 per kg.

Chemical contents – Fruits contain umbelliferone, scopoletin, rutin, coumarins, quercetin, kaempferal and aflatoxins. Essential oil from seeds contain geranyl content.

Properties and uses – The fruits are used as condiment in confectionery. Used as anthelmintic, aphrodisiac, aromatic, astringent, carminative, digestive, diuretic, stimulant and tonic. Useful in burning sensation of the body; in colic, chronic conjunctivitis, dropsy, diarrhoea, dysentery, dyspepsia, headache, as heart tonic and as a poultice for indolent swellings, nose bleeding and scrofula. The oil is used chiefly as a flavouring agent for spirituous liqueurs, in cocoa and chocolate industries. It is also employed in medicines. It is stable and retains its agreeable odour for a long time as compared to other oils. Also used in perfumery. The green leaves are relished as flavouring agents in cookeries, eaten as pot herb and chewed to correct foul breath (Wealth of India, II: 348, Dey, 42-43; Martingale; 1017). Extract of the seed has been found antimicrobial, antibacterial, sedative and is frequently used in lotions and shampoos. Essential oil is also antimicrobial.

Preparations – Dhanyapanchak-kwath, Abhayarishta, Guduchyadi-kwath, Lavanbhaskar-churna etc.

Cultivation – In Himachal Pradesh, it is cultivated as a winter crop. Fruits are crushed to expose the seeds which facilitates proper germination.

Corydalis govaniana Wall.

Family – PAPAVARACEAE

Local/Common name(s) – Bhutkes, Bhutjata, Bhutbishi.

Distribution – Pakistan to E. Nepal, 2400-4800 m in shrubberies, open slopes; a common alpine species. In H.P. it has been collected from Dhamkhri Gajiyani (4000-5000 m), Bijahi-Katha bawli (2500 m), Chounti-Kalga (4500 m), Chansil-Dhamkhari (3470 m). Rohtang slopes, Churdhar and Lahaul-Spiti. (Chauhan, N.S. 1134, 1263, 1564, 2341).

Description – A soft, slender, herb with long, thick, woody, finely knit type, dark brown rootstock, crowned with withered leaf sheaths. Radical leaves are long stalked, 2-pinnate, cut to the base and as long as stem. Leaflets-wedge shaped and deeply lobed. Flowers are yellow, crowded, 2.5 cm long and arising from 5-10 cm long individual axis. Capsule is about 1-1.5 cm long.

Flowering/Fruiting – June-September/August-October.

Parts used – The root and the juice.

Corydalis govaniana Wall. (Bhutjata)

Current market rate – Not available.

Main ingredients – A great variety of alkaloids derived from isoquinoline have been isolated from *C. govaniana* and allied species, *C. cornuta.* They contain protopine and stylopine. Further detailed investigations need to be done on all Himalayan species of this genus.

Uses – The root is considered tonic diuretic, alterative and antiperiodic. It is prescribed in syphilitic, scrofulose and cutaneous affections. The yellow juice of *C. ramosa* is employed in eye diseases. (Kirtikar and Basu: 136-37; Wealth of India. II: 358). *C. cornuta, C. ramosa* and *C. rutaefolia* have also been collected from the area.

Preparations – Not available.

Cultivation notes – Can be propagated by seeds or by divisions of underground stem having vegetative buds. Seeds should be sown immediately after ripening.

Costus speciosus Smith

Family – ZINGI-BERACEAE

Local/Common name(s) – Kemuk, Keu, Kushtha.

Distribution – Distributed more or less throughout India upto 1500 m elevation in forests and shrubberies. In Himachal Pradesh, it has been collected from Bharwain, Jwalamukhi, Khaltoo, Dilmanjori etc. upto an elevation of 1200 m. Recently put under domestication in Herbal garden, Nauni (Chauhan, N.S., 3174).

Description – An erect stout, rhizomatic, perennial herb with 15-30 cm × 5.7-5 cm; subsessile, spirally arranged, oblong, acute or acuminate leaves;

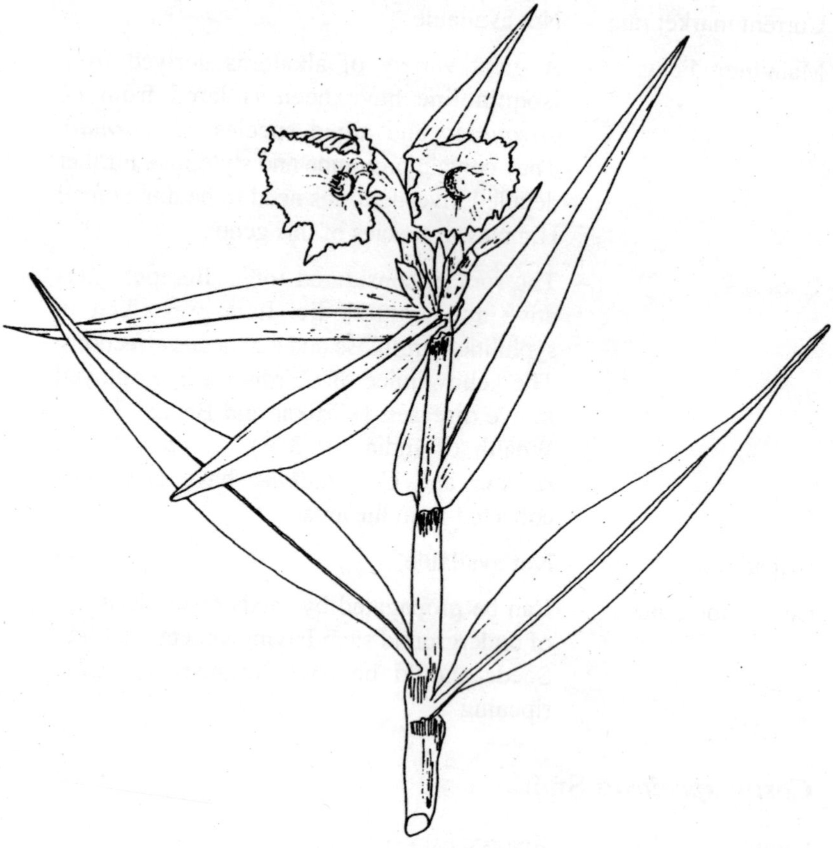

Costus speciosus (Konig) Smith (Pokhara)

glabrous above and silky pubescent beneath. Flowers white, funnel shaped, with contrasting, large, bright red bracts, numerous in a very dense cylindrical spike 5-12 cm long by 3.8-7.5 cm in diameter borne on a stout red leafy stem ranging from 1-2 m. Flowers 4 cm, bracts ovate, 1.5-4 cm long. Calyx with 3 oval lobes; corolla tube shorter than calyx, with unequal petals 2.5-4 cm; lip white with an orange red centre, rounded 5-8 cm, forming a funnel with margins incurved and meeting. Capsule red, globosely 3-gonous, crowned with persistent calyx and about 2 cm in diameter.

Flowering/Fruiting – June-August.

Parts used – Rhizome.

Current market rate – Not available.

Main ingredients – Contains cortico-steroids and starch.

Uses – Root is bitter, astringent, purgative, depurative, tonic, anthelmintic, used in snake bite; useful in bronchitis, fever, Kapha and Vata, dyspepsia, anaemia, rheumatism, hiccough and inflammation. Root is rich in starch. The plant is often cultivated as an ornamental.

Preparations – Eladiganum. In modern pharmaceutical preparations, it is employed for corticosteroidal preparations.

Cultivation notes – The plant is propagated vegetatively by rhizomes. Stem cuttings have also been found successful. The plant requires cool, shady and moist place for a luxuriant growth. Sandy loam or sandy clay loam has been found most suitable for optimum growth. Rhizomes are used as such or cut into pieces in such a way that each piece has 1-2 viable buds. The cut ends are smeared with fungicide to check fungal infection. Planting is done during May-June, before the onset of monsoon in the well prepared field supplemented with FYM and at a depth of 10 cm, keeping a plant to plant and row to row distance to be at 45-50 cm. FYM has been recommended to be added once in 3 months followed by occasional weeding. Harvesting can be done in 1½ to 2 years time (Anonymous, 1990).

Crocus sativus Linn.

Family – IRIDACEAE

Local/Common name(s) – Keshara, Kesar, Kunkuma, Jafran, Zabran, Saffron.

Distribution – Native of South Europe and West Asia, cultivated in Kashmir valley, especially in Pampore area upto an altitude of 1600 m. In Himachal Pradesh, it is cultivated in Sangla, district Kinnaur under Himachal Pradesh Krishi Vishva Vidyalaya, Palampur.

Description – A bulbous, perennial herb with long, linear, channelled leaves; flowers deep violet, funnel shaped, stalkless, and appearing in autumn with the leaves. Petals, narrow elliptic, equal, fused below into long slender corolla tube; styles orange, stigma brick red.

Flowering/Fruiting – October-December.

Parts used – Dried stigma and top of the style.

Current market rate – Rs. 35,000/- to 55,000/- per kg.

Main ingredients – Saffron contains glucoside, crocin, crocetin, picro-crocin essential oil, also contain β and R carotene and lycopene (Chopra et al. 1956), riboflavin and thiamine (CCRAS, 1990:72).

Uses – Saffron is used principally for its colouring and flavouring properties. It is used for colouring butter, cheese, puddings, pastry and confectionery. It is credited with various medicinal properties. It is used as a nerve sedative, stimulant, stomachic, abortifacient and as emmenagogue and is used in fevers and snake bite, melancholia and the enlargement of liver and spleen and is a good remedy for catarrhal affections of children. Saffron has been employed as an abortifacient and is frequently adulterated with styles, anthers and parts of corolla of saffron as an adulterant. Saffron corms are toxic to young animals and stigmas in overdoses are narcotic. It is mixed with other drugs to normalise menstruation, urine troubles; is very effective against diabetes, strengthens heart and brain, stops vomiting, expels worms, cures head-ache and heals wounds. It is also a useful remedy for epilepsy.

Preparations – Kumkumadi ghrita, Kumkumadya taila. Besides this, it is in demand for colouring, flavouring and for other ceremonial purposes.

Cultivation notes – The plant survives well in cold regions with warm or sub-temperate climate between 1500-2500 m above m.s.l. and receive snow during winter. It requires a rich, well drained sandy or loam soil. The plant is propagated vegetatively by bulbs. In Pampore, near Srinagar, bulbs are trans-planted in August-September, in raised plots (1.5 sq. m) surrounded by drains 22.5 cm × 15 cm deep, laid out in well pulverized soil. The plots are hoed and weeded. No manure is applied or irrigation given once the plants are established. In Kishtwar, in the Jammu valley, where the soil is lighter and well drained, the bulbs are planted on flat ground. Bulbs once established, continue to live for 10 to 15 years. New bulbs are being produced annually and the old ones rotten away. About 18 quintals of bulbs are required to plant one acre. In good soils, the bulbs multiply to about 22-23 quintals in 3 years and to double their weight in 10 years.

For collecting saffron, the flowers are collected daily, early in the morning, after the dew disappears. Styles and stigmas are separated and dried either in the sun or over artificial heat. In the latter case, the material is placed on sieves in the earthen kilns or pots containing low fire.

Cuscuta reflexa Roxb.

Family – CONVOLVULACEAE

Local/Common name(s) – Amarvela, Akasvela, Akasbel, Sutrawlo.

Distribution – Throughout, India and Ceylon. In H.P., this parasitic plant grows upto an elevation of 2800 m on shrubs and has been collected from Kandaghat, Lohardi forest, Winch Camp, Nalagarh, Bagriad Kandha, Girth-aurdi-Mari, Palah-da-Jungle, U.G. Hostel area. (Chauhan, N.S., 826, 1293, 1348, 1394, 1349, 1591, 1817, 1866, 2118).

Description – A leafless, twinning /spreading parasite throwing profuse branches and often covers the host shrub or other plants completely. Stems/branches are fleshy; flowers small, white, pedicelled and arranged in loose racemes or often in short clusters. Sepals acutish, ovate-oblong; corolla tubular, lobes minute, acute, externally reflexed. Fruits small, globose, depressed containing 2-4 seeds embedded in mucilaginous pulp.

Flowering/Fruiting – April-May/June-July.

Parts used – Whole plant & seeds.

Current market rate – Not available.

Main ingredients – Amerbelin, cuscutin and wax. Seeds yield semi-drying oil.

Uses – The plant is purgative and used internally in protracted fevers, as astringent and anthelmintic. Its decoction made with aromatics is given in flatulence, induration in liver,

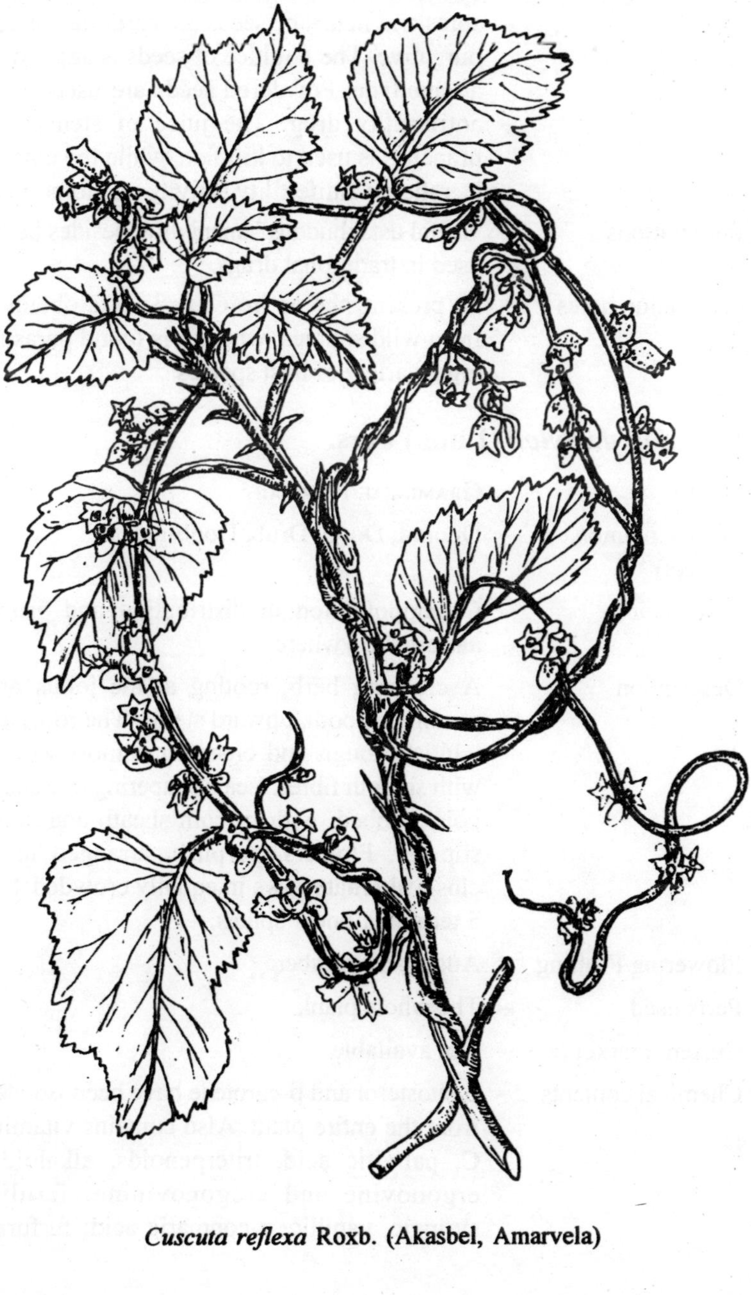

Cuscuta reflexa **Roxb. (Akasbel, Amarvela)**

diarrhoea. The infusion is used as a wash for sores and itch. The seeds are carminative and purgative. The poultice of seeds is applied as an anodyne. Powdered seeds are used as an antifertility drug. The juice of stems and branches is used to kill lice, while the extract has shown antifertility activity on albino rats.

Preparations – "Itrifal ustakhuddus" unani drug besides being used in traditional drugs.

Cultivation notes – At present, the entire *C. reflexa* is obtained from wild sources, where it is found parasiting a variety of host species.

Cynodon dactylon (Linn.) Pers.

Family – GRAMINEAE/POACEAE

Local/Common name(s) – Doorva, Doob, Drub, Doob-grass.

Distribution – Cosmopolitation in distribution and grows almost everywhere.

Description – A creeping herb, rooting at the joints and sending smooth, upward stems. The roots are whitish, tough and creeping, almost woody with smooth fibres. Leaves tapering to a sharp point, ribbed, with smooth sheath and hairy stipules. Flowers, purplish, arranged in 2, close, alternate rows in equally crowded 4 or 5 terminal, linear spikes.

Flowering/Fruiting – August-September.

Parts used – The whole plant.

Current market rate – Not available.

Chemical contents – β-sitosterol and β-carotene have been isolated from the entire plant. Also contains vitamin-C, palmitic acid, triterpenoids, alkaloids-ergonovine and erogonovinine; ferulic, syringic, vanillic, p-coumaric acid; furfural, alcohol, glucose, fructose, sucrose and starch. Also contains lignin, hydrocarbons and esters.

Properties and uses – Decoction of the root is used as diurctic, in dropsy and in secondary syphilis. Infusion of the root is useful in stopping the bleeding from piles. Crushed root mixed with curd are used in chronic fleet. Juice of the plant is astringent, used as application to fresh cuts and wounds, diuretic; useful in dropsy, anasacra, in hysteria, epilepsy, insanity; astringent in chronic diarrhoea and dysentery and catarrhal in ophthalmia. It is also regarded anti-abortive, conceptive, demulcent, haemostatic and suppurative. Useful in marks and blemishes, ulcers and wounds (Chopra *et al.*, 1958: 504; Wealth of India, II: 420; Dey, 45). Locally, the leaf juice is dropped into nostrils to stop nose bleeding. Roots are taken with mishri as memory sharpner for students and scholars.

Preparations – Doorvadi kwath, Doorvadi-ghrita, Doorvadi-taila, Chandrakala-rasa.

Cultivation – Doob is a very hardy plant and can grow in any type of soil. Its branches spread out on the ground, root at the nodes and bear new plants. Splits of the plant with root or branches are planted mostly in lawns, play fields and courtyards.

Cyperus rotundus Linn.

Family – CYPERACEAE

Local/Common name(s) – Nagarmotha, Motha, Mustak, Nut grass.

Distribution – Distributed throughout Himachal Pradesh in moist areas, ricefields and along water courses upto 1500 m elevation.

Cyperus rotundus Linn.

Description	– A troublesome weed in rice fields. Culms are erect, triangular, with rounded angle. Leaves-radical, sheathing, shorter than the culms. Roots-tuberous, sweet smelling, rounded black and hard like a knot; often crowded together. Tubers are black externally, white internally and smell like *Acorus*. Umbels are terminal, compound and 3-leaved.
Flowering/Fruiting	– July-September.
Parts used	– Tuberous roots.
Current market rate	– Rs. 8-10/- per kg.
Chemical contents	– Tubers contain terpenes, olealonic acid and its glycosides; beta-sitosterol and essential oil. Essential oil contains pinene, traces of cineole, monoterpens, sesquiterpenes and patchoulenone. Also contains fatty oil (2.7%).
Properties and uses	– The roots are considered stomachic, tonic, diuretic, demulcent, stimulant and galactagogue; given in fevers, dyspepsia, diarrhoea and cholera; also in urinary calculi and amenorrhoea. As a galactagogue, the fresh tubers are applied to the breasts. Useful in appeasing of thirst, disorders of stomach, irritation of bowels, febrile and dyspeptic affections; heals wounds and ulcers, pain in the abdomen and in scorpion stings (Chopra, *et al.*, 1956: 89, Khory and Katrak: 632; Dey, 110-11). Root paste is applied for healing wounds and sores. The extract is anti-inflammatory, antipyretic and analgesic. It's tea is taken in colds and against jaundice. Essential oil is insecticidal, anti-microbial and anthelmintic.
Preparations	– Jawarish Jalioos (Unani), Mustakadi-Kwath, Mustakadichurna, Panchbhadra-kwath, Gokshuradi-guggulu, Chandraprabha-vati etc.

Cultivation – It grows as a gregarious weed in moist places during rains and can meet the requirement from its natural population.

Dactylorhiza hatagirea **(D. Don.) Soo**

Syn: *Orchis latifolia* L.

Family – ORCHIDACEAE

Local/Common name(s) – Panja, Salampanja, Salep, Hathpanja.

Distribution – The plant is found in the Western Himalayas ranging between 2800 m - 4200 m.

In Himachal Pradesh, the plant is found growing between 3000-4200 m in Chhota Bhangal and Bara Bhangal of Kangra, Parvati Valley of Kullu, Pangi-Bharmour area of Chamba; Shego, Mudh, Mane and Rangrik areas of Spiti; Chansal and Maral Danda areas of Rohru Forest Division. It has also been found in Chounti-Thatch (3400 m), Shagali thatch (3200 m), Chansal (3800 m), (Chauhan, N.S. 168, 1210, 1221, 2048, 2342, 2751 and 5344).

Description – The plant is a terrestrial orchid in alpine meadows with fleshy tuberous roots. The tubers are slightly flattened, palmately lobed like a hand with fingers open, hence the name 'Panja'. Stem is usually 30-50 cm tall, leafy and with a few sheathing scales in the lower portion. Leaves are erect, oblong-lanceolate, 7-15 cm long, obtuse and with a sheathing base. Flowers are pink-purple to dull purple in colour, crowded in terminal, spicate racemes; are spurred and the darker lip is spotted. The market sample is a dried tuber, pale-white or greyish in colour; rounded, ovate or digitate in form and has hard and horny consistency.

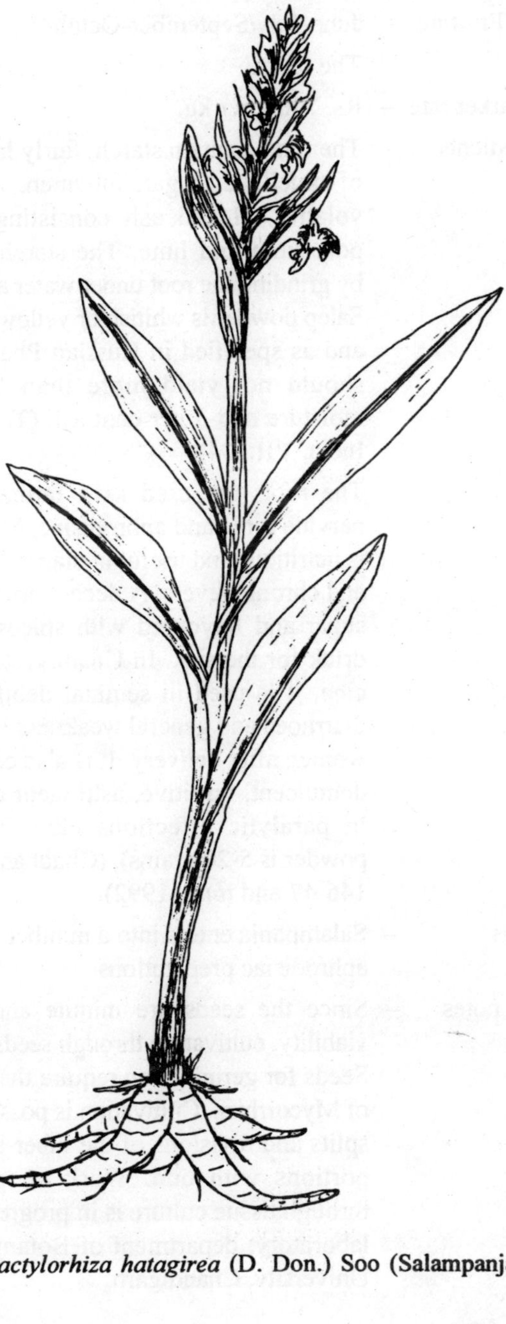

Dactylorhiza hatagirea (D. Don.) Soo (Salampanja)

Flowering/Fruiting – June-July/September-October.

Parts used – The roots.

Current market rate – Rs. 700/- per kg.

Main ingredients – The roots contain starch, fairly large quantity of mucilage, sugar, albumen, a trace of a volatile oil and ash consisting chiefly of potassium and lime. The starch is obtained by grinding the root under water and straining. Salep powder is whitish or yellowish in colour and as specified in Russian Pharmacopoeia, should not yield more than 14 per cent moisture and 3 per cent ash (The Wealth of India, VII:104).

Uses – The roots are used as a farinaceous food, nervine tonic and aphrodisiac, Mucilage jelly is nutritious and useful in diarrhoea, dysentery and chronic fevers. A decoction of salep with sugar and flavoured with spices is a useful drink for the sick. In Unani system of medicine, it is used in seminal debility, chronic diarrhoea and general weakness in debilitated women after delivery. It is also considered as demulcent, nutritive, astringent and is given in paralytic affections etc. (Dose of the powder is 5-20 grains), (Chauhan, N.S. 1984: 146-47 and *ibid.*, 1992).

Preparations – Salampanja enters into a number of tonic and aphrodisiac preparations.

Cultivation notes – Since the seeds are minute and have low viability, cultivation through seeds is difficult. Seeds for germination require the association of Mycorrhiza. Cultivation is possible through splits and divisions of the tuber having stem portions with bud. Work on propagation through tissue culture is in progress in orchid laboratory, department of Botany of Panjab University, Chandigarh.

Datura fastuosa **Linn.**

Family – SOLANACEAE

Local/Common name(s) – Kala-dhatura, Krishmadhastura, Siahdhatura.

Distribution – Sporadically distributed in Shivalik belt areas on Waste grounds, roadsides and debris dumping sites. Domesticated at Herbal Garden, Nauni-Solan (HP).

Description – A robust, erect, 60-120 cm tall herb having woody stem at the base and purplish to dark-purplish upwards. Leaves 7.5 cm to 20 cm long and 5 to 10 cm broad, ovate or ovate-lanceolate acute or acuminate, sinuate, dentate with cuneate base, unequal sided; petiole 3 cm to 7 cm long. Flowers pedicilate, solitary, calyx tube upto 7 cm long having 5 triangular lanceolate, acuminate tooth, base reflexed in fruit. Corolla funnel shaped 10-17 cm long, 10-12.5 cm in diameter of the mouth, violet or purplish outside, white within, 5 angled with long cuspidate folds. Stamens 5, attached near the base of the corolla tube, style filiform having 2 lobed stigma. Capsule globose 2.5-3.5 cm in diameter covered with blunt tubercles. Seeds flattened, greyish white.

Under domestication in the herbal garden, flowers with 1, 2 and 3 funnel shaped tubes in the same flower are commonly seen.

Flowering/Fruiting – May-September.

Parts used – Whole plant.

Current market rate – Not available.

Main ingredients – Leaves contain scopolamine, hyoscine and hysocyamine. Seeds contain mainly hyoscine and hyoscyamine. Seeds possess daturaolone and fastuic acid.

Properties and uses – Leaves are smoked to cure coughs. Seeds are used in asthma, hydrophobia, malaria and making liquors more intoxicating. Roots are used to cure toothache and for brushing teeth. Kala dhatura is mostly demanded by tantrics for various processes under Jantra-mantra therapy.

Preparations – As in *D. stramonium*.

Cultivation notes – It can be grown on a variety of soils but prefer rich clay loam soil and sunny situations. The land is plagued two to thrice times followed by planking. Farm yard manure mixed with a basal dose of fertilizers (NPK: 25; 50; 25 Kg/ha) is applied in the beginning. *D. metel* is grown as summer crop and seeds are sown in March. It can be grown directly by seed or by transplanting of seedlings. Germination of seed is low and irregular due to the presence of an inhibitor. Seed germination can be enhanced by soaking the seed for overnight in water before the sowing. In case of direct sowing, seeds are sown in rows 1 m apart and covered with soil. 7-8 kg seed is required per hectare. The seed starts germinating within a fortnight. Weeding and thinning is done when plants are 10-12 cm high, keeping plant to plant distance of 70-90 cm within the rows. After sowing of seed, irrigations is given if there is no rain within a week.

Datura metel Linn.

Family – SOLANACEAE

Local/Common name(s) – Dhustura, Dhatura, dhastur, Kanak.

Distribution – Throughout India, occasionally grown in gardens. Found as a weed of waste places in

Nalagarh, Parwanoo, Una, Kangra, Nahan, Paonta etc. upto 1100 m.

Description – Herbs or shrubs, the whole plant densely clothed with greyish tomentum. Stem erect, stout, herbaceous, terete. Leaves 15-20 cm long, broadly ovate, acute, unequal at the base and often cordate, densely tomentose on both surfaces and generally glandular. Flowers, pale-white, large, erect. Fruit is a capsule; globose, covered with long, rather slender spines. Seeds numerous, compressed, whitish grey.

Flowering/Fruiting – April-September.

Parts used – Whole plant including seeds.

Current market rate – Whole plant Rs. 10-12/- per kg.
Seed Rs. 250/- per kg.

Main ingredients – Plant contains scopolamine, hyoscyamine, hyoscine and atropine. Leaves contain mainly atropine and small amount of hyascyamine. Seeds contain upto 12 percent fixed oil and allantoin. Leaves also contain Vit. C.

Uses – The plant is violently narcotic. Leaves and root are a good palliative in spasmodic asthma. Seeds, leaves and roots are considered useful in insanity, fever with catarrhal and cerebral complications, diarrhoea, skin diseases, lice infestation etc. Dried leaves are used in medicines in the same manner as the leaves of belladona and stramonium.

Preparations – Kanakasava, Dhaturadi-pralep, Dugdha-vati, and Laghuvishgarbha taila.

Cultivation notes – It is generally propagated by seeds. Vegetative propagation by stem cuttings were tried and found to be successful. Seeds are collected from the mother plant when the fruits turn

Datura metel Linn. (Kala dhatura)

yellow. These are washed immediately to remove the fruit pulp and dried in the shade. Dried seeds are sown in sand beds during March-April. Frequent watering and shade is necessary for proper germination. Dormancy of 30-40 days has been noticed in the case of seeds. Pre-soaking the seeds for 24 hours reduces the dormancy period to 20 days. The seedlings are transferred to the field directly during the monsoon. The percentage of success is reported to be 70 percent.

For planting suitable field (with proper drainage that ensures 'no water logging' during the rains) should be selected and ploughed. All weeds are to be removed. Pits measuring 30 cm × 30 cm × 30 cm with a spacing of 0.5 meter filled with a mixture of topsoil, sand and cowdung in the ratio of 1:2:1 should be prepared. The plant starts yielding from the sixth month.

Scale insects are noticed on the lower surface of the leaves and young branches during dry season. They suck the sap leading to the drying up of plants. Application of pesticides is called for in such an eventuality.

Datura stramonium Linn.

Family – SOLANACEAE

Local/Common name(s) – Dhatura, Safed dhatura, Dhouturo, Thorn-apple.

Distribution – Native of Tropical America, naturalized in the Himalaya, upto 2700 m, in cultivated areas, waste lands and near village habitations. Collected from Bagi-Odari, Jakha (2750 m), Dodra-Punjarli, Kupri, Jogindernagar, Nauni and Solan (Chauhan, N.S., 392, 1018, 1203, 1291, 2271 and 2698).

Datura stramonium Linn. (Safed dhatura)

Description	–	An annual, erect, nearly glabrous, tall, robust herb with broad, ovate, coarsely and irregularly lobed and toothed leaves. Flowers are white, single on short, usually axillary stalks; 7.5-15 cm long, tubular. Capsules are ovoid, 3.5 × 2.5 cm covered with rigid, sharp prickles, surrounded at the base by enlarged and reflexed lower part of the calyx. Seeds-black, wrinkled.
Flowering/Fruiting	–	March-September.
Parts used	–	All parts of the plant.
Current market rate	–	Seed Rs. 250/- per kg. Panchang Rs. 10/- per kg.
Main ingredients	–	The dried leaves and flowering tops contain 0.3 to 0.5 per cent of alkaloids chiefly as hyoscyamine with smaller amount of hyoscine and atropine (Martindale, 1977: 248). The dry seeds contain 0.033 per cent and fruits 0.038 per cent of total alkaloid of which 25 per cent is scopolamine and 75 percent hyoscyamine. Leaves contain vitamin C and D. (Chopra et al., 1956:91, Dey, 1980:73). This species is more suitable for the extraction of scopolamine.
Uses	–	The drug is narcotic and anodyne. It affects the sympathetic nervous system, but does not affect the motor or the sensory nerves. Its effect is similar to that of atropine. The plant and its preparations are given to relieve spasm of the bronchioles in asthma and for this purpose, it is given in tablets or in mixtures. It is also smoked in cigarette or burnt in powders and the fumes are inhaled in bronchitis. It is useful in dysmenorrhoea, neuralgia and sciatica; is given with benefit in cases of nymphomania and in purpura mania with a tendency to suicide. Useful in boils, breast

inflammation caused by excessive formation of milk; asthma, salivation, travel sickness (sea or air), diarrhoea, enlarged testicles, insanity and itch. Oil medicated with dhatura is applied to head in headache, to boils and skin diseases. The local application of dhatura ointment gives relief in haemorrhoids, fissures and other painful diseases of the rectum. It has also been used to control the salivation and muscular spasm in paralysis and in post encephalitic parkinsonianism. The paste or hot fomentation of dhatura, bhang and nirgundi leaves is a good household remedy for tonsilitis (Martin-dale, 248-49; Khory and Katrak, 442; Chopra et al. 1956: 91 and Dey, 72-73).

Preparations – Kankasava, Dugdhavati, Laghu vishgarbha-taila, Dhathuradi-pralep etc. Salts of its esters are used in antiperspirant preprations.

Cultivation notes – Cultivation practices are same as in case of *Datura metel*. Naturally regenerating plants are more robust, profusely branched and bear bold capsules and profuse quality of seeds. Under domestication, it suffers from severe viral disease wherein all leaves dry up, drop and ultimately killing the plant.

Dioscorea deltoidea Wall. ex Kunth.

Family – DIOSCOREACEAE

Local/Common name(s) – Singli-mingli, Dioscorea, Kins, Ganj.

Distribution – North-Western Himalayas from Kashmir and Punjab eastward to Nepal and China at 1000-3500 m (Chopra et al. 1956). In Himachal Pradesh, about three decades back, it was frequently found in Kullu, Chhuar valley in

Mandi, Rohru and Chopal areas of Shimla district between 1500 m-2800 m elevation. Due to over exploitation, its population has dwindled alarmingly. Now it has been found in sporadic to scattered form in Manekhdhar, Kundrala (Rohru) 2660 m, Lohardi to Dogra (Polling) in Mandi and Kangra districts. Achnetugarh, Gaurigarh etc. (Chauhan, N.S., 1023, 1489, 2006, 2469, 2668).

Description – An extensive climber with stems twining to the left. The leaves are stalked, cordate, acute, often triangular. Fruits-triangular, trigonous, membranous, resembling the nose ornament (Nath) worn by ladies in the hills. Rhizomes are horizontal, borne close or often deep in the soil (depending on the site), greyish brown in colour, with rigid scattered roots on them.

Flowering/Fruiting – May-July.

Parts used – The dried rhizomes (best collection season for the drug is from November-March, when the plant is in dormant stage and the shoot portion is withered).

Current market rate – Rs. 10-15/- per kg.

Main ingredients – A rhizome of good quality is reported to contain from 4.8-8 percent of diosgenin content, which is used in the partial synthesis of modern drugs like cortisone and other steroids. The main intermediate isolated from diosgenin is 16-DPA (16, dihydropregenolone acetate) which can further be synthesized to any desired steroid hormone used in various formulations including oral contraceptive pills. The plant is one of the richest sources for this valuable steroid sapogenin (Chopra and Handa, 1961; Chopra et al.; 1958; Sood and Kalia, 1982).

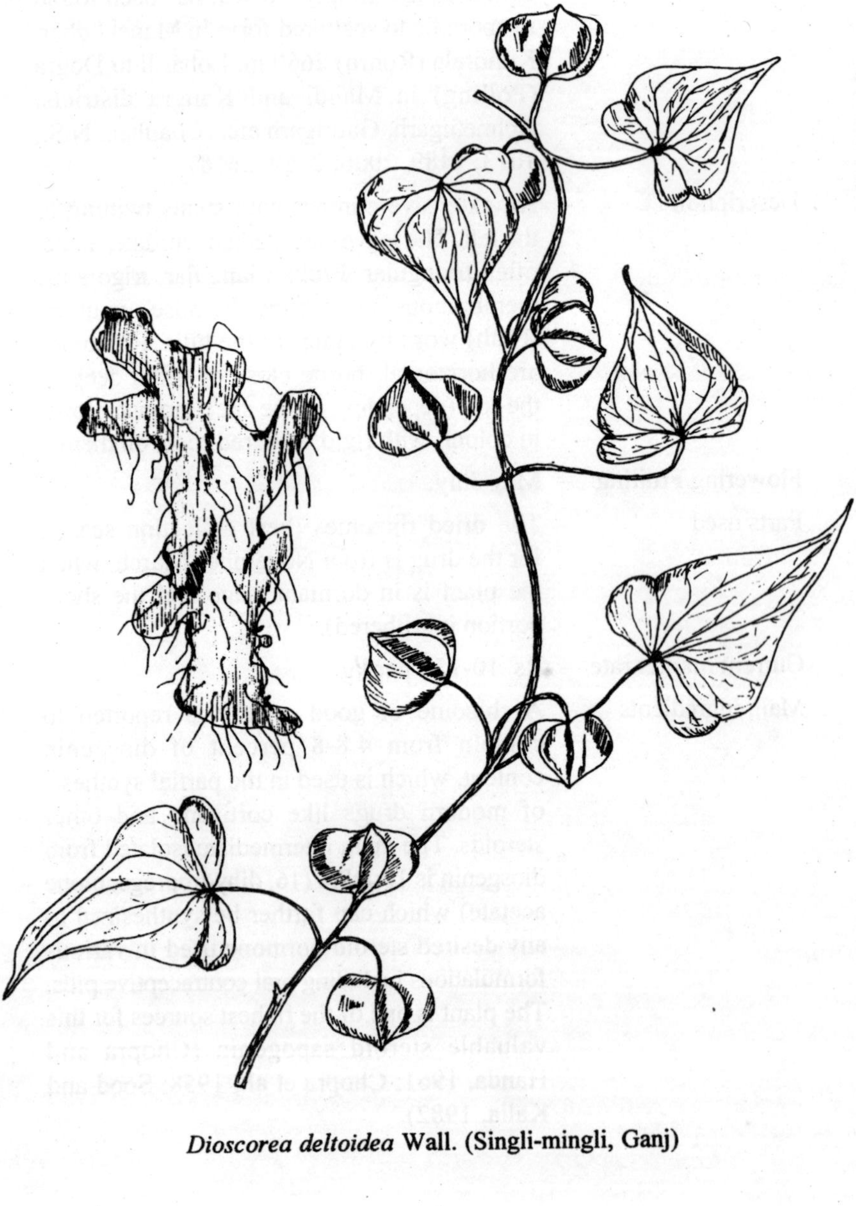

Dioscorea deltoidea Wall. (Singli-mingli, Ganj)

Uses – Cortisone has proved of great value in the treatment of a large variety of diseases, particularly in rheumatic diseases; certain ophthalmic disorders, allergic states and in idiopathic thrombocytopenic purpura. Being rich in saponin, the rhizomes are used for washing silk, wool and hair and also in dyeing. They are reported to kill lice. Locally, the crushed rhizomes are given with kneaded flour and salt to oxen for general gastric complaints. The plant contents are utilized in the preparation of various injections and tablets for use in modern medicines including birth control pills.

Preparations – Diosgenin enters into various modern formulations involving cortico-steroids; sex-hormonal preparations; tablets and injections including oral contraceptive pills and also the anabolic agents.

Cultivation notes – Dioscorea is one of the most important plants for the Pharmaceutical industry. The state had dominated its production for about 15 years as a result of which the entire population of the herb in its natural habitat was exploited resulting in its fast depletion and endangering its existence. State forest department had tried its cultivation in the plantations of deodar, fir and spruce in Kullu and Sundernagar forest divisions. Rhizome pieces weighing 50-100 g were sown by dibbling and yielded around 500 g. fresh weight of rhizome after 10 years.

It requires sandy loam to humus rich loamy soils for its proper growth. Rhizome pieces weighing 40-70 g are sown in well dug pits during February-March at a spacing of 60 cm × 45 cm. About 15-18 quintals of rhizomes are required for planting one hectare of land. FYM @ 20-25 tonnes are required for plantation. Experiments carried out by the scientists of Dr. Y.S. Parmar University of Horticulture & Forestry have given a yield of 177.16 quintals of fresh rhizomes per hectare when harvested after 44 months.

It is desirable to grow the herb through seed to save the valuable rhizomes for drug manufacturing. 3-4 seeds should be sown in well prepared pits during June-July. Harvesting can be done after 3-5 years. Attempts for its captive cultivation have been initiated by Dr. Y.S. Parmar University of Horticulture & Forestry and State Forest Department.

Dolichos uniflorus Lank.

Syn. *D. biflorus* Linn.

Family – PAPLIONACEAE/FABACEAE

Local/Common name(s) – Kulthi, Kulath, Horsegram.

Distribution – Cultivated in the entire sub-Himalayan tract upto 1800 m in sunny exposed, pebbly soils.

Description – An annual, branched, sub-erect or trailing, hairy cultivated herb with 3-foliolate leaves and yellow flowers. Pods are narrow, flat, compressed, sword shaped, about 5 cm long, containing 5-6 flattened, ellipsoid, dark grey-brownish seeds.

Flowering/Fruiting – August-September.

Parts used – Seeds

Current market rate – Rs. 20-30/- per kg.

Chemical contents – Seeds are a rich source of urea. Also contain lectins, phytoagglutinin, globulin, coumesterol and psoralidin.

Properties and uses – The seeds are considered anthelmintic, antilithiatic, astringent, diaphoretic, diuretic, emmenogogue, expectorant, febrifuge, ophthalmic and tonic. Useful in asthma, cold, cough, colic, hiccough; effective against stones, leucorrhoea and menstrual disorders in the form of decoction. Also effective in eye disease, piles and worms. It is being studied for its use in urinary tract stone dissolution (Anonymous, 1980). The plant is useful in tumors, adenites, sores, burns, smallpox, dysurea, prolapsus ani, fistula ani, dysentery, neuralgia, pleurisy, pneumonia, leprosy and rinderpest disease.

Preparations – Several formulations in Indian medicines.

Cultivation – It is a commonly grown rainy season crop in the upper Gangetic area and sub-Himalayan tract. As an important pulse crop, it is cultivated in Tamil Nadu, Karnataka, Maharashtra and Andhra Pradesh in large scale and to a small scale in H.P.

***Drocera peltata* Sm.**

Syn. *D. lunata* Buch-Ham.

Family – DROSERACEA

Local/Common – Mukhajali, Sundew.

Distribution – Temperate Himalaya between 1300-2800 m. Found growing in scattered form in pasture lands in shimla hills i.e. Khabal and Churdhar areas between 2400-2800 m. (Chauhan, N.S. 2282).

Description – A perennial, insectivorous, erect, delicate herb; 7-20 cm tall with small brown bulbous roots (0.5 cm in diameter). Leaves are peltate, half moon shaped, glandular, viscid. Flowers bisexual, regular, white, in terminal, branching racemes.

Flowering/Fruiting – August-September.

Parts used – The herb (The entire dried plant).

Current market rate – Not available.

Chemical contents – Resin and plumbagin. Also contains a proteolytic enzyme of the pepsin type and a yellowish-brown crystalline pigment.

Properties and uses – The yellowish brown crystalline pigment dyes the silk a fast rich brown. The plant is reported to be used by Ayurvedic practitioners in the preparation of *gold bhasma* used as antisyphilitic, alterative and tonic. The air dried entire plant is reported to be used in chronic bronchitis, asthma and whooping cough, given usually as a tincture in the dose of 0.5-2 ml. The bruised leaves are used as a blister in Kinnaur. Since the plant is bitter and caustic, it is not browsed by cattle (Dymock *et al.*, 1: 593; Nadkarni, 314; Martindale: 1753; Kirtikar and Basu: 1005).

Preparations – Swaranbhasma

Cultivation – No attempt has so far been made to cultivate this insectivorous plant.

Eclipta prostrata Roxb.

Syn: *E. alba* Hook.

Family – Compositae (Asteraceae)

Local/Common name(s) – Bhringraja, Kesaraja, Bhangra, Bhangru.

Distribution – Common weed in moist situations throughout India, ascending upto 2000 m on the hills (Chopra et al., 1956). In Himachal Pradesh, the herb is growing in sporadic to moderate quantity in moist places and along water channels, kuhls, ditches etc. upto 1300 m elevation (Chauhan, N.S., 110, 1447, 1615).

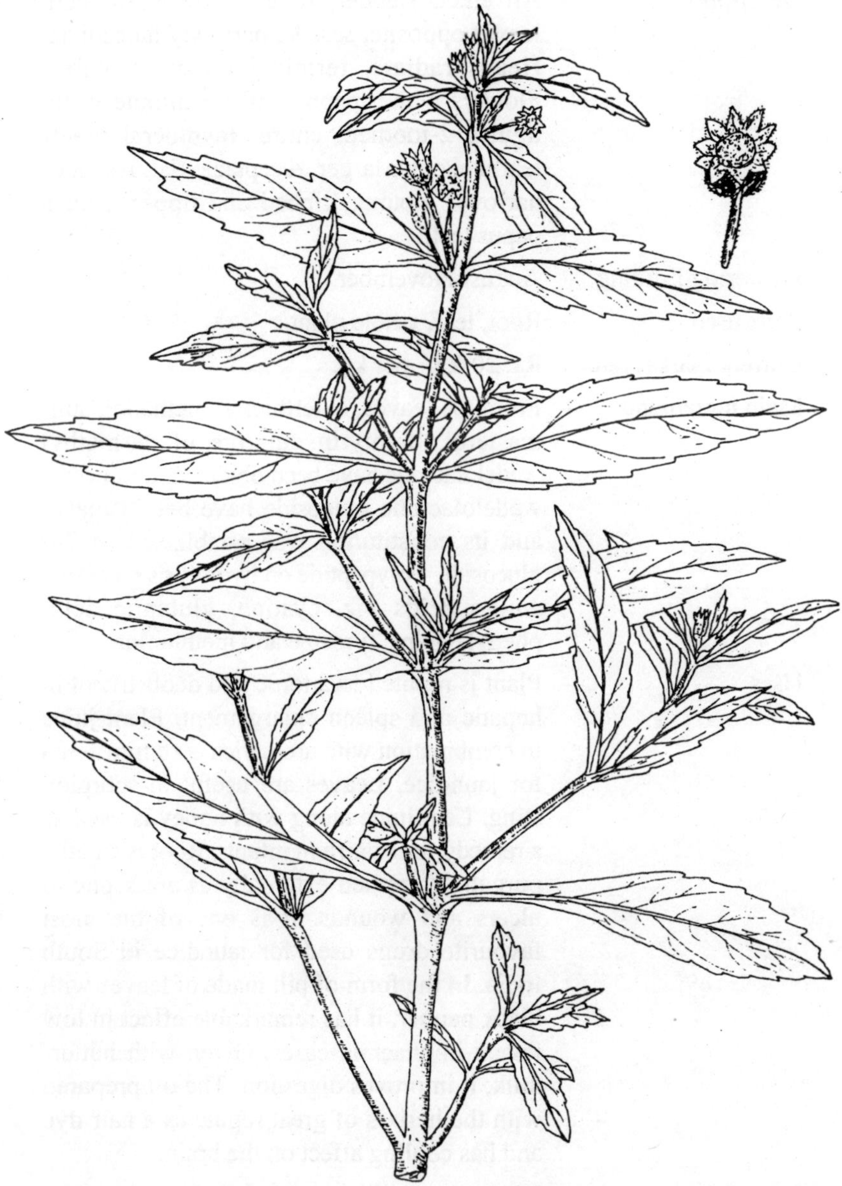

Eclipta prostrata (L.) Mant. (Bhringraja)

Description – An erect, slender, roughly pubescent herb leaves opposite, sessile, narrowly lanceolate. Heads radiate, terminal on erect stalks. Flowers white; pappus of 2-5 minute teeth; ligules 2-toothed, entire. Involucral bracts leaflike, outer larger, receptacle flat. Achenes narrowly oblong, ribbed and tipped with a papus teeth.

Flowering/Fruiting – August-November.

Parts used – Root, leaf, whole plant.

Current market rate – Rs. 20-25/- per kg.

Main ingredients – From the leaves, α-terthienyl-methanol, stigmasterol, β-amyrin, mixture of desmethyl wedelolactone have been obtained. Desmethyl wedelolactone glucoside have been isolated and its constitution was established as 7-0 glucoside. Polypeptide on hydrolysis gave five amino acids viz. cystine, glutamic acid, phenylalanine, tyrosine and methionine.

Uses – Plant is reputed as a tonic and deobstruent in hepatic and spleen enlargement. Plant juice in combination with aromatics is administered for jaundice. Leaves are useful in scorpion sting. Leaf juice along with honey is used as a remedy for catarrh in infants. Root is emetic, purgative; applied externally as antiseptic to ulcers and wounds. It is one of the most favourite drugs used for jaundice in South India. In the form of pill made of leaves with black pepper, it has remarkable effect in low fevers in anaemic cases. Given with buttermilk, it improves digestion. The oil prepared with the herb is of great repute as a hair dye and has cooling affect on the brain.

Preparations – Bhringarajasava, Maha-bhringraja-taila, Gandhaka-rasayana, Shadabindu-taila,

Ashvakanchuki-rasa, Sutashekhara-rasa, Ananda-bhairva-rasa.

Cultivation notes – It is easy to cultivate the plant through seeds. Seeds have shown good germination capacity. The seedlings so produced are planted in a well moistened field at 30 × 45 cm distance.

Emblica officinalis Gaertn.

Family – EUPHORBIACEAE

Local/Common name(s) – Amla, Dhatriphala, Amalaki, Aonla, Emblica or Emblic Myrobalan.

Distribution – Throughout tropical and sub-tropical India, chiefly in the dry deciduous forests ascending upto 1350 m in the Himalaya. In Himachal Pradesh, it is frequented in dry deciduous forests in Nahan, Nalagarh, Kunihar, Mandi, Bilaspur, Kangra, Hamirpur, Solan and Rajgarh divisions (Chauhan, N.S., 344, 369, 1352, 1411, 1701 and 1928).

Description – The plant is a deciduous tree and is nearly 12 m tall. Leaves are 1 cm × 0.2 cm, small narrowly linear and resemble pinnate leaves. Flowers are yellowish and are axillary, fascicled on lower leaves. Fruits are fleshy, globose and pale yellow in colour. Fresh fruits are globular, smooth, six-striated, of yellow-green colour with an obovate-obtusely triangular, 3-celled nut, each cell containing 2 seeds. The dried fruit is of grey-black colour, small sized, sub-hexagonal, wrinkled. If collected at proper maturity, the fruit is of yellowish-brown colour.

Flowering/Fruiting – March-May/December-February.

Parts used – Fruits and seeds.

Current market rate – Fresh fruits Rs. 4/- per kg.
Dried and without seed Rs. 12-14/- per kg.

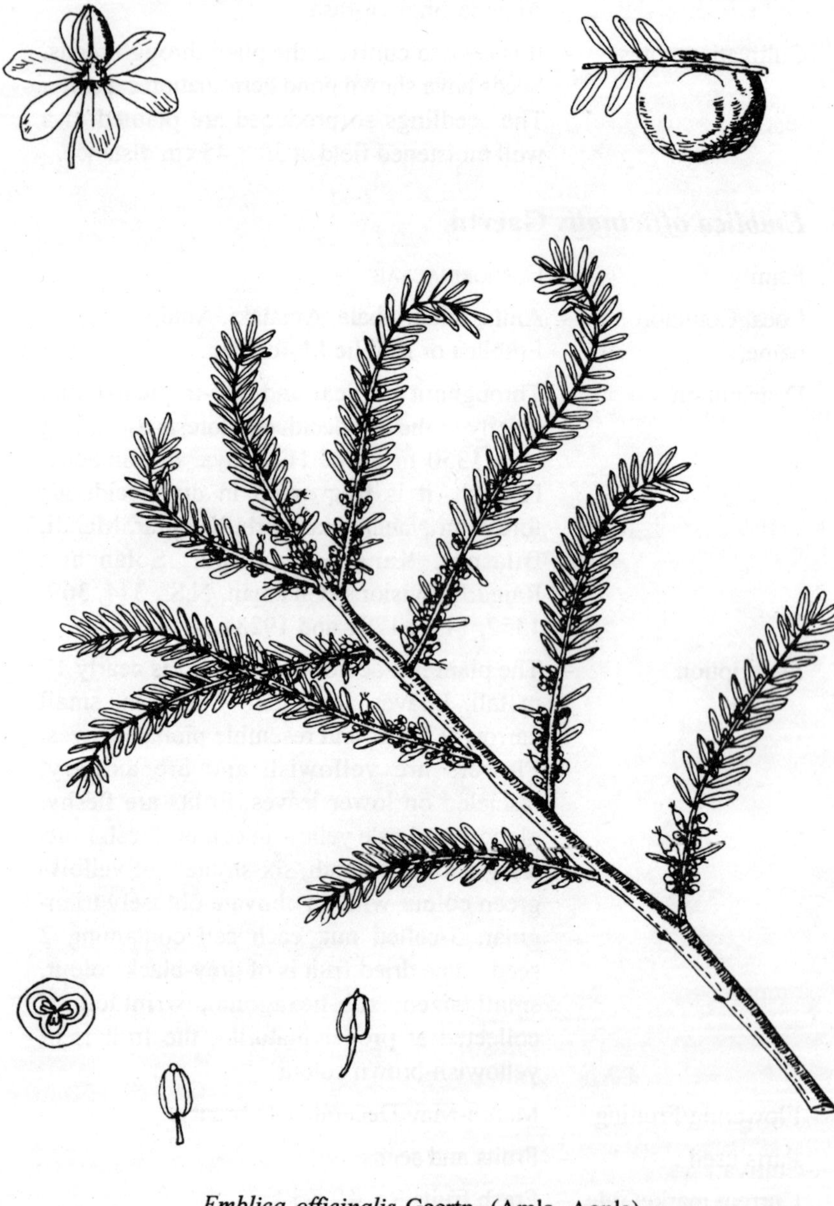

Emblica officinalis Gaertn. (Amla, Aonla)

Main ingredients – It is a rich source of vitamin C. Seeds contain fixed oil, phosphatides and an essential oil. Fruits and leaves contain tannins, polyphenolic compounds, 1,3, 6, trigalloylglucose, terchefin, corialgin and ellagic acid. Also contain allealoids phyllantidine and phyllantine. Leaves and stems yield lupeol and β sitosterol. Phyllemblin and an antimicrobial substance have been isolated from suspension cultures.

Uses – Fresh fruit is a rich source of vitamin C. It is diurctic, laxative, cardiac, astringent and liver tonic. Useful in anaemia, diarrhoea and dysentery, dyspepsia, haemorrhage, inflammation of the eyes, irritability of the bladder, jaundice, leucorrhoea, menorrhagia and discharge of blood from the uterus; affects blackening of hair. The fresh fruits are edible. The dried fruits are used as a remedy in diarrhoea. Fruit is a prized expectorant and has anti-cancerous properties. Fruit, juice, sediment and residue act as antioxidant due to gallic acid. Juice in ghee is used for abdominal and glandular tumors and in cancer. Decoction of pericarp cures peptic ulcers. Liquor fermented from fruit is good for indigestion, anaemia, jaundice, certain heart complaints, bad colds and for promoting urination. It is a chief ingredient of Chyavanprash Avleha, a tonic and restorative in Indian medicines.

Preparations – Chyavanprash, Brahmarasayan, Dhatrilauha, Triphala, Sanjivani-vati, Agnitundi-vati, Ambalyadi-churna, Phalatrikadi-kwath, Raktapittantak taila, Arogyavardhini, SG-1, Switradilepa etc.

Cultivations notes – Amla has many varieties both wild and cultivated, of which Banarasi, Chikaiya,

Hathighul and Desi are more prominent. Though, it can grow on any type of soil, yet it thrives best in well drained loamy soil. Propagation is done through seed, through budding and top working followed by grafting and budding. Seeds are sown in nursery after immersing them in water for 24 hours. Seedlings are transplanted during rains in well dug pits (1 × 1 × 1 m) at a spacing of 7.5 × 7.5 m. A full grown tree on an average yields 187-299 kg fruits per year.

Ephedra gerardiana Wall.

Family – GNETACEAE

Local/Common name(s) – Asmania, Budagur, Rachi, Khanda phag, Tutgantha, Somlata, Soma, Chhe, Chhedang.

Distribution – Drier regions of temperate and alpine Himalayas, from Kashmir to Sikkim at 2800-5100 m frequently met with at Pangi, (Chamba), Lahaul-Spiti, Chini, Kilba-Kailash ranges of Kinnaur (Bushahr), Shali hills (north of Shimla), Kashmir and Ladakh. It is a common scrub to under shrub in Lahaul-Spiti and Kinnaur, attaining a bushy growth in Pooh division and Spiti between Samdoh and Tabo. (Chauhan, N.S. 2390, 5512).

Description – A slow growing, rigid, tufted shrub, 30-60 cm, often having variable size with numerous, densely clustered, erect, slender, smooth green, jointed branches arising from a branched woody base in whorls. Internodes of branchlets are 1-4 cm long and 1.2 mm in diameter. Branches have scales at the joints. Male cones are ovate, 6-8 mm, solitary or 2-3, with 4-8 flowers, each with 5-8 anthers with fused filaments and rounded fused bracts. Female cones are usually solitary. Fruit is

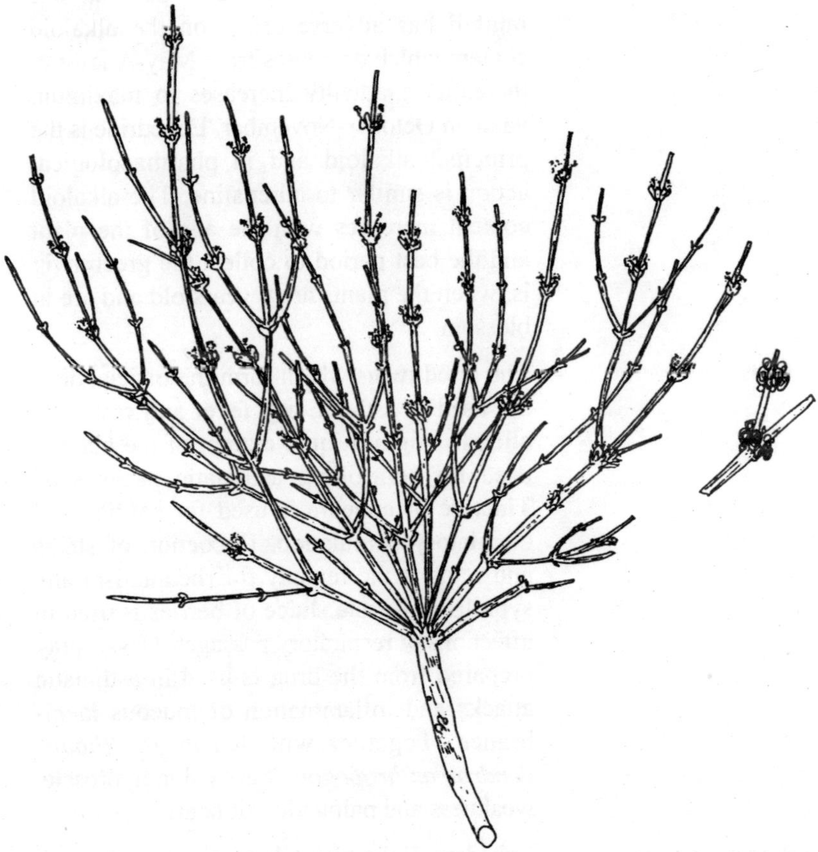

Ephedra gerardiana Wall. (Somlata)

ovoid, 7-10 mm, with fleshy red succulent bracts enclosing the 1-2 seeds within.

Flowering/Fruiting – May-June/August-September.

Parts used – Aerial shoots/twigs.

Current market rate – Rs. 7/- to 15/- per kg.

Main ingredients – Alkaloids ephedrine and pseudo-ephedrine. Total alkaloid content in Indian ephedras varies from 0-28 percent–2.79 percent. The rainfall has adverse effect on the alkaloid content which decreases from May-August & thereafter gradually increases to maximum value in October-November. Ephedrine is the principal alkaloid and in pharmacological action is similar to adrenaline. The alkaloid content increases with the age of the plant and the best period to collect the green twig is, when the plants are 4 years old and are in blossom.

Uses – The dried twigs which form the official drug are used in asthma, hay fever and rashes of allergic origin. Liquid extract of the plant is used for controlling asthmatic paroxysms. Tincture of ephedra is used for cardiac and circulatory stimulation. Decoction of stems and roots is the remedy for rheumatism and syphilis in Russia. Juice of berries is used in affections of respiratory passages. Nasal spray prepared from the drug is used in asthmatic attacks and inflammation of mucous membrance. Together with leaves of *Rhododendron anthropogon*, it is used in hydrosele, weakness and palpitation of heart.

Preparations – Ephedrax, Ephedrin. A large number of cough syrups under different proprietary names are also being manufactured.

Cultivation notes – Propagation is through seeds, suckers and layers. It is a most desirable for plantation in dry, sandy banks or rocky slopes in Pooh, Spiti, Lahaul and Bharmour areas. Harvesting of the twigs should be done during October-November and dried in the sun. Artificial drying at high temperature must be avoided. The dried twigs when stored in dry, closed containers, protected from light, retain their activity for a long period. Best material is obtained from plants after 4 years of age. Other species available are *E. intermedia* & *E. major*.

Fagopyrum esculentum Moench

Family	– Polygonaceae
Local/Common name(s)	– Ogla, Kathu, Bhares, Buckwheat.
Distribution	– Cultivated as a rainy season crop in mid and high hill, tribal areas of the state upto an elevation of 2600 m. (Chauhan, N.S. 1021, 1306 and 2674).
Description	– An annual purplish herb 1-1.5 m, tall with alternate, hastate, acute leaves; 2.5-7.5 cm long. Flowers in axillary or terminal cymes, pinkish white, fragrant, dimorphic and highly self sterile. Fruit is a 3 cornered achene approximately 0.5 cm long with keeled edges, varying in colour from silvery-grey to brown or black.
Flowering/Fruiting	– July-September/October.
Parts used	– Fruits for general purpose and leaves for rutin.
Current market rate	– Seeds Rs. 50/- per kg.
Chemical contents	– Leaves contain 3.5-4.7%, while fruits yield about 1.3% rutin. Also contain fagopyrin, flavanol, fagomine, alanine, rhamnodiastase, enzymes hydrolyse and various flavanone glycosides.
Properties and uses	– The plant has assumed a great medicinal importance due to the presence of the glucoside rutin used in modern medicines as antihaemorrhagic. The fruits ground into flour are used as a food by the natives on fast days and also used to fill the intestine of goat

during Bakrid to make a special preparation. Biological value of the protein from this plant is superior to that of other plant foods and nearly equal to that of the eggs. Decoction of the root is used in rheumatic pains, lung diseases and typhoid; while the juice is useful in urinary disorders. Rutin is valued as a cure for capillary fragility of the heart.

Preparations – Rutin based formulations.

Cultivation – In hills, the crop is sown during May-June and harvested during September-October. In plains, it can be grown in any season as a pot herb.

Other useful species: *F. dibotrys* (D.Don.) Hara syn. *F. cymosum* Meissn and *F. tataricum* (Linn.) Gaertn.

Fumaria parviflora **Lam.**

Family – FUMARIACEAE

Local/Common name(s) – Pitpapra, Papra, Shahtera, Bansulpha, Shahteraj, Baklat-el-malik, Parpat, Fumitory.

Distribution – Common over the greater parts of India, mostly in wheat field growing in the winter season and all over Himachal Pradesh in lower warmer areas upto 1400 m. (Chauhan, N.S., 839, 1463, 1762).

Description – A pale green, diffuse, much branched annual herb. Leaves much divided, segments very narrow, flat. Flowers are small, purplish, in lax flowered racemes. Sepals lanceolate, much smaller than the corolla tube, pedicels exceeding the bracts. Ovary 1 celled, style filiform; stigma entire or short lobed, ovules 2, on placentas. Fruit globose, 1 seeded, pin head sized capsule bearing dark brown seeds.

Flowering/Fruiting – February-March.

Parts used	–	Whole plant.
Current market rate	–	Rs. 10-20/- per kg.
Main ingredients	–	The major alkaloids protopine along with phenolic and non phenolic alkaloids have been isolated. Protopine has relaxant, non specific antispasmodic activity (CCRAS, 1986). Alkaloid cryptopine, sanguinarine, fumaridine, fumaramine, parfumidine, parfumine, fumariline etc. have been isolated. Also contain non-acosanol and sitosterol.
Uses	–	Dried plant is anthelmintic, diuretic, diaphoretic in low fever and is used to purify blood in skin diseases (Chopra et al. 1956).
Preparations	–	Parpatadi-kwath, Parpatadyarista, Sadangapaniya, Panchabhdra-kwath, Satyadi-kwath, Kirattiktadi-kwath, Sudersan-churna, Amritarista etc.
Cultivation notes	–	Grows profusely as a weed in wheat fields, where its density often outnumbers the population of the main crop and in such cases has to be uprooted manually. Propagation is through seed harvesting is done during February-March.

Gentiana kurroo Royle

Family	–	GENTIANACEAE
Local/Common name(s)	–	Karu, Kaur, Trayamana, Indian Gentian.
Distribution	–	Pakistan to Uttar Pradesh, 1800-2700 m in temperate to alpine grassy slopes. In Himachal Pradesh, it has become sporadic in sub-alpine to alpine meadows. It can be collected from Kalga Pattan (4200 m); Dainasar rocky slopes (4500-5000 m); Pathrala Thatch, Gajyani Catchment (4000 m), Dhamkhari-Marvi, Karol tibba in Solan and Mangarh area of

Gentiana kurroo Royle (Kaur, Karu)

Sirmour district it lower altitudes at 2000 m elevation. (Chauhan, N.S., 2023, 2329, 6459).

Description – A small, decumbent, tufted herb with perennial root stock, stout, giving out numerous arching stems, 5-30 cm. Basal leaves are long lanceolate, usually 10-12 cm long; stem leaves 2-3 pairs, narrow, linear about 2 cm long. Flowers-handsome, blue, often spotted with white. Usually 2 or more on each stem, but usually solitary, funnel shaped.

The drug consists of brownish, sub-cylindrical entire or longitudinal split pieces of roots. The root is wrinkled and the rhizome has transverse annulations. The drug is brittle and breaks with a short fracture. Taste is sweet at first intensely bitter afterwards.

Flowering/Fruiting – August-October.

Parts used – Root and rhizome pieces.

Current market rate – Rs. 150-165/- per kg.

Main ingredients – The root contains gentiopicrin, gentianic acid, a pectin and a sugar.

Uses – Karu is used as a bitter tonic, stomachic and febrifuge. It is considered to be an anthelmintic, blood purifier, carminative, diphoretic, digestive, useful in hepatic disorders and urinary affections and is a well known remedy for fattening the horses (The Wealth of India, 4:124; Watt III, 486; Dey, 150).

Preparations – It enters into most of the stomachic and tonic preparations like Sudarshan churna, Arogyavardhini-vati, Punarnava-mandoor etc.

Cultivation notes – It requires a cool temperature and well drained moist soils. Propagation is through seed, by cuttings or by root divisions. Seeds should be sown as soon as ripe. Attempts have recently been made to propagate the plant through tissue culture. *Gentianella moorcroftiana* and *G. paludosa* are used as substitute as Tikta in Amchi system of medicine.

Geranium nepalense Sweet

Family	– GERANIACEAE
Local/Common name(s)	– Laljari, Ratanjot (substitute), Laldori.
Distribution	– Frequent in temperate Himalaya between 1500-3000 m. Common in temperate forests and pasture lands in H.P. in Churdhar of Sirmour, many parts of Mandi, Kangra, Shimla, Kullu, Chamba and Kinnaur between 2000-3200 m. (Chauhan, N.S. 720, 788, 1228 & 2173).
Description	– A perennial, pubescent, diffuse herb, 15-45 cm long, prostrate, often ascending and branches rooting at the joints. Leaves-orbicular, 3.5-7.5 cm across, palmately 3-5 lobed. Flowers-about 1.5 cm across, purplish red. The root is fleshy, branched, red.
Flowering/Fruiting	– May-September.
Parts used	– The root.
Current market rate	– Rs. 35-40/- per kg.
Chemical contents	– Abundance of red colouring matter, tannins, gallic acid, quercetin and citric acid. Kaempferol 7-rhamnoside, kaempferitrin and geranin have also been reported from the plant.
Properties and uses	– The plant is considered as an astringent and also used in the disease of the kidneys. Due to the red colouring matter, it is used to colour various medicinal preparations and oils. The

roots are also reported to be used for tanning (Chopra et al., 1956: 124; Wealth of India, IV: 126). Ethyl alcohol (50%) extract is reported to be hypotensive.

Preparations – Not known.

Cultivation – Can be cultivated through seeds and root slips bearing stem portions. Allied species: *Geranium wallichianum* D.Don ex sweet.

Ginkgo biloba Linn.

Family – GINKGOACEAE

Local/Common name(s) – Bal kunwari, Fossil tree, Maiden hair tree.

Distribution – Native of China, occasionally cultivated in Indian Gardens, particularly on hills. A few trees, deformed and mutilated are found in Kalpa in Kinnaur, Mashobra in Shimla; a handsome tree at Manali and one small plant in Nauni-Solan, UHF Garden.

Description – It is a handsome, straight tree, upto 30 m high, spreadingly branched when young, bearing clusters of fan-shaped leaves. Flowers dioecious; fruit a drupe consisting of a foul smelling pulp surrounding a cream coloured thin shelled, sweet kernel (stone).

Flowering/Fruiting – Has not been observed so far in H.P.

Parts used – Leaves and roots (seeds); Root bark.

Current market rate – Not available.

Main ingredient – Leaves contain shikimic acid and flavonoids. Leaves also constitute phenolic components; sesquiterpene, bilobalide and beta-sterol. Ginkgoloides, diterpenes have also been extracted from leaves and root bark. Phenolic, toxic substances can be extracted from fruit pulps and sensitizers from seed. Ginnol, identical with celidoniol have also been found. Fruit extracts contain anacardic acids. Seeds contain alkaloids-ginkgotoxin and amino-acid.

Properties and uses – Grown as an ornamental. Nuts are used against cancer in China. Leaves are used in cardiovascular disorders, increased cerebral blood circulation and parkinson's disease. Biflavonoids of leaves are active in peripheral vasodilation. The leaves are recommended as being beneficial to the heart and lungs. Inhalation of decoction of the leaves is useful in treating asthma. The seed is used as an anti-tusive and expectorant in Japan and China. It is an CNS improver, best brain tonic, memory sharpener in old ages, but contra-indicated in young people aged below 12 years.

Preparations – Tebonin (standardized preparation containing procyanidin flavon mixt.) used in cardiovascular disorders; preparation used in patients suffer-ing from parkinson's disease and increased cerebral blood circulation. Ginkocer, Bilovas, Tanakan and Rokan are the standard concentrated extracts of its leaves marketed in France and Germany respectively.

Cultivation – The tree can be raised through seeds and vegetative cuttings. Cuttings take a long time to root which is facilitated by the application of rooting hormones.

Gloriosa superba Linn.

Family – LILIACEAE

Local/Common name(s) – Langli, Methonni, Kalihari, Agnishikha, Glory Lily.

Distribution – Throughout tropical India. Frequently distri-buted in Sal forests. The plant has been

		collected from Jwalamukhi, Bharwayin, Nauni, Kunihar division upto 1250 m elevation in H.P. (Chauhan, N.S., 1140, 1246).
Description	–	A beautiful, herbaceous, tall, glabrous, branching leaf tip climber, about 1-3 m tall. Leaves alternate, opposite or ternately whorled, lanceolate, strongly nerved, nerves parallel, leaf-tip ending in a tendril like spiral. Flowers large, showy, axillary, solitary on long pedicels, nodding. Perianth 7.5-10 cm across, with crisply waved margins, greenish at first, then yellow, passing through orange and scarlet to crimson. Fruit is a large oblong capsule upto 5 cm long. Seeds subglobose, testa spongy, wing like.
Flowering/Fruiting	–	Rainy season during September-October.
Parts used	–	Tuberous rootstock and seeds.
Current market rate	–	Rs. 18/- to 20/- per kg; seeds Rs. 300/- per kg. approx.
Main ingredients	–	Contains superbine, gloriosine and 1.3 percent colchicine, Gloriosine promises to be more effective than colchicine in plant breeding work for inducing polyploidy (Dey, 1980:95). Colchicine is the chief alkaloid found in roots, seed and flowers alongwith its other derivatives. Flowers contain luteolin.
Uses	–	Tuber is pungent, bitter, acrid; heating; anthelmintic, laxative, abortifacient, useful in chronic ulcers, leprosy, inflammation, piles, thirst etc; removes thorns, spines, nails, expels skin parasites and intestinal worms. The leaf juice is used for killing lice in hair. Root paste is an antidote to snake bites, while root powder is given in rheumatic fevers.
Preparations	–	Kasisadi-talia, Lughu-vishgarbha-talia, Langulirasayana.

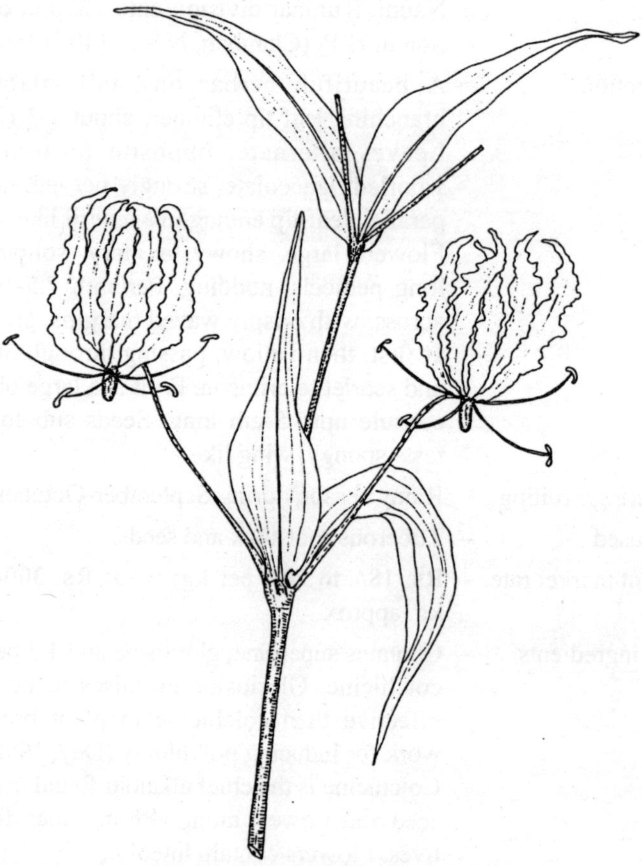

Gloriosa superba Linn. (Kalihari, Langli)

Cultivation notes – Planting is done during monsoon. It grows in a wide variety of soil conditions, but healthier growth has been noticed in soils having an optimum pH of 5.5 to 6.5. Fields are ploughed and harrowed several times, levelled, weeds removed and care has to be taken to avoid water logging. During the onset of monsoons,

30 cm × 30 cm pits are prepared mixed with soil, FYM and sand in the ratio of 1:1:1. Rhizomes of 2 months old seedlings are planted on the raised beds, about 15-20 cms apart. The plant being a leaf climber, supports have to be provided. The plant reaches a height of about 2.5 mts. in 6 months time. After fruit setting, the aerial shoot dies and the rhizomes hibernate. The plant starts yielding within 6-8 months time. Rhizome of each plant weighs around 50-100 gms under optimum conditions. The crop can be raised through seeds, but the germination is found to be low. Pre-soaking of seeds in water for 24-36 hours has increased the germination rate from 20 to 40 percent. Harvesting is done during autumn, by which time fruits ripen and the aerial parts die down.

Hedychium acuminatum Rosc.

Syn. H. spicatum Buch. -Ham. ex Smith

Family – ZINGIBERACEAE

Local/Common name(s) – Shati, Kapurkachari, Kachur, Van-haldi.

Distribution – Found in the entire Himalayan region between 800-2800 m. In Himachal Pradesh, it is found in isolated patches elsewhere and in frequent groups in Chail and Shimla catchment forests between 1600-2800 m elevation (Chauhan, N.S. 931, 1050, 2196, 2324 and 4400).

Description – A perennial, rhizomatic, robust herb upto 1 m tall. The rhizomes are quite thick upto 7.5 cm in diameter, aromatic, knotty, spreading horizontally under the soil surface, grey-brown in colour and have long, thick fibrous roots. Leaves large with long clasping sheaths, oblong-lanceolate, acuminate. Flowers pinkish white in terminal spikes.

The rhizomes come in the market in circular slices, covered by a rough reddish-brown bark, marked with numerous scars and circular rings, white and starchy within. Odour is strongly camphoraceous and the taste is pungent, bitter and aromatic.

Flowering/Fruiting – July-September.

Parts used – Cut, dried rhizomes in slices.

Current market rate – Rs. 15-60/- per kg.

Chemical contents – Rhizomes contain about 4% of an essential oil containing P-methoxy cinnamic acid, ethyl cinnamate, d-sabinine sesquiterpene, alcohols, pentadecane and traces of cinnamaldehyde. Rhizomes also contain starch (52%), organic acid, glycosides, ash (4.6%), fixed oil and an odorous material.

Properties and uses – Kapur kachri is employed in the preparation of 'Abir', a fragrant coloured powder used during the holi-festival and in religious ceremonies in plains. Due to insect repellent properties, the rhizomes are used to preserve clothes. Powdered rhizomes can be used for perfuming tobacco. The rhizome is aromatic, anti-arthritic, appetiser, cardiac, stimulant, carminative, deodorant, dentrifuge, hair tonic, stomachic and tonic. Useful in colic, cough and asthma; diarrhoea, dysentery, dropsy, headache, hair falling, liver complaints, rheumatism skin diseases and vomiting. Acts as a very good deodorant and the oil is used in the manufacture of incense, for perfuming soaps, hair oils and face powders. Rhizomes are much used in veterinary medicines. (Nadkarni, 1: 608, Chopra *et al.*, 1956: 121,

Hedychium acuminatum **Rosc.**

Wealth of India, V: 14, Dey: 137-38). Ethyl alcohol (50%) extract of the rhizome is hypoglycaemic anti-inflammatory, vasodilatory, mild hypotensive, antiseptic in animals, useful in tropical pulmonary eosinophilia and as insect repellent. In Unani medicines, it is regarded as aphrodisiac.

Preparation – Sudarshan-churna, Chandraprabha-vati, Satyadi-churna etc.

Cultivation – The plant is easily propagated through rhizome pieces and prefers rich organic matter, partial shade and moist situations.

Helicteres isora **Linn.**

Family – STERCULIACEAE

Local/Common name(s) – Marorphali, Jonkaphal, Avartani, Mrigashinga. East Indian screw tree.

Distribution – The plant occurs often gregariously, throughout India, from Jamuna eastwards to Nepal, Bihar and Bengal and South-wards in central, western and southern India and Andaman Islands. In Himachal Pradesh, it grows in Jwar forest, Dada Siba, Nalagarh and Tatta pani, Paonta, Mubarakpur areas in dry forests, Neri, Sarkaghat, Hamirpur, Kangra and Nahan (Chauhan, N.S., 319, 482, 1836, 1871 and 1929).

Description – The plant is a sub-deciduous shrub or a small tree with stem 2.5-12.5 cm in diameter, reaching a height of 1.5-4.5 m. Bark is grey in colour and in young parts is covered by stellate hairs. Leaves are simple, palmately nerved or obliquely cordate with serrated margin, scabrous above and pubescent beneath. Flowers are solitary or in sparse

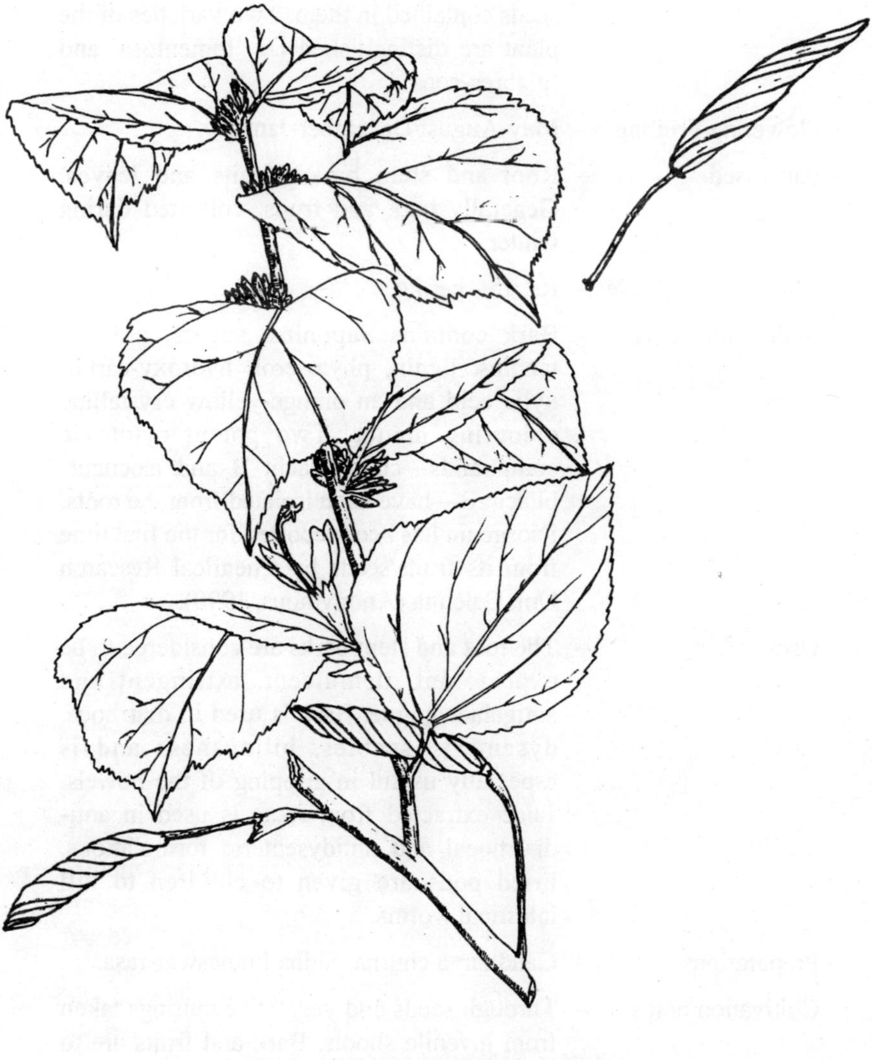

Helicteres isora Linn. (Marorphali)

clusters, 2.5-5 cm long, with red petals turning pale blue when old. Fruits are about 5 cm long; greenish brown, beaked, cylindrical, composed of 5 siprally twisted follicles which on ripening untwine and scatter the small seeds contained in them. Two varieties of the plant are distinguished i.e. 'tomentosa' and 'glabres-cens'.

Flowering/Fruiting – May-August/December-January.

Parts used – Root and stem barks, fruits and leaves. Generally bark and fruits, collected during winter.

Current market rate – Rs. 15/- per kg.

Main ingredients – Bark contains Saponins, sugars, phlobotannins, lignin, phytoserol, hydroxy-carboxylic acid and an orange-yellow crystalline colouring matter. Two potent cytotoxic compounds—cucurbitacin B and isocucurbitacin B—have been isolated from the roots. Diosgenin has been reported for the first time from its fruits/seeds by Chemical Research Unit Calcutta (Anonymous, 1990).

Uses – The root and stem barks are considered to be expectorant, demulcent, astringent and antigalactagogue. Bark is used in diarrhoea, dysentery, scabies, biliousness and is especially useful in gripping of the bowels. Juice extracted from roots is used in antidiarrhoeal and antidysenteric formulations. Fried pods are given to children to kill intestinal worms.

Preparations – Gandharva churna, Sidha Praneswar-rasa.

Cultivation notes – Through seeds and vegetative cuttings taken from juvenile shoots. Bark and fruits are to be collected during winters.

***Heracleum lanatum* Michx.**

Syn: *Heracleum candicans* Wall. ex DC.

Family	– UMBELLIFERAE/APIACEAE
Local/Common name(s)	– Padara, Patrala.
Distribution	– Pakistan to S.W. China 1800-4300 m in open slopes and drier area. In Himachal Pradesh, the plant is frequently found to occur in Pangi Bharmour, in Kullu (Malana), Rohtang Pass; Lahaul, Kinnaur and common in Spiti division, where it has not been extracted because of inaccessible and distant location. In Rohru forest division of Shimla district, it has been collected from Dhola slope water-falls, Khabalgoti 2800 m, Kalapani forest, Chhajpur etc. (Chauhan, N.S., 739, 787, 945, 2225, 2402 and 2489).
Description	– Rootstock stout, perennial, odorous. Stem upto 1.75 cm tall, densely pubescent. Leaves 15-45 cm, pinnate or pinnatifid; rarely 2-pinnate, leaflets oblong, 15-30 cm, pinnately lobed and irregularly toothed, lower surface pubescent. Rays 10-40, upto 15 cm long in fruit. Flowers small, white in compound umbels. Fruit pubescent, obovate 0.75 cm-1.25 cm × 0.4 cm-0.75 cm.
Flowering/Fruiting	– July-August/September-October.
Parts used	– The roots.
Current market rate	– Rs. 20-25/- per kg. (roots).
Main ingredients	– Roots are the chief source of xanthotoxin, a photosensitising coumarin and bergapten. Seeds also are reported to contain coumarin (cf. Mehra et al., 1969:313). The roots of the plant contain 6-9 percent of the total coumarins, the furanocoumarins, of which on

Heracleum candicans **Wall.**

		hydrolysis and methylation, the yield of xanthotoxin is 1.5 percent (Sood and Kalia, 1982). Furocoumarin, heraclenin, heraclenol, xanthotoxol, spondin and cnadicanin have also been reported from the roots.
Uses	–	The xanthotoxin isolated from the plant is highly efficacious in the treatment of leucoderma and psoriasis. It is also used in the preparation of Sun-tan lotions.
Preparations	–	Sun-tan lotions.
Cultivation notes	–	Planted in wild gardens for bold effects and sometimes as specimen plants. They thrive on rich, moist soil. Propagated by seed or division. The foliage has been known to produce a skin rash in susceptible person (BN Bailey and Bailey, 1977). The plant can be successfully grown in temperate forest soils between 2000 m—3200 m. Seedling can be raised in a nursery during March-April and transplanted in June. However harvesting is done after 3 years in October-November. Collection from the wild is done during September-November. Experiments in herbal gardens have shown a high rate of seed germination and survival rate after transplantation is also promising.

Hippophae rhamnoides Linn.

Family	–	ELAEAGNACEAE
Local/Common name(s)	–	Chharma, Sarla, Tirku, Seabuckthorn.
Distribution	–	Pakistan to Himachal Pradesh and Central Asia between 2100 and 3600 m on river sides. Most common at Keylong, Pooh division and Baspa valley in Kinnaur between 2800 m and

4200 m. Shego, Attargu, Kumling, Rangrik etc. in Spiti. (Chauhan, N.S. 2701, 5360).

Description – A varying sized shrub, shoot often modified into spines. Leaves lanceolate-linear, obtuse, with peltate and stellate scales on lower surface. Male flowers in clusters at bases of the shoot; tepals free, suborbicular. Female flowers pedicellate in clusters 2-4; tepals 3-4 mm long, united. Utricles subglobose, succulent, red or orange coloured. Seeds solitary, unequally bilobed, light-black, stony.

Flowering/Fruiting – June-September.

Parts used – Fruits

Current market rate – Fruit Rs. 10/- per kg.

Chemical contents – Alkaloids derived from the plant are harmol and harman. Polyphenols, quercetin and isorhamnetin have also been found. Berries yielded isorhamnitol, while the leaves contain flavones. Serotonin has been extracted from its bark. Plant contains steroids, terpenoids, alkaloids and sugars. Fruits are rich in Vitamin-C. Seeds yield fatty oil.

Uses – Fruits are used for preparation of jelly, syrup and used against pulmonary complaints. In Siberia, the plant is used in tumors, in sunburn preventing preparations; as emollient in prevention of eye region skin wrinkles and in other cosmetic preparations.

Preparations – In China, various parts/products of the plant are utilized in medicines, skin creams, oil, jams, jellies squashes, wines etc.

Cultivation – This is a shrub very easy to propagate. Can be propagated through seeds, suckers and vegetative cuttings.

Hippophae salicifolia D. Don.

Family	– ELEAGNACEAE
Local/Common name(s)	– Seabuckthorn, Chuk, Amli, Chhuchha, Chuma, Kalabis.
Distribution	– An erect shrub or a small tree. Leaves 5-10 cm long, linear lanceolate, softly white-pubescent above when young, densely tomentose or stellately-hairy beneath; margins recurved. Fruit ovoid, yellow in axillary clusters.
Flowering/Fruiting	– June-September.
Parts used	– Fruits, bark.
Current market rate	– Fruit Rs. 10/- per kg.
Chemical contents	– Beta-sitosterol and 2 alkaloids have been detected from the bark.
Properties and uses	– Bark has given promising results as a cure for certain types of cancer. Fruits are highly acidic, the source of vitamin C and made into chutneys. The paste of its bark is used for curing wounds and ulcers. The fruit juice and jellies are prescribed in lung complaints and tuberculosis. Plant is a medicine in tumors, used in sun-burn preventing, medicinal and cosmetic preparations for skin. Bark extract has shown tumor inhibition activity.
Preparations	– Facial creams, ointments, chutneys, syrups, wines, oils etc.
Cultivation	– Like *H. rhamnoides* it can also be propagated through seeds, suckers and shoot cuttings.

Holarrhena antidysenterica Wall. ex DC.

Family	– APOCYNACEAE
Local/Common name(s)	– Kurchi, Kura, Kutaja, Tiptaka, Indrajau (seed).

Distribution – The species occurs almost throughout India upto an altitude of 1250 m often gregariously in deciduous forests and open waste lands and is especially abundant in the sub-Himalaya tract. It is a pioneer plant in newly cleared forests.

In Himachal Pradesh, it is sporadic to common in Nahan Forest division (especially in Paonta Range), Una, Hamirpur, Nurpur division of Kangra district, Bilaspur, Kunihar of Solan district and Sarkaghat area of Mandi district (Chauhan, N.S., 397, 1409). Planted under the university arboretum at Dhaula-Kuan, Distt. Sirmour.

Description – The plant is a tall shrub or a small deciduous tree nearly 2-6 m tall with exfoliating bark. Leaves are stalked, opposite, ovate-oblong, entire and pale-green in colour. Flowers are white in colour and are fragrant. Fruits are upto 30 cm long in pairs, cylindrical, dark grey in colour and with white specks all over. Seeds are 1 cm long having a tuft of 2-2.5 cm long, brown hairs at the top. Bark is grey and nearly smooth outward and of cinnamon colour from inside, quite bitter in taste.

Flowering/Fruiting – May-June/September-October (Maturing in winter).

Parts used – Bark and seeds.

Current market rate – Bark Rs. 16/- per kg. Seeds Rs. 28/- to 30/- per kg.

Main ingredients – Bark contains a large number of alkaloids, the chief among them being conessine, kurchicine, kurchine, conamine, conkurchine, holarrhenine, holarrhinine, holarrhine and lettocine (I.P.C.; 135).

Uses – Bark is used in amoebic dysentery and is prescribed in piles, anaemia, asthma, bron-

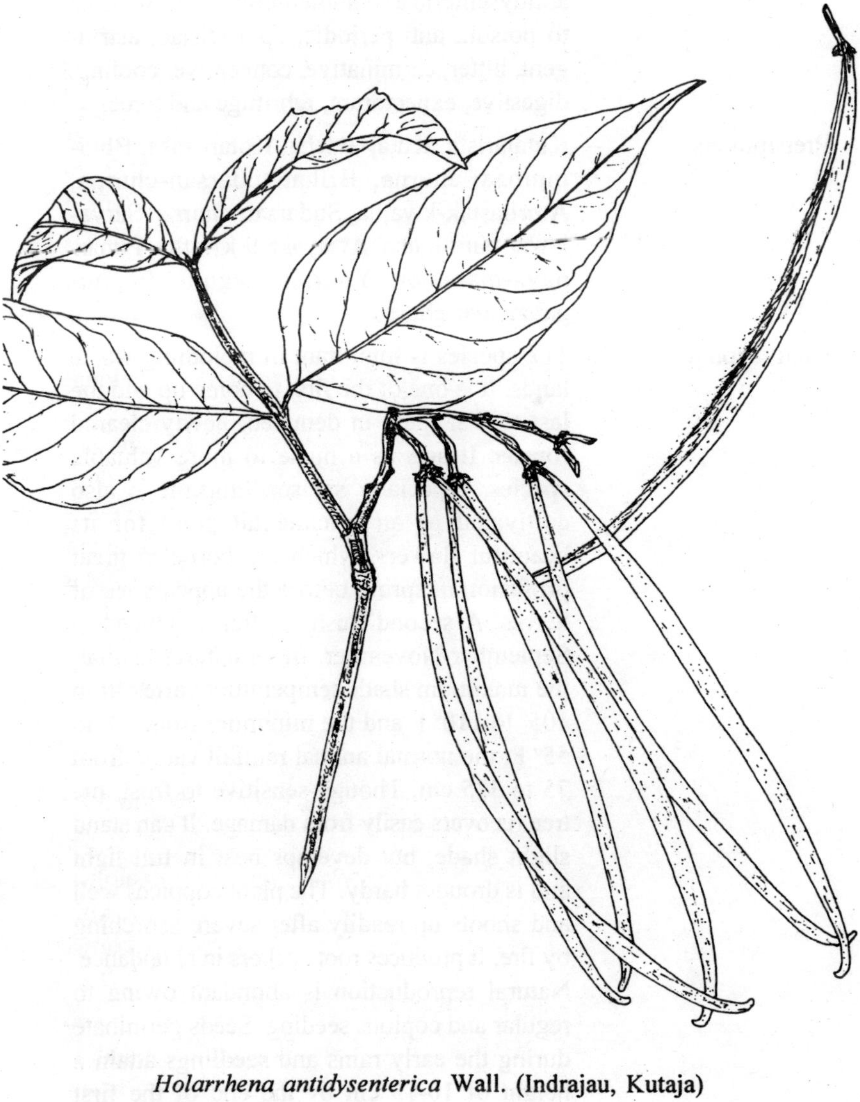

Holarrhena antidysenterica Wall. (Indrajau, Kutaja)

chopneumonia, dropsy, dysurea, influenza, rheumatism, tooth-ache, vomiting and nausea, intestinal worms, dyspepsia, chest complaints and in the diseases of the skin and the spleen. Seeds are astringent, carminative, febrifuge, antidysenteric and anthelmintic. It is antidote to poison, anti-periodic, aphrodisiac, astringent, bitter, carminative, conceptive, cooling, digestive, expectorant, febrifuge and tonic.

Preparations – Kutajarista, Kutaj-avleha, Kotari-rasa, Bhuinimbadi-churna, Brihat-sudarsan-churna, Amritastak-kwatha, Sudarsan-churna, Sarvajawar-hara-louha, Atsayakadi-kwatha, Punarnava-mandoor, Yograjaguggulu, Laghugangadhar-churna.

Cultivation notes – The species is important in reclothing waste lands. It is one of the first to come up and the last to disappear in denuded, newly cleared forests. It acts as a nurse to more valuable species, especially sal seedlings. It is also cultivated as an ornamental plant for its beautiful flowers, which are borne in great profusion in spring before the appearance of leaves. A second flush is often produced in September-November. In its natural habitat, the maximum shade temperature varies from 105^0 to 110^0 F and the minimum from 30^0 to 55^0 F; the normal annual rainfall varies from 75 to 375 cm. Though sensitive to frost, the tree recovers easily from damage. It can stand slight shade, but develops best in full light and is drought hardy. The plant coppices well and shoots up readily after severe scorching by fire. It produces root suckers in abundance. Natural reproduction is abundant owing to regular and copious seeding. Seeds germinate during the early rains and seedlings attain a height of 10-15 cm by the end of the first year. Weeding and watering stimulates growth in subsequent years. Growth is rapid and the mean annual girth increment is recorded to be at 1.80-2.25 cm. Artificial reproduction can be secured both by direct sowing and by transplanting. Fresh seeds have a high percentage of germination, but if stored for more than a year, the viability is low.

Humulus lupulus Linn.

Family	– MORACEAE
Local/Common name(s)	– Hop, Hops.
Distribution	– North-West Himalayas, a native of North America and perhaps of North Asia. Cultivated in commercial scale in Lahaul valley and has to some extent been extended to Kinnaur and scarce in Spiti.
Description	– A perennial, twining, scabrid herb, 4-6 m in height. It is generally dioecious; monoecious plants are occasionally found. Aerial stems are oppositely branched, angular. Leaves are opposite, stipulate; lower larger, 3-5 lobed; upper smaller; ovate or cordate, serrate. Male flowers are small, greenish yellow, arranged in paniculate cymes. Female flowers are in a cone like inflorescence resulting into scaly bracteate fruiting body.
Flowering/Fruiting	– June-July/August-September.
Parts used	– Mature female cones.
Current market rate	– Rs. 35 to 51/- per kg. (dried hops).
Main ingredients	– Essential oil, choline asparaginer and lupulin (Chopra et al., 1956); bitter substances like lupulones, humulinones and hulupones.

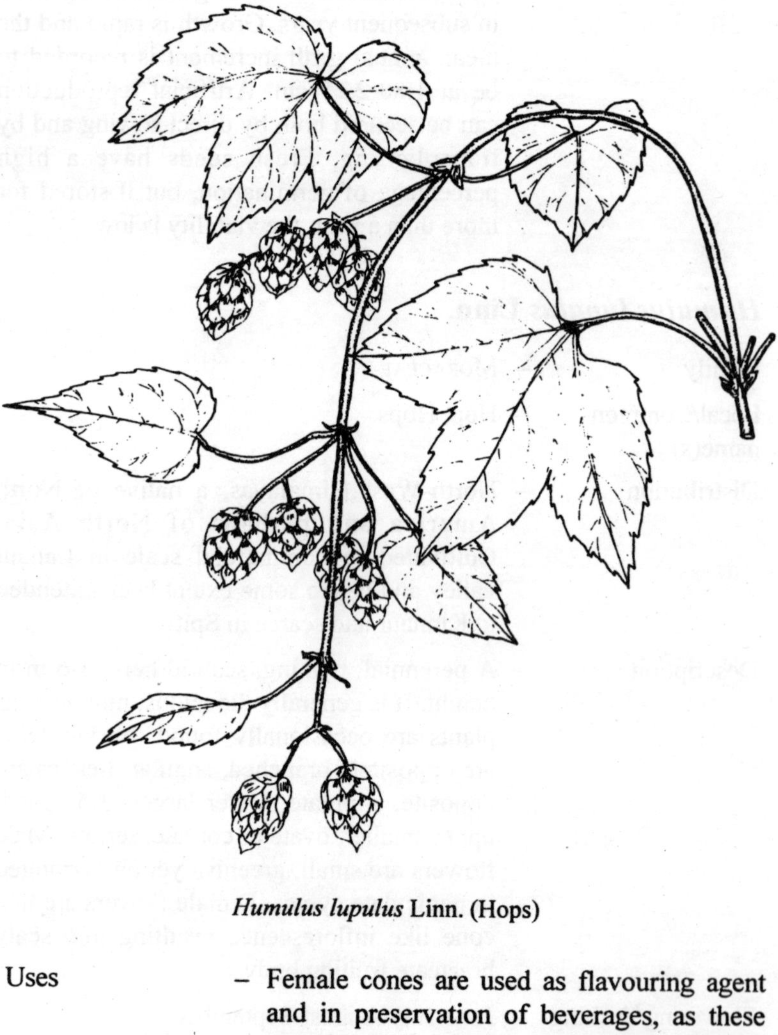

Humulus lupulus Linn. (Hops)

Uses – Female cones are used as flavouring agent and in preservation of beverages, as these contain alpha-acids, hop oil and hop extract. It is also used as tonic, stomachic and diuretic. It prevents worms and allays the nervous tension and prevents indigestion. Infusion is regarded as tonic and sedative (Chopra et al., 1956).

Preparations	–	Chiefly used in beer industry.
Cultivation notes	–	Hop plant can be raised both from seed and by vegetative method, but commercial propagation is by vegetative means and the seed is used only for breeding purposes. It is propagated by cuttings, layering and by way of runners and lateral branches. Planting should be done in autumn before the soil becomes cold so that plants may establish and have a good start in spring. Another good planting period is 2nd half of February.

Hyoscyamus niger Linn.

Family	–	SOLANACEAE
Local/Common name(s)	–	Dipya, Parasikayavani, Khurasani ajwain Tukhlang, Thuklang
Distribution	–	The plant is distributed in the Western Himalayas from Kashmir to Kumaon at altitudes of 1600-3000 m (The Wealth of India, Vol. V, 1959:151); frequent in Spiti near villages and roadsides and sporadic elsewhere in Lahaul and Kinnaur regions (Chauhan, N.S., 2005, 2268).
Description	–	The plant is an erect, viscidly hairy, foetid annual or biennial, upto 1.5 m tall. Leaves are radical and cauline, coarsely dentate, to pinnately lobed. Flowers are yellowish green, having dark purple centre, sessile or subsessile and are in terminal scorpioidal cymes. Capsules encircled by globular base of the enlarged calyx. Seed are knee shaped and are about 1.5 mm long, brown in colour and marked with fine, but conspicuous reticulations.
Flowering/Fruiting	–	July-August.

Hyoscyamus niger Linn. (Khurasani ajwain)

Parts used	–	Dried leaves and flowering tops, seeds.
Current market rate	–	Seeds Rs. 35/- per kg. Leaves Rs. 15/- per kg.
Main ingredients	–	Alkaloids, hyoscyamine, hyoscine or scopolomine, volatile-bases; a bitter glucoside, hyoscypicrin, choline, mucilage and albumin and potassium salts have been isolated from the plant.
Uses	–	*Hyoscyamus* has anodyne, narcotic and mydriatic properties. It is principally employed as a sedative in nervous affections and irritable conditions, such as asthma and whooping cough and is substituted for opium in cases, where the latter is inadmissible. It is also used to counteract the gripping action of purgatives and to relieve spasms in the urinary tracts. In veterinary practice, it is used as a urinary sedative. The leaves have been employed, externally, to relieve pain, but their utility for the purpose is not well established.
Preparations	–	Vedantak rasa, Parasiyadi-churna etc. in Ayurvedic medicines in addition to its use in modern pharmaceutical preparations.
Cultivation notes	–	The plant requires a well drained, fertile, sandy loam or silt loam. Forest soils and waste lands, rich in organic matter and free from weeds are suitable. Propagation is by seed and pre-treatment with concentrated sulphuric acid for about 75 seconds followed by washing or exposure to freezing temperature are reported to facilitate the germination. Seeds are sown broadcast or they may be sown in the nursery and the seedlings transplanted in the field. The distance between the plants being 10 cm. When sown broadcast,

the seed rate is about 1 kg per acre. On account of their small size, seeds are mixed with fine dry earth before sowing. Shallow sowing (0.5 cm below surface) is usually recommended. The seeds germinate in 2-4 weeks, depending upon the weather.

When rosettes of radical leaves appear, the plants are thinned out and the field hoed and weeded. During transplantation, the root is likely to be injured; better results are obtained by transplanting fair sized plants. Frequent irrigation in the early stages is necessary. Application of inorganic nitrogenous fertilizers and ample sun-shine are favourable for the development of the active principles. The plant has also been grown as a mixed crop.

The main pests are potato beetle and its larvae. Spraying of plants with rotenone and pyrethrin and rotation of crops are effective in warding off the pest. The plant is affected by few virus diseases.

In the case of the annual crop, the leaves and tops are picked at the time of flowering while in the case of the biennial, the first year leaf is collected late in summer and the second year leaf at the time of flowering. Leaves are dried in the sun for 3-4 days or in the shade for 7-10 days before packing. The yield of the dried leaves is about 14 percent of the green material. A yield of 2-2.25 quintals per acre is reported under favourable conditions.

Hypericum perforatum Linn.

Family	–	HYPERICACEAE
Local/Common name(s)	–	St. John's Wort, Bassant

Hypericum perforatum Linn.

Distribution	–	Shimla in shady, damp forest, Western Himalaya, 1000-2600 m. Sporadic between Larot-Chansal dhamkhari; Hiuni-nala and Shogra thatch between 2500-2800 m in Pabbar valley (Chauhan, N.S. 1114).
Description	–	A perennial herb. Stem 30-60 cm tall, 2-angled. Leaves oblong, 0.8-2.5 cm; obtuse veins pellucid, lower surface pale. Flowers upto 2.5 cm diameter in terminal corymbs. Sepals narrowly lanceolate acute, margins often black-dotted. Petals black-dotted on the margins. Stamens in 3 bundles; anthers black-dotted. Ovary 3-celled, styles 3, twice as long as the ovary. Capsule 0.75-0.85 cm.
Flowering/Fruiting	–	April-October.
Parts used	–	The whole plant.
Current market rate	–	Rs. 50-60/- per kg.
Chemical contents	–	It's aerial parts contain anthraquinone derivatives, hypericin, usigtoercin and protohypericin. Also contains kaempferol and quercetin and rutin. Quercetin found in Hypericum can enhance or potentiate effects of hypericin. Also contains phenolic acids and tannins.
Properties and uses	–	An oil (oleum hyperici), obtained by infusing fresh flowers in olive oil, is used externally for sores, wounds, ulcers, swellings, and sometimes for rheumatism and lumbago. Also employed as sunburn oil and as cosmetic application to skin. An aqueous extract of the plant is used for preparing an ointment which is a hair restorer. The plant is stated to possess sedative and astringent properties. It has been used for excitability neuralgia, fibrrositic, sciatica, wounds and specifically for menopausal neucrosis.

Preparations – It is used extensively in homoeopathic preparations as well as several herbal products including alcoholic beverages.

Cultivation – The plant is cultivated through seeds which are quite minute and need utmost care while sowing these in the nurseries and watering afterwards.

Hyssopus officinalis Linn.

Family – LABIATAE/LAMIACEAE

Local/Common name(s) – Hyssop, Juffa, Juffa-yabis, Tengu, Tyangu.

Distribution – The plant is found in the Himalayas from Kashmir to Kumaon at altitudes of 2600 m-3500 m. In Himachal Pradesh, the plant is found in Chamba, Kinnaur and Lahaul & Spiti (Chowdhery & Wadhwa, 1964). It has been collected by the author from Pin Valley where it is more prominent, moderate in Kishori bridge below Mani Mahesh, Attargu and sporadic in Kaa loops (Chauhan, N.S., 1227, 2812). Some plants have recently been spotted at Kaza, which have come up as a result of seed shattering from the specimens collected by the author during the process of drying in the rest house compound.

Description – The plant is an aromatic, shrubby perennial, 30-60 cm high. Branches are erect or diffuse and the leaves are sessile, linear, oblong or lanceolate, obtuse and entire. Flowers, blue purplish in colour, aggregated in axillary tufts and are arranged unilaterally on terminal branches. Nutlets are dark brown, narrow, trigonous and smooth.

Morphology of the official parts used: Several, upright, ascending more or less

Hyssopus officinalis Linn. (Tengu)

branched stems are arising out of vertical tap root. A sharp, aromatic smell is emitted out by numerous sunken glandular scales present on the leaves and the shorter shoots emerge at the nodes giving a whorled appearance. Inflorescence is spike like 5-15 cm long, bearing violet blue flowers; 3-7 flowers grouped together in leaf axils; corolla is tubular with prominent, exerted, 4 stamens. The aerial flowering shoots constitute the drug of commerce.

Flowering/Fruiting – July-October.

Parts used – The leaves and the flowering tops.

Current market rate – Rs. 90/- per kg.

Main ingredients – The sample collected from Pin valley in Spiti yielded 1 percent essential oil. The volatile oil consists of the betone 1-pinocamphene, β-pinene, L-pinene, camphene, 1-pinocampheol, an aldehyde, bicyclic sesquiterpenes and sesquiterpene alcohols of the cardinen type and small amounts of primary and secondary alcohols.

Besides the volatile oil, the herb contains fat, sugar, choline, tannins, carotin and xanthophyll. The tops contain ursolic acid and flavonoid glucosides-hesperidin, diosmen, besides the bitter principle marubion, resin, sugar and gums.

Uses – Hyssop is considered stimulant, carminative and pectoral and used in such medicines. An infusion of the tea, prepared from the plant is said to be effective in nervous disorders, toothache; in pulmonary, digestive, uterine and urinary troubles. The leaf juice is employed for the expulsion of the round worms. The crushed herb is applied as a resolvent and vulnerary. Steeped in hot water, it is used as a fomentation for wounds, sprains and sprains; muscular rheumatism and for clearing blackness due to the blows. It is also used as a salve in catarrhal ophthalmia. The hyssop oil is used as a flavouring agent in bitters and tonics and is also used in perfumes. The flowering tops are used as a substitute for saffron by the Tibetan Medical Centre, Dharamsala. It is recommended for use as a gargle in throat inflammations and hoarseness and internally for chronic bronchitis and bronchial asthma. (The Wealth of India, V. 1959:160; Ambasta, 1986; Chauhan, 1989; Dofler & Hosett, 1989).

Preparations – Used in several Ayurvedic and Unani preparation in the name of 'Juffa and Juffa-yabis' respectively. As Tengu/Tyangu, it is an important ingredient of Tibetan medicines.

Cultivation notes – The cultivations can be done through seeds and divisions of several stems arising from the common rootstock.

Inula racemosa Hook f.

Family – COMPOSITAE/ASTERACEAE

Local/Common name(s) – Poshkar, Poskarmul, Pushkarmool, Manu.

Distribution – The plant is found in the North-Western Himalayas at altitudes ranging from 1600-4200 m (The Wealth of India, V. 1959:236). In Himachal Pradesh, it is found in Parwati-valley of Kullu and found in captive cultivation in Lahaul division. (Chauhan, N.S., 1381, 6632).

Description – The plant is a stout herb, upto 1.5 m tall. Leaves are leathery, rough above, densely hairy beneath and toothed. Radical leaves are

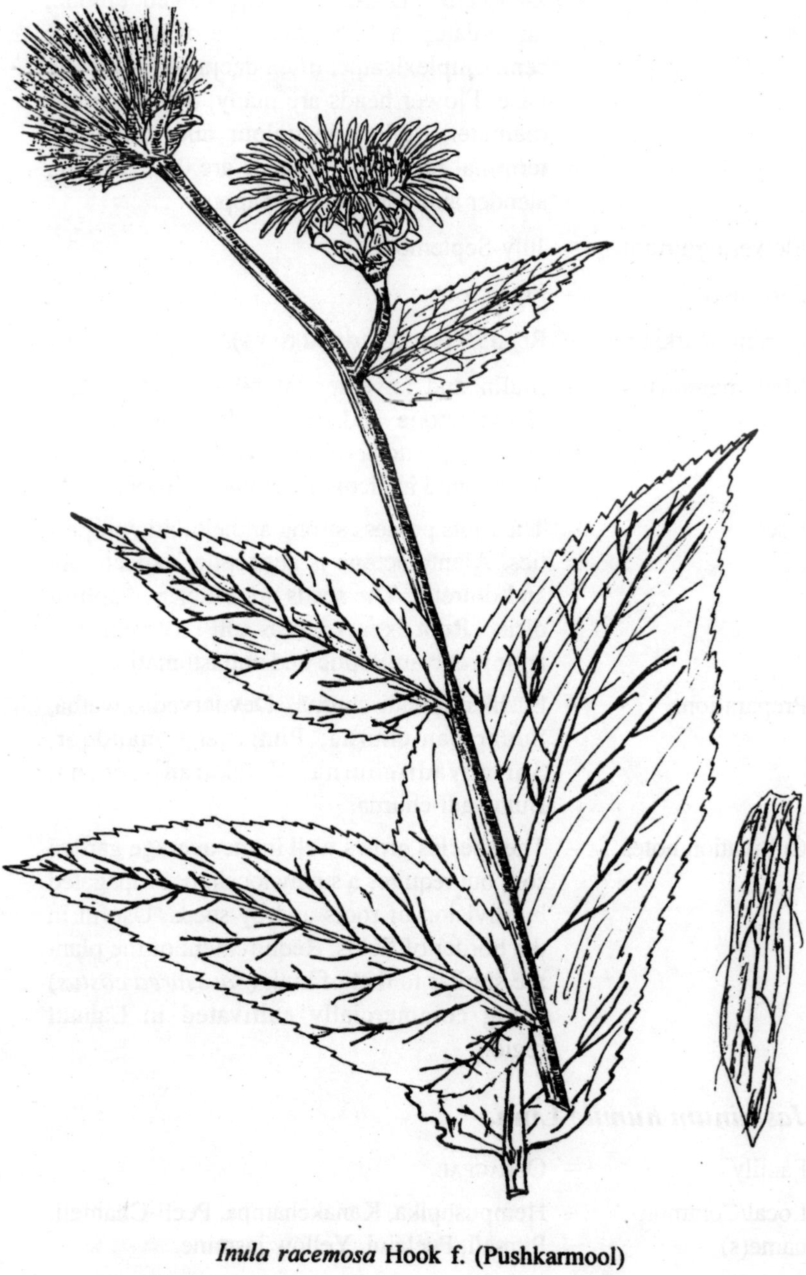

***Inula racemosa* Hook f. (Pushkarmool)**

		20-45 cm × 12.5-20 cm long, stalked, elliptic, lanceolate; cauline leaves are ovate-oblong, semi-amplexicaule, often deeply lobed at the base. Flower heads are many; 3.5-5.0 cm in diameter, yellow in colour and arising in terminal racemes. Achenes are 0.4 cm long, slender and with reddish pappus.
Flowering/Fruiting	–	July-September.
Parts used	–	Fresh roots.
Current market rate	–	Rs. 35/- per kg. (dried roots).
Main ingredients	–	Inulin and an essential oil containing iso-alantolactone and di-hydralactone. Alanto-lactone, β sitosterol and its glucosides have been found in alcoholic extract of roots.
Uses	–	The roots possess strong anthelmintic proper-ties. Alantolactone is antiseptic, expectorant and diuretic. The seeds are bitter and aphro-disiac. Root extract acts as anti-inflammatory, antipyretic/antiseptic and antiasthmatic.
Preparations	–	Pushkarmuladi-churna, Devdarvadi-kwatha, Sudarshan-churna, Punarnana mandoor, Shringyadi-churna, Puskaradi-churna, Bukuvadi-churna.
Cultivation notes	–	The species grows well in the average garden soil, but requires a sunny location. Propagated by division of roots and by seeds. Useful in the border of fields. Requirement of the plant are similar to that of kuth (*Saussurea costus*) being commercially cultivated in Lahaul region.

***Jasminum humile* Linn.**

Family	–	OLEACEAE
Local/Common name(s)	–	Hempushpika, Kanakchampa, Peeli-Chameli, Pitmali, Peelajui, Yellow jasmine.

Distribution – The plant is common in the outer Himalayas, Jaunsar and Tehri Garhwal between 1200 m-2800 m. It is also found throughout the sub-tropical Himalayas from Kashmir to Nepal upto an altitude of 2800 m and in Mount Abu and the hills of South India above 1500 m. (The Wealth of India, Vol. V, 1959:283 and Gupta, 1969).

In Himachal Pradesh, the plant is found in the districts of Mandi, Kullu, Chamba, Kangra, Shimla, Solan and Sirmour (Chauhan, N.S., 2430: Tikbri Khabol (1800-2400 m); 2634: Gourie-Bagi (2200 m).

Description – An erect, diffuse shrub with profuse, angular branches. Leaves are alternate, 5-10 cm long; leaflets are 3-7 in number and are oblong or ovate; end one 2.5-5 cm. Flowers are yellowish in colour in terminal corymbose cymes. Fruits are black, when ripe with crimson juice.

Flowering/Fruiting – September-December.

Parts used – Leaves and flowers.

Current market rate – Not available.

Main ingredients – Flowers yield a volatile oil, whereas the leaves yield salicylic acid.

Uses – The paste of the leaves is applied locally in skin diseases, wounds and ulcers. Leaves are chewed in stomatitis and toothache. Oil prepared out of the leaves is used in earache. Flowers are used in the form of a paste in the diseases of the eye, headache etc. It is also considered as an antidote in the poisonous cases, besides being used as an astringent, tonic to the heart and bowels. Roots are used in ringworm infections and the juice of the plant is useful in chronic sinuses and fistulas.

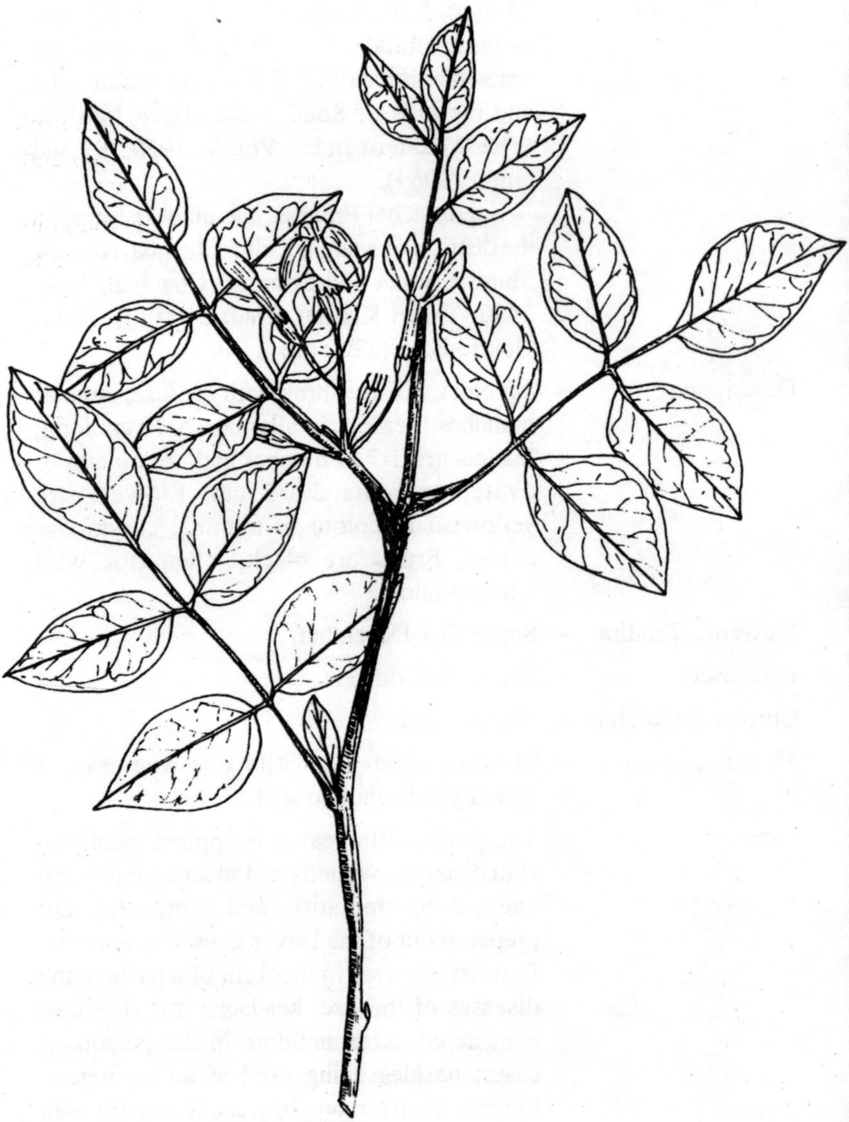

Jasminum humile Linn. (Pitmali)

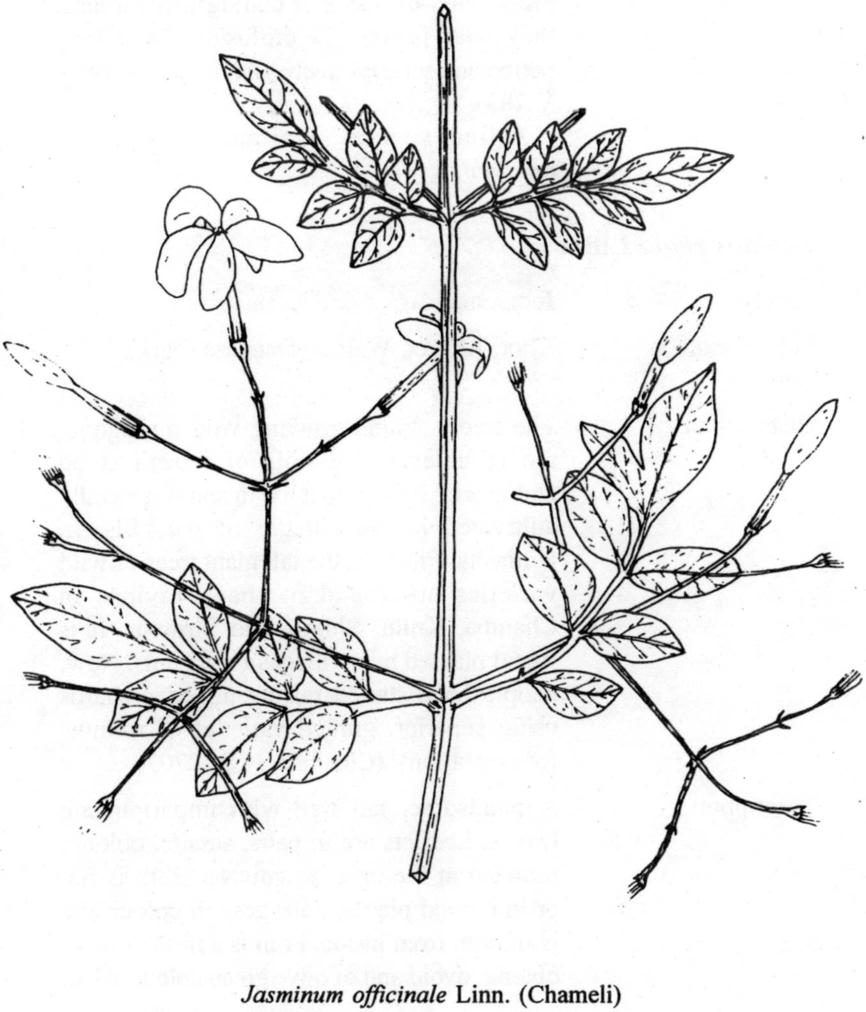

Jasminum officinale Linn. (Chameli)

	(Chopra et al., 1965:144). Useful in the treatment of skin eruptions, wounds and ulcerations. Oil from the flowers is used in perfumery and cosmetics.
Preparations	– Jatyaditaila and Jatyadimalhar.
Cultivation notes	– The plants are grown in gardens for their fragrant yellow flowers. They are easily propagated by seeds or cuttings. In gardens, they bear flowers in profusion for a long period and set seed freely. (Anonymous, 1959, V:283). Allied species available: *Jasminum officinale* L.

***Juglans regia* Linn.**

Family	– JUGLANDACEAE
Local/Common name(s)	– Khor, Akhrot, Walnut, Dandasa (bark).
Distribution	– The tree is found growing wild throughout the Himalayas and hills of Assam at an altitudes of 1000 m to 3300 m and is generally cultivated in the villages of the hills. In Himachal Pradesh, the tall giant trees of wild varieties are found in shady ravines in Chamba, Kullu, Shimla and Sirmour. It is found planted near villages for its nuts. Now, people have started raising mixed orchards using superior, grafted varieties of walnuts for plantations. (Chauhan, N.S., 2707).
Description	– A handsome, tall tree with imparipinnate leaves. Leaflets are in pairs, sessile, oblong, rounded at the base, acuminate. Bark is flat or in curved pieces, dark grey in colour and is smooth from inside. Fruit is a fleshy drupe, oblong, ovoid and of olive green colour, when

Juglans regia Linn. (Akhrot)

immature. Shell is hard and thick. Kernels are pale white, thick and oily. (Chauhan, N.S., 1984).

Flowering/Fruiting – March-May/Fruit are harvested during September-October.

Parts used – Root bark, leaves and fruit. Bark of root and stem is popularly known as Dandasa and is illegally traded.

Current market rate – Dandasa is a restricted item and sells at a very high price. Nuts (fruits) sell Rs. 30-75/- per kg depending upon the quality.

Main ingredients – Juglandic acid, Juglonone; Barium and Arsentic in traces. Fruits contain the oxalic acid, resin and a fixed oil of a drying nature, like the linseed oil. Kernels contain B group of vitamins and several mineral elements. Unripe fruits are the richest source of vitamin C (Ascorbic acid).

Uses – The leaves are alterative and given in scrofula, rickets and leucorrhoea. The oil is a sure and mild laxative, cholagogue and is given in torpid liver. The decoction of the bark is galactafuge and is used to stop mammary secretions. It is used as an astringent to check diarrhoea and menorrhagia and for gargling in case of a sore throat. The kernel is supposed to possess aphrodisiac and vermifuge properties. It is given in general debility in children, for intestinal worms, and as an alterative in rheumatism. The plant is of immense value in confectionery and ice creams and is used as an article of food, when dried. The oil is used as a drying agent for paints, oil paints, printing inks, varnishes and for making soaps. (Chopra et al., 1956:145, Khory and Katrak, 561-62). Locally, the juice of the unripe fruit is considered to be useful in alopacia, whereas the root bark juice mixed in mustard oil is used in leucoderma, (Chauhan, N.S., 1984).

Bark is used as a dye, for cleaning teeth; whereas the leaf juice (water extract) of leaves has strong bactericidal activity against microorganisms. Pulverised walnut can be used for the manufacture of decolourising charcoal and activated carbon. (Umrao Singh et al., 1983).

Preparations – Cotyledons are added in tonics. Dandasa is used in dyeing and tannin industries.

Cultivation notes – Under natural conditions, the fruits fall to the ground under and around the tree. The exocarp cracks and rottens off. The nuts, however, are subject to attack by birds, monkeys and rodents and large quantities of them are destroyed. A covering of earth or debris over the nuts and a fair degree of warmth and soil moisture are necessary for germination. Natural reproduction takes place on gentle slopes with loose, but deep, fairly moist soil. Where boulders and rock fragments are plentiful, they act as a protection to nuts. Good crops of walnut are found on detrital accumulations, formed by the weathering of cliffs or erosion of hill sides and on deep loose rubble on landslides in the open. Walnut is a light demander, though it withstands slight shade in young stages. It coppices well and the plantation for timber purposes may be raised by direct sowing or transplanting nursery seedlings. In the former case, the nuts are sown at a depth of 5 cm. For transplanting, nursery seedlings are raised in the prepared fields of rice mould in warm situations. Nuts are usually sown in December-February in drills 20-22 cm apart and the distance between nuts being 7.5-10 cm. The beds are watered regularly till the rainy season. Seedlings may be transplanted in the following cold season, after trimming the tap root, if necessary. If larger seedlings are required for transplanting, the plants are pricked out with a spacing of 30-45 cm and tended till the second winter. Planting in the field should be somewhat close in order to prevent branching. On gentle slopes or level ground, a spacing of 2 m × 2 m is considered suitable; on steeper slopes, the spacing allowed is 1.5 m; and in contour lines, it is kept at 2.5-2.7 m apart.

Juniperus communis Linn.

Family – CUPRESSACEAE

Local/Common name(s) – Hapusha, Hauber, Bethar, Bitharh, Theleru, Chhershup, Chhershi, Pethri, Juniper.

Distribution – In Himachal Pradesh, the plant grows between 3000 m-4500 m elevation in Chhota and Bara Bhangal, Rorag thatch in Kullu, Chounti, Kalga Pattan, Chansil, Chheradhank in Rohru Forest Division, Chitkul and Sangla Kanda in Kinnaur, Mani Mahesh in Chamba and Churdhar in Sirmour district (Chauhan, N.S., 746, 2029, 2038 and 2066).

Description – The plant is a dense, dwarf, procumbent, bushy, evergreen shrub, or a near bush 60-150 cm high with upright branches, or a small tree reaching upto 9 m high. Bark is chocolate brown, peeling off in papery shreds; leaves are numerous and in whorls of 3, persistent, spreading at right angles and rigid; young twigs are divaricate; fruit is subglobose. The plant is very variable with a number of geographical varieties and garden forms. It becomes prostrate, not more than 60-70 cm in height at higher altitudes in the Himalayas.

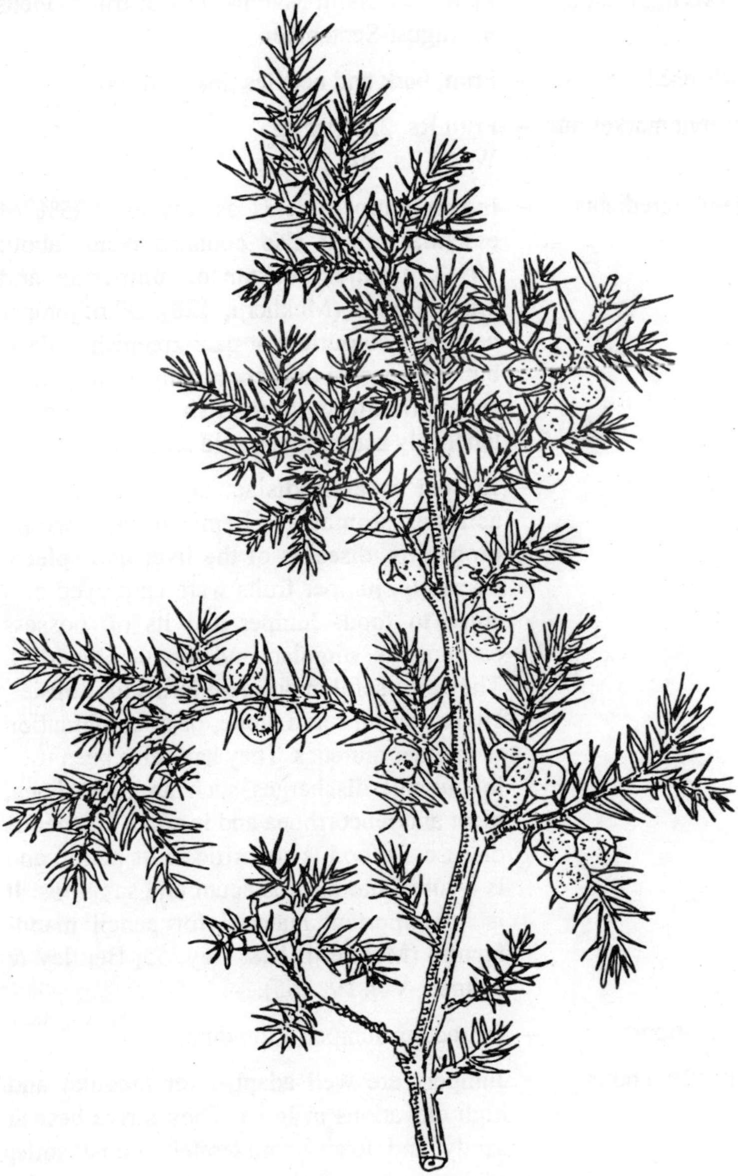

Juniperus communis Linn. (Hauber)

Flowering/Fruiting – Plant flowers in March-April & fruits ripens in August-September.

Parts used – Fruit, bark and needles (leafy twigs).

Current market rate – Fruit Rs. 22/- per kg.
Wood Rs. 50/- per kg.

Main ingredients – Indian juniper yields as low as 0.25% of essential oil. It also contains resin (about 10%), a bitter substance, juniperine and organic acids. (Mukherji, 128). Oil of juniper berries is colourless or pale-greenish yellow, but on exposure thickens rapidly, turns yellow and ultimately to reddish brown simultaneously acquiring an acid reaction.

Uses – Juniper is aphrodisiac, styptic, useful in asthma, stomatitis, hemicrania, chronic bronchitis, diseases of the liver and spleen. Formerly, juniper fruits were employed as a spice to food. Juniper and its oil possess carminative, stimulant and diuretic properties. They are useful in different forms of dropsies, either administered alone, or in combination with other diuretics. They have also been used in mucous discharges such as gonorrhoea, gleet and leucorrhoea and in some cutaneous diseases. Wood is sudorific in its action and is a substitute for guaiacum and sassafras. It is an important material for pencil manufacture (Mukherji, 128; Dey, 55; Bentley & Trimen, Vol. IV, 265).

Preparations – Extractum Juniperi Liquidum.

Cultivation notes – Junipers are well adapted for medium and high elevations in India. They thrive best in sandy and loamy, moderately moist soils; rocky and gravelly grounds. They prefer sunny, open conditions and can be propagated by seeds, cuttings and layers. Viability of seeds when stored in a cool dry place can be maintained for several years. When sown, they often take one year to germinate. Germination may be hastened by soaking the seeds in hot water for a few minutes, before sowing. (The Wealth of India, V:306).

***Juniperus macropoda* (Boiss.)**

Family	–	CUPRESSACEAE
Local/Common name(s)	–	Indian junipers, Chalai, Dhup, Leur.
Distribution	–	Pakistan to Uttar Pradesh, 2400-4300 m. Dry river valleys, gregarious; forming open forests in Lahaul (Temperate and Cold regions of Northern Hemisphere), sporadic in Pooh (Kinnaur), a dilapidated patch at Giu and Attargu hill in Spiti (Chauhan, N.S. 1625, 2244, 3585).
Description	–	A tree upto 18 m tall; with light bark, open foliate with spreading, sharp pointed leaves on lower branches. Scale leaves are on the upper branches and branchlets. Ripe fruits are bluish-black; resinous having 2 or more seeds.
Flowering/Fruiting	–	May-August/September.
Parts used	–	Wood and berries (Fruits).
		Most valuable for furniture, building and pencil making, used for fuel and charcoal. Berries are useful in colic, cough and pectoral affections, diarrhoea, impotency, indigestion; leucorrhoea, paralysis, piles and skin affections. Fruit and resin used in ulcers and nasal polyps.
Current market rate	–	Rs. 25-30/- per kg.
Chemical contents	–	Berries yield essential oil constituting Sugiol, (10-non-acosanol, beta-sitosterol glucoside, flavonoid, junipodin and its glycosides-

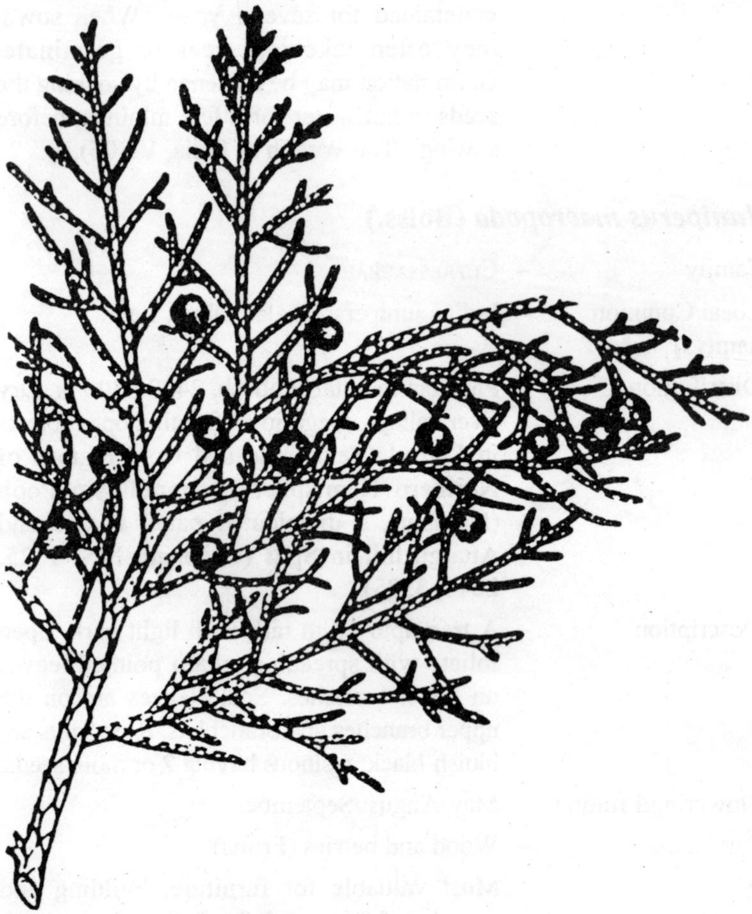

Juniperus macropoda (Boiss.)

junipin. Also contain hypolaetin. Essential oil constituents from leaves include stilbenes; 2 new isoflavons, junipegenin B & C. Essential oil from berries contains resin-juniperin and oxalic acid; alpha pinene with small amount of camphor, cardinene, juniper-camphor and terpenenol etc.

Preparations – Essential oil is mostly used in gin flavours and cordials. Berries as 'Hauber' are used in ayurvedic formulations.

Cultivation – Cultivation through seeds raised in polybags. But, the germination being very-very poor. It has become difficult to raise it in large scale. Seeds take 1½ years to germinate. Natural regeneration is good in protected areas.

***Juniperus recurva* Buch. Ham. (*J. squamata* Buch.-Ham.)**

Family – CUPRESSACEAE

Local/Common name(s) – Phillu, Thelu, Bhedara, Wetyar, Betharh, Shingpillu, Dhoop.

Distribution – The plant is distributed almost throughout the temperate and alpine Himalayas and Assam, at altitudes of 2200 m-4500 m. (The Wealth of India, Vol. V: 309). The plant is found in the districts of Kinnaur and Shimla, Sirmour, Chamba, Kangra, Lahaul & Spiti in H.P. upto an elevation of 4000-4500 m (Chauhan, N.S., 746, 2029, 2038 & 2066 and 2066).

Description – A prostrate or decumbent shrub or a small tree of graceful, drooping habit, often attaining a height of 10-12 m in favourable localities at lower elevation. Stems creep over ground and rock boulders. Leaves are owl shaped; flowers monoecious or dioecious, fruit ovoid and wood is light red in colour.

Flowering/Fruiting – June-September.

Parts used – Wood, leaves, twigs and the fruits.

Current market rate – Fruit Rs. 20-22/- per kg.

Dhoop Lakkar (branches and twigs for incense Rs. 50/- per kg.

Main ingredients – Wood yields resin and the fruits give essential oil (0.46-0.88%).

Uses – Wood is locally used as a fuel. It is suitable for use as a pencil wood and is used as a coffin wood in Burma. The wood, leaves and twigs are used as incense. The smoke from the green wood is reported to be emetic. Ethyl alcohol extract of aerial parts has been reported to be anticancer.

Preparations – Mostly oil is extracted from berries for various uses and is largely used in compounding gin flavours and cordials. Most important use in Europe is for the preparation of gin and gin type alcoholic beverages. (Dey, 1980).

Cultivation notes – The plant is propagated by nursery seedlings, raised from cuttings of prostrate stems and planted out early in the rainy season. (The Wealth of India, Vol. V, 310).

Other species – *Juniperus macropoda* Bois.

Jurinea dolomiaea Boiss.

Syn: *Jurinea macrocephala* Benth.

Family – COMPOSITAE/ASTERACEAE

Local/Common name(s) – Dhoop, Jari-dhoop.

Distribution – Pakistan to East Nepal between 3000-4300 m in open slopes. In Himachal Pradesh, dhoop is commonly found in alpine pastures/slopes between 3500 m-4500 m in Pangi-Bharmour (Mani-Mahesh), Dainasar and Thamsar of Kangra, Rorag thatch in Kullu, Chitkul, Sangla and Manjiban Kandas of Kinnaur; Chandranahan, Chansal, Gorju, Kalgapattan and Muraldanda of Shimla district. (Chauhan, N.S., 134, 799, 981, 1140, 1226, 1292, 2326,

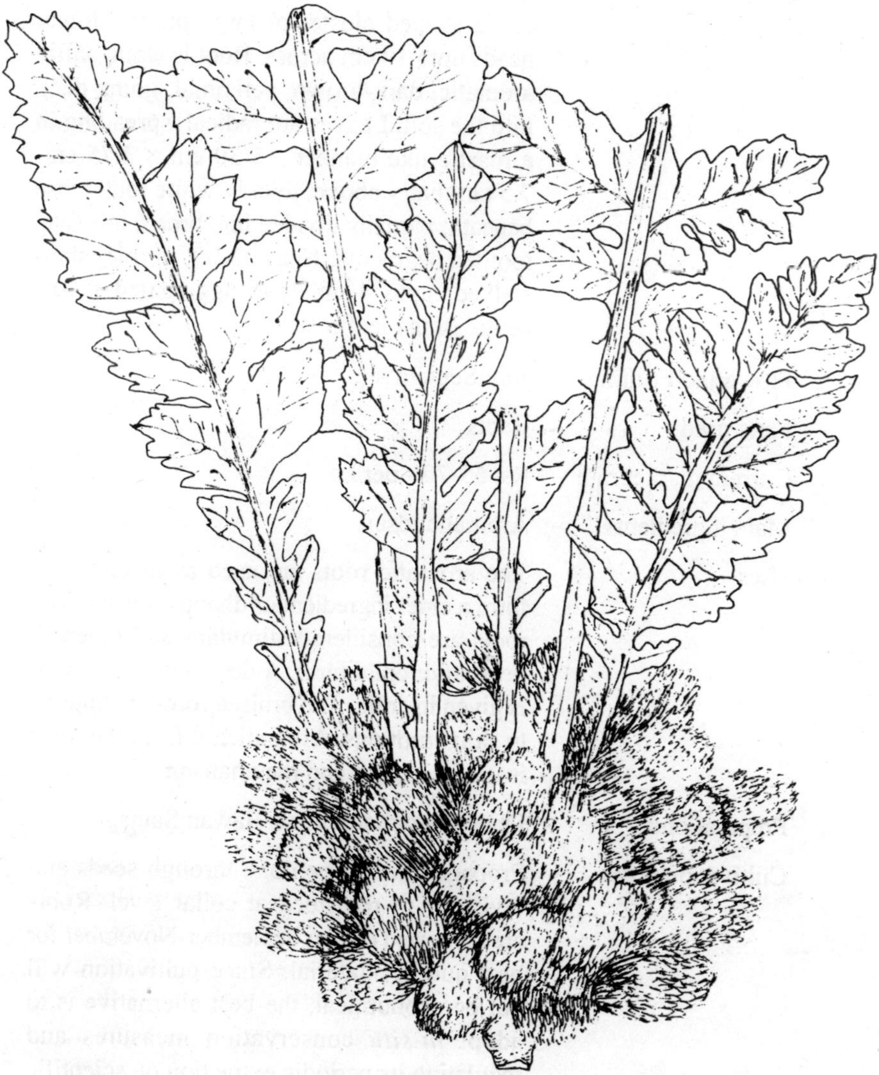

Jurinea dolomiaea **Boiss. (Dhoop)**

2349 and 2567) Dhoop of Chamba is more in demand because of its high resin content.

Description – A prostrate, perennial herb with a central dome shaped cluster of large purple flower heads upto 10 cm across. Root is stout, softly aromatic, dark-brown, perennial, going deep into the soil. Leaves are radical, spreading in a rosette like manner, 15-30 cm × 3.75 cm-7.5 cm; hairy above, thickly white tomentose beneath; pinnally divided into broad, lobulate, toothed segments. Heads 3-30, purple, short stalked, upto 4 cm long aggregated in the centre of the leaf rosette.

Flowering/Fruiting – July-September.

Parts used – Roots

Current market rate – Rs. 55-70/- per kg.

Main ingredients – Aromatic resin

Uses – The aromatic roots are used as incense and form a chief ingredient of dhoop industry. The roots are considered stimulant and given in fever after child birth. A decoction of the root is given in colic. The bruised roots are applied to skin eruptions. Aromatic oil from the roots is useful in gout and rheumatism.

Preparations – Dhoop, agarbatties and Hawan Samagri.

Cultivation notes – Propagation can be done through seeds and divisions of rootstock at collar level. Roots are harvested during September-November for best quality material. Since cultivation will not be economical, the best alternative is to adopt *in-situ* conservation measures and regulating its periodic extraction on scientific bases by employing skilled labour.

***Justicia adhatoda* Linn.**
Syn: *Adhatoda vasica* Nees.

Family	– ACANTHACEAE
Local/Common name(s)	– Vasa, Adusa, Vasaka, Basuti, Vasika, Banshte, Bashti.
Distribution	– Throughout in plains of India, and in the Sub Himalayan tracts upto 1450 m. In Himachal Pradesh, it is common in patches at various places in Shivalik belt especially at Mandi-Drang, Nahan-Paonta, Oachghat, Neri-Langna, Ahju, Jogindernagar, Mahanta-di-Rakh, Jwar Forest (Tatta-Pani), Machhyal, Mohand khad and other part of the state below 1300 m. (Chauhan, N.S., 295, 671, 868, 1412, 1453, 1824, 1907).
Description	– An evergreen shrub with yellowish brown bark. Leaves are elliptic, acuminate; base narrowed, minutely pubescent, especially when young; lateral nerves, 8-14 pairs. Flowers arise in dense, axillary pedunculate spikes. Bracts and bracteoles are foliaceous green, minutely pubescent. Calyx lobes are equal; corolla is white throughout or with branched pale purple vein on the inside of lower lip, pubescent outside. Capsule pubescent dehisces with a cracking noise and scattering the seeds to a considerable distance.
Flowering/Fruiting	– December-April/February-May.
Parts used	– Leaves, flowers, roots.
Current market rate	– Leaves Rs. 7.50 to Rs. 10/- per kg.
Main ingredients	– Contains alkaloids vasicine. Vasicinone, β-sitosterol, Kaemp-ferol have been isolated from the plant and vasicine has been oxidised to vasicinone. Vasicoline, betaine, adhotodine, vasicalinone, anisoline and 6-hydroxy peganine have also been isolated. Fatty oil of

the seeds contain arachidic, behenic, lignoceric, oleic, linoleic acid and sitosterol (CCRAS, 1990:255).

Uses – Leaves and roots are useful in cough, asthma, bronchitis, rheumatism and as insecticidal. Flowers, leaves and roots are antiseptic antispasmodic, allevative, blood purifier and tonic. It is a common remedy for cough, lung complaints anaemia and asthma. For immediate expectoration in children and spasmodic attack of asthma, fresh juice gives immediate relief, which may be followed by other preparations of the drug. The twigs of the plant when used as tooth sticks offer a cure for pyorrhoea. Vascicine has been reported to have oxytocic effect on uterus.

Preparations – Vasavaleha, Vasarista, Vasapanaka, Chandanadi-taila, Punarnavarista, Asokarista, Chyavanaprasa, Pradarari-rasa, Kanakasava, Kasamrita.

Cultivation notes – Natural regeneration is through seeds. Ripe capsules split with a force to sprinkle seeds upto a considerable distance and as a result the plant has been found to colonize vast tracts. It can be propagated through seeds, cuttings and divisions.

Leucas cephalotes (Roth) Spreng.

Family – LABIATAE/LAMIACEAE

Local/Common name(s) – Dronpushpi, Dhurpi sag, Deladona, Goma, Gumabuti, Barakotha.

Distribution – The plant is found throughout the greater part of India, ascending upto 1800 m in the Himalayas. (The Wealth of India, V: 80). The plant is found as a weed in all parts of the H.P. state below 1800 m (Chauhan, N.S., 1979).

Leucas cephalotes Spreng. (Dronpushpi)

Description – An erect, scabrous or pubescent, stout annual, 30-100 cm high, found as a weed in common wastelands. Leaves are shortly petioled, flowers small and white in colour, aggregated in axillary globose whorls.

Flowering/Fruiting – Almost throughout the year, but generally during summers and rainy season.

Parts used – Entire herb.

Current market rate – Not available.

Main ingredients – Seeds & Flowers yield an essential oil. β-sitosterol and a glycoside which yielded glucose and β-sitosterol on hydrolysis, have been isolated from the whole plant and an alkaloid has also been detected. Laballenic acid has been reported from seeds. (CCRAS, 1990).

Uses – The plant is used as a pot-herb. The seeds yield an oil, used for illumination purposes; and are used as a auxiliary in the extraction of the dye. The plant has rennet like properties. It is considered to be a stimulant diaphoretic, laxative, anthelmintic, antiseptic and insecticidal. A syrup of the flowers is used as a domestic remedy for cough and colds.

Preparations – Vishmajvaradi-kshar, Sudarshan-Churna, Balarogantaka-rasa.

Cultivation notes – Easily cultivated through seeds. The plant comes out during rainy season in field borders and waste places.

Mallotus philippinensis Muell. Arg.

Family – EUPHORBIACEAE

Local/Common name(s) – Kamala, Sindur, Rohini, Kampillak, Kamila.

Distribution – The tree is found throughout India, occasionally ascending to 1500 m in the outer Himalayas. (The Wealth of India, Vol. VI, 229). In H.P., it is found in the districts of Chamba, Kangra, Kinnaur, Mandi, Solan, Sirmour and has been collected from Dhaulakuan, Nauni Khaltoo, Palah-da-Jungle, Nahan Paonta and Jwar Forest (Tattapani), (Chauhan, N.S., 841, 1345, 1369, 1669, 1878 and 1921).

Description – It is a much-branched, evergreen tree, 6-9 m high, with a trunk of 90-120 cm round. Young branches and inflorescence are rusty-pubescent. Leaves are alternate, long stalked, variable in shape and size; 7.5-15 cm long. Flowers are small, in panicled racemes on different trees. Fruits are globose with red powdery bloom. Seeds are smooth and black in colour.

(Mukerji, 132; Bentley & Trimen, IV:236). Kamala is a red powder and like Lycopodium is inflammable and resists admixture with water (Dey, 1980:71).

Flowering/Fruiting – October-April.

Parts used – Fruit powder known as Kamila powder.

Current market rate – Rs. 40-65/- per kg.

Main ingredients – Contains 80 percent resin, tannic acid, gum and volatile oil. The most important constituent is a brownish resin composed of crystalline substance called rottlerin. Besides rottlerin, isorottlerin is also present alongwith a small amount of homo-rottler and octacosanol. It contains traces of wax, volatile oil, starch, sugar, tannins, oxalic and citric acids. Alcoholic extract gave bergenin. (Mukerji, 132; CCRAS, 1990:48). Seeds yield a fixed oil.

Mallotus philippinensis Muell. Arg. (Rechanaka, Kamala)

Uses – Kamala is anthelmintic, antioxidant, aphrodisiac, blood purifier, purgative in full doses, occasionally causing nausea, but seldom vomiting. It is used as an anthelmintic against tapeworm. It is used, extensively, in the dyeing industry. Kamala is also used in parasitic affections of the skin, particularly scabies, pimples and ringworms, weeping eczema, boils and is also useful in bile trouble. (Mukerji; 132, Dey, 1980:72).

Preparations – Tinctura Kamala; Krimighatini-vati.

Cultivation notes – The tree can withstand considerable shade and is frost hardy and drought resistant. It coppices well. Natural reproduction takes place by seeds, which fall to the ground in the beginning of the hot season and germinate in the ensuing rainy season. Artificial propagation is done by sowing fresh seeds by April. Germination is rather uncertain and it is advisable to sow seeds quite close; i.e. 5 cm apart, and to thin out, during the first rainy season. The seedlings are ready for transplanting during the first year. Smaller ones can be kept for another year in the nursery. Line sowing with field crops has proved to be successful. Regular weeding and loosening of the soil should be carried out during the first two years and as often as may be necessary, afterwards. (The Wealth of India, VI:229).

Mentha longifolia **(L.) Huds.**

Syn. *M. sylvestris* L.

Family – LABIATAE/LAMIACEAE

Local/Common name(s) – Podina, Jungli pudina, Horsemint, Safed-pudina, Chachri.

Distribution	– The herb is found in the temperate Himalayas and Western Tibet at the altitudes of 1200-3600 m and in Kashmir, Himachal Pradesh, Garhwal and Kumaon (The Wealth of India, VI, 1962:341). In Himachal Pradesh, it is found growing in abundance in moist nalas, ravines in almost all the districts of the state above 1200 m going upto 4800 m in Spiti; especially in Sagnam, Gulling, Sheego, Rangrik top, Gue and Kibber (Chauhan, N.S. 2148, 2200, 2271, 2358, 2493, 2608, 2707 & 5569).
Description	– *Mentha longifolia* is an erect or diffuse, aromatic, tomentose herb with white pubescent, lanceolate and toothed leaves. The flowers are pink in slender terminal spikes.
Flowering/Fruiting	– June-September.
Parts used	– Leaves and top-shoot.
Current market rate	– Not available.
Main ingredients	– Essential oil, obtained from the plant contains piperitone oxide, diosphenol, piperitenone, piperitinone oxide and diosphenolene.
Uses	– Horsemint is considered to be carminative, antiseptic and stimulant. The leaves are astringent and are used for rheumatic pains. A decoction of the plant is used in fever and heat apoplexy. The plant is taken in the form of chutney and is, widely, used for culinary purposes. Its oil can be used as a substitute for imported peppermint oil for flavouring confectionery. The oil is used in U.S.A. for flavouring powder and is used as a snuff to remove headache. A poultice placed over the lumber and renal region acts as a good diuretic and emollient. As a liniment, it is applied in headache. It is much used as an incense, as

Mentha longifolia (L.) Huds. (Jungli pudina)

masala in washing the hair and is useful in the stomach disorders, dyspepsia, vomiting, and in case of the pain in the liver or womb. The plant is stated to be useful in scorpion sting and snake bites. (Chopra et al. 1956:186; Khory and Katrak:652).

Preparations – Arka-pudina, Pudina-sat, Amritdhara and Amritanjan. Also useful for flavouring candy, cough mixtures etc.

Cultivation notes – It is easy to propagate the plant through suckers in a moist soil. Suckers can be transplanted between March-May depending upon the altitude and climate. Agrotechniques applied to other species of Mentha under commercial cultivation are equally suitable for this species as well.

Other species – *M. spicata* & *M. arvensis*.

Morchella esculenta (L.) Pers.

Family – HELVELLACEAE

Local/Common name(s) – Guchhi, Chiayun, Honey combed mushroom.

Distribution – It is a valuable mushroom growing naturally in humus rich forest soils in Shimla, Kullu, Mandi, Sirmour and Kangra districts during March-April between 2200-3200 m. Also found growing occasionally along field border during monsoons in the lower ranges.

Description – Guchhi is an ascomycetes having pale-white stalk which supports the thick, erect, fruiting body, bearing a rounded or conical pileus covered with a number of prominent ridges. It is yellowish brown to olive coloured, often dark brown also. It appears in the decaying organic matter in forests during March-August

		especially following the thunder storms during March-April.
Flowering/Fruiting	–	March-April
Parts used	–	The whole body excluding the base of the stalk which contains soil humus.
Current market rate	–	Rs. 1500-3500/- per kg.
Main ingredients	–	It is reported to contain moisture (89.07%), protein (3.59%), carbohydrates (5.73%), fat (0.27%) and ash (1.34%). The ash contains calcium (3.92%), iron (6.8%) and phosphorus (22.82%).
Properties and uses	–	The mushroom is eaten both fresh and dry as vegetable and as a rich man's food, it is considered a delicacy of the table. It sells at the average rate of Rs. 1500.00 to 4000.00 per kg. It is sold in the market in the dried state. Medicinally, it is considered as tonic, aphrodisiac and narcotic (Chopra et al., 1956: 169). As a domestic remedy, the paste of the mushroom is applied to cure boils that have become chronic.
Preparations	–	In tonic and aphrodisiac formulations.
Cultivation	–	No success has been achieved till date to raise it artificially. It grows automatically in nature in a very short season during March-April especially in humus rich soil. Deodar forests, fir and spruce, Himalayan polars and wild bird cherry trees provide better sites for its regeneration.

Moringa oleifera Lamk.

Family	–	MORINGACEAE
Local/Common name(s)	–	Sanjana, Sahinjan, Soanjana, Drum stick tree, Horse-radish tree.

Distribution – The tree is found in the outer Himalayas at low elevation and is cultivated throughout the hotter parts of India and most tropical countries (Ahuja, 51). In Himachal Pradesh, it is found growing in lower warmer areas in Kangra, Una, Hamirpur, Mandi, Solan, (Nalagarh division) and Nahan divisions.

Description – It is a small tree, 9-12 m, tall with corky bark. Wood is soft and tastes pungent. Leaves are usually 3-pinnate and leaflets are in 6-9 pairs with an odd leaflet elliptic-ovate or obovate. Flowers are pedicelled, white and honey scented in spreading, paniculate racemes; bracts linear; capsule is 20-50 cm long, pendulous and 9-ribbed.

Flowering/Fruiting – January-July.

Parts used – Root, root bark, stem bark, leaves, fruit, seeds and gum.

Current market rate – Not available.

Main ingredients – Root contains spirochin and pterygospermin (an antibiotic). Root bark contains alkaloids moringine and moringinene. Pterygospermin from the plant is antibiotic against both gram positive, gram negative and acid fast bacteria.

Uses – The plant is used as a tonic, abortifacient, stimulant, aphrodisiac and in rheumatism. Oil obtained from the seeds is a useful remedy for gout and acute rheumatism. Native practitioners prescribe the fresh root as a stimulant in paralysis and intermittent fevers and also in epilepsy and hysteria. Leaves, bark and roots are considered spasmodic. Leaf juice mixed with pepper is applied to hasten suppuration in boils. Warmed leaves are applied in hydrocele. Bark offers a cure for gum boils and toothache. Spirochin enhances

Moringa oleifera Lamk. (Sahinjan, Shigru)

		the growth of epithelial cells and suppresses the irregular contraction of the uterus and provides strength to it. Pteryospermin is antifungal and antibacterial.
Morphology of the official parts used	–	The roots are light greyish brown in colour and lenticelled. The wood is very soft, porous and yellowish in colour and is surrounded by outer layer of bark and cortex. A gum exudes from wounds in the roots and stem and becomes pink on exposure to air. On drying, the colour turns to purplish-red to reddish-brown. The gum swells into a jelly, when placed in water. Seeds are three angled and the angles expand into wings.
Preparations	–	Various formulations in Indian medicines.
Cultivation notes	–	The drug is collected during March-July. It grows in all types of soils except stiff clays and thrives best under tropical climate. Can be propagated through seeds or cuttings which is generally preferred as the plants raised from seed produce inferior quality of fruits. Cuttings of fairly large size are planted in moist soil, strike roots easily and attain a sizeable growth in a field. Under our conditions, the tree sheds its leaves in December-January and new leaves appear in February-March onwards.

Mucuna pruriens **(L.) DC.**

Syn: *Mucuna prurita* Hook.

Family	–	LEGUMINOSAE (Papilionaceae)
Local/Common name(s)	–	Kiwanch, Kaunch, Goncha, Drougalbel, Atmagupta, Cowhage.
Distribution	–	The herb is found almost all over India. It is cosmopolitan in the tropics. (The Wealth of India, VI:422 and Mukerji, 150).

It is found in the foothills of Himachal Pradesh and has been found in Dehra, Nahan, Bharwain, Jwalamukhi, Nurpur, Nalagarh, Nauni-Khaltoo, Tattapani, Bilaspur and Sarkaghat areas (Chauhan, N.S. 1238, 1287 and 1483; Uniyal and Chauhan; 285).

Description – It is an annual, half-woody, twining herb with long cylindrical branches. Leaves are alternate, 3-foliate; petioles are 6.3-11.3 cm long; are ovate or rhomboid, membranous and glabrescent above, 15-20 cm long and are on short, thick, hairy stalks with stipullae at their base. Flowering racemes are 15-30 cm long and are solitary or 2-3 together bearing purplish flowers. Pod is 5-7.5 by 1.2 cm; 4-6 seeded, turgid, falcately curved on both ends, densely clothed with persistent irritant bristles, which are at first pale-brown and afterwards turn steel grey.

Seeds are 4-5, separated by cellular partitions, about 0-5-0.75 cm long, ovoid somewhat compressed, smooth, brownish often mottled with black, Hilum is large and oblong.

Flowering/Fruiting – July-September/October-November. Seeds are collected on ripening in January-February.

Parts used – Seeds.

Current market rate – Seeds Rs. 12-15/- per kg.

Main ingredients – The seeds contain alkaloids mucuadine mucunadinine, mucuadininene and prurienidine alongwith mucunin, mucunadine and small amount of nicotine. Seeds also contain L-dopa, glutarthione lecithin, gallic acid, a glucoside and amino acids. Trichomes from pods yield serotonin.

Uses – Kaunch is anthelmintic (hairs), aphrodisiac, astringent, diuretic and nervine tonic. It is useful in cholera (in the form of a strong infusion), dropsy, general debility, leucorrhoea, seminal weakness and as a tonic, (Dey, 76 and Bentley & Trimen, II:78).

Preparations – Vanari-gutika and masbaladi-pachan, L-dopa.

Cultivation notes – The plant is occasionally cultivated. It is hardy, grows vigorously and forms a thick soil covering. It is also grown for its pods and young leaves, which are used as vegetable and as fodder. For use as fodder the crop must be cut, when in flower. The plant is found in bushes and hedges, damp places, ravines and scrub-jungles throughout the plains of India where it regenerates itself. Experimental cultivation in UHF herbal garden has proved successful.

Murraya koenigii Spreng.

Family – RUTACEAE

Local/Common name(s) – Kathnim, Mitha neem, Curry patta, Gandhela, Barsanga, Curry leaf.

Distribution – The plant is found, almost, throughout India, upto an altitude of 1500 m. It is found in warmer areas in Himachal Pradesh. Commonly found to occur in Zimzima, Jogindernagar, Gallu, Doeg-Dhaniatar (1100-1350 m), most common between Tanda and Shahpur in Kangra, Nalagarh and Nahan-Paonta Ranges. (Chauhan, N.S., 108, 693, 980, 1456, 1500).

Description – It is a handsome, aromatic, more or less deciduous shrub or a small tree, upto 6 m in height and 15-40 cm in diameter. Bark is dark brown in colour and the leaves are imparipinnate. Flowers are white in colour, fragrant and arising in terminal, corymbose panicles.

		Fruit is sub-globose, 0.5 cm in diameter, rugose, black when ripe.
Flowering/Fruiting	–	April-May/July-September.
Parts used	–	Leaves, fruit, shoot and bark.
Current market rate	–	Not available.
Main ingredients	–	Essential oil and glucoside-koeinigin. (Chopra et al., 1956:171).

Murraya koenigii, Spreng. (Mitha neem)

Uses – Plant is used in the preparation of a tonic. Bark and roots are used as stimulants and applied externally to cure eruptions and bites of poisonous animals. Green leaves are eaten raw as a cure for dysentery. As decoction with bitters, it acts as a febrifuge and is also used in snake bite. Leaves are frequently used in curries. Fruit are edible.

Preparations – Not available.

Cultivation notes – Propagation is by seeds, which germinate freely under partial shade and a spacing of 3 to 5 m has been suggested. Natural regeneration through seeds has been found quite high. Vegetative propagation through cuttings is also successful.

Myrica esculenta Buch. Ham.

Family – MYRICACEAE

Local/Common name(s) – Kaiphal, Kahi, Katphal, Kaphal, Box-myrtle.

Distribution – The tree is found in the sub-tropical Himalayas from Ravi Eastwards to Assam and in Khasi, Jaintia, Naga and Lushai hills at altitudes of 900-2100 m. (The Wealth of India, VI, 472). It is found in the districts of Chamba, Mandi, Solan, Sirmour and Shimla in Himachal Pradesh. (Chauhan, N.S., 1492). Common in Gallu forest at Jogindernagar; Shilly-Jaunaji and Dochi-Chail in Solan districts.

Description – It is a small moderate-sized evergreen tree, 3-15 m high. The bark is grey or brownish grey and the leaves are lanceolate. Flowers are minute and the fruits are ellipsoid, reddish in colour and sweet-acidic in taste. Bark, which is an official drug is 10-12 mm thick,

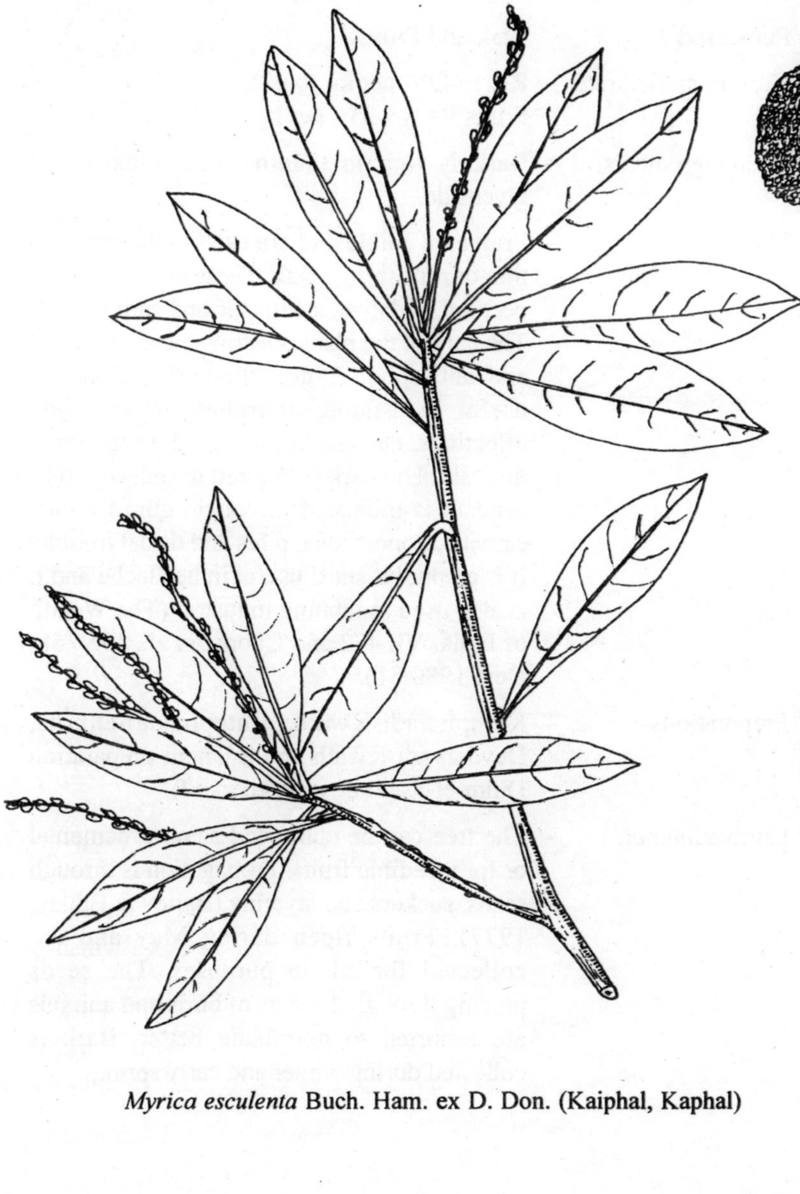

Myrica esculenta Buch. Ham. ex D. Don. (Kaiphal, Kaphal)

		externally scabrous, rusty-brown and dirty white; inner bark has a deep, dull-red colour, gives a deep red solution when dipped in water and has a highly astringent taste.
Flowering/Fruiting	–	August-December/April-May.
Parts used	–	Bark and Fruits.
Current market rate	–	Rs. 15-20/- per kg (bark). Fruits Rs. 15-25/- per kg.
Main ingredients	–	Bark is rich in tannins and contains the glycoside.
Uses	–	Fruits are edible and are used in the preparation of refreshing drink. These are considered pectoral, sedative, stomachic and carminative. The bark of the plant is astringent, carminative and antiseptic. A decoction of the bark is useful in asthma, diarrhoea, fevers, lung affections, chronic bronchitis, dysentery and diuresis. The bark is chewed to relieve toothache. It is indicated in cough, gum trouble, earache, gonorrhoea, piles and throat trouble. It is a popular snuff useful in headache and it is also used in tanning industry. (The Wealth of India, VI, 472 and Chopra et al., 1956:58; Dey, 1980:81).
Preparations	–	Kathphaladi-Kwath, Kathphaladi-churna, Devdarvadi-Kwath and a Unani preparation Dulm-el-Kanduh, (cooled in oil).
Cultivation notes	–	The tree can be cultivated as an ornamental or for its edible fruits. Propagation is through seeds, suckers and layering (Bailey & Bailey, 1977). Fruits ripen during May and are collected for edible purposes. The seeds passing through the guts of birds and animals are reported to germinate better. Bark is collected during winter and early spring.

Myrsine africana Linn.

Family	– MYRSINACEAE
Local/Common name(s)	– Kakhum, Shamshad, Berang, Bidang, Bai-bidang.
Distribution	– The shrub is found in the outer Himalayas from Kashmir to Nepal and Khasi hills at altitudes of 300-2700 m. The plant is found in the districts of Chamba, Kinnaur, Kullu, Mandi, Solan, Sirmour and Shimla in Himachal Pradesh. The shrub is also found in Rohru forest division between 2000 m-2500 m (Chauhan, N.S., 1979).
Description	– It is an erect shrub; 0.6-1.2 m high. Leaves of the plant are sharply toothed and the flowers are minute and white in colour. Fruit is small, ovoid, deep purple on ripening and contains a single seed.
Flowering/Fruiting	– April-October.
Parts used	– Fruit (Dried).
Current market rate	– Rs. 18/- per kg (Used as a substitute for seeds of *Embelia ribes*).
Main ingredients	– Fruit contains embelin upto the extent of 3 percent. Quercitol is also present in the fruits to some extent. Seeds contain traces of emobellin. The plant is also known to yield a gum. (The Wealth of India, Vol. VI, 482 and Chauhan, N.S., 1984).
Uses	– *Myrsine africana* is considered to be suitable for hedges. The fruit is edible and has anthelmintic properties and is particularly effective for the expulsion of the tape worms. Fruit is used as a substitute for Embelia (From *Embelia ribes*). The fruit is, also, used as a laxative in dropsy and colic. The gum is considered as a 'Warm remedy' and is

employed for dysmenorrhoea. It is used with liquorice for strengthening the body and preventing the affects of age. Decoction of the leaves is used as a blood purifier. (Chauhan, N.S. 1984).

Preparations – Vidangadi-Churna, Bidangarishta, Sudarshan-Churna, Agnitundivati, Sanjivani-vati.

Cultivation notes – The plant is readily propagated by seeds or cuttings.

Nardostachys grandiflora DC.

Syn: Nardostachys jatamansi DC.

Family – VALERIANACEAE

Local/Common name(s) – Jatamansi, Balchharh, Mansi, Machhi, Spikenard.

Distribution – The herb is found in the alpine Himalayas from Himachal Pradesh to Sikkim and Bhutan, at altitudes between 3000-5000 m. In Himachal Pradesh, this plant is found in alpine rocky slopes in Manjiban, Jakha-Kanda, Tangankhai between 3000-4000 m. (Chauhan, N.S., 2421). Has also been collected from the Great Himalayan National Park area in Kullu district.

Description – It is an aromatic, erect, perennial herb, 10-60 cm high with long, stout, woody rhizomatic rootstock covered with tail like brown fibres left over from the withered leaves towards the stem, while the root continues to penetrate deep in the soil. Leaves are radical in nature long, narrow and the flowers are creamy white, often rosy or pale pink in appearance arising in terminal corymbose cymes. Fruit is small, 4 mm long, covered with minute hairs. Market drug consists of a short portion of rhizome, as thick as little finger, dark-grey,

Nardostachys grandiflora DC. (Jatamansi)

covered by a tuft of fine, reddish brown fibres and gives an appearance of a tail. Fibres are malted together as a network. It has a heavy aromatic odour and tastes bitter.

Flowering/Fruiting – June-July/September-October.

Parts used – Rhizomes.

Current market rate – Rs. 90-160/- per kg.

Main ingredients – The hairy roots contain essential oil having jatamansone, jatamansic acid, lupeol, jatamansinone, jatamansinol, jatamansin, angelicin, dihydrojatamansin, oroselol, oroselone, angelic acid, seselin, α and β pinene, Δ carene, b eudesmol, lemol, b-sitosterol, valeranone, valeranal, nardostachone and nardosinone etc. Rhizome also contains alkaloid-actinidin. (CRAS-1990).

Uses – The rhizome is considered as tonic, stimulant, anti-spasmodic, diuretic, deobstruent, emmenagogue, stomachic and laxative. An infusion of the rhizome is reported to be useful in epilepsy, hysteria, leprosy, respiratory disorders, nervous excitement, palpitation of the heart and a tincture is given in intestinal colic and flatulence. The rhizome has the property of promoting the hair growth. The roots have remarkable properties to tone up the brain. Spikenard oil possesses antirhythmic activity with possible therapeutical usefulness in cases of auricular blutter. (Chopra, 1958:679; The Wealth of India, VII:3; Dey, 1980:63). Jatamansone is more potent than the oil and is also more active than quinidine in ventricular trachycardia, resulting from acute myocardial infarction and has also an anti-convulsant action. Prevalently used as a hair and heart tonic. (Chauhan, N.S., 1984).

Preparations – Mansadi-kwath, Rakshoghna-ghrita, Sarvousadhi-snan, Dasanga-lep, Laghu-visgarva-taila, Pippalayasav and Becamin. (Chauhan, N.S., 1984: Dey, 1980:63).

Cultivation notes – The plant is propagated by cuttings, divisions of under ground parts and sometimes by seeds. Rawat et al., (1985) have reported that in the absence of its cultivation know how, the plant is difficult to establish. The plant being of great importance as a drug and having been enlisted as endangered species needs special attention for cultivation taking support from micropropagation techniques.

Nerium indicum Mill.

Family – APOCYNACEAE

Local/Common name(s) – Kaner, Karveer, Kuruvira, oleander, Lal kaner.

Distribution – The shrub is found in the Himalayas from Nepal Westwards to Kashmir upto 1950 m, in Gangetic plain and Madhya Pradesh. (The Wealth of India, VII:16). The plant is found in the districts of Kangra, Kullu, Mandi Bilaspur, Chamba, Sirmour and Shimla in Himachal Pradesh and is also cultivated as an ornamental shrub all over Himachal Pradesh upto Shimla. (Chauhan, N.S., 1603).

Description – It is a large, evergreen shrub, with milky juice. Leaves are mostly in whorls of 3. Flowers are red or white in colour fragrant in terminal cymes and the fruit is a 15-30 cm long follicle. The seeds are numerous with woolly hairs.

Flowering/Fruiting – April-June, Fruits in cold season.

Parts used – All parts.

Current market rate – Not available.

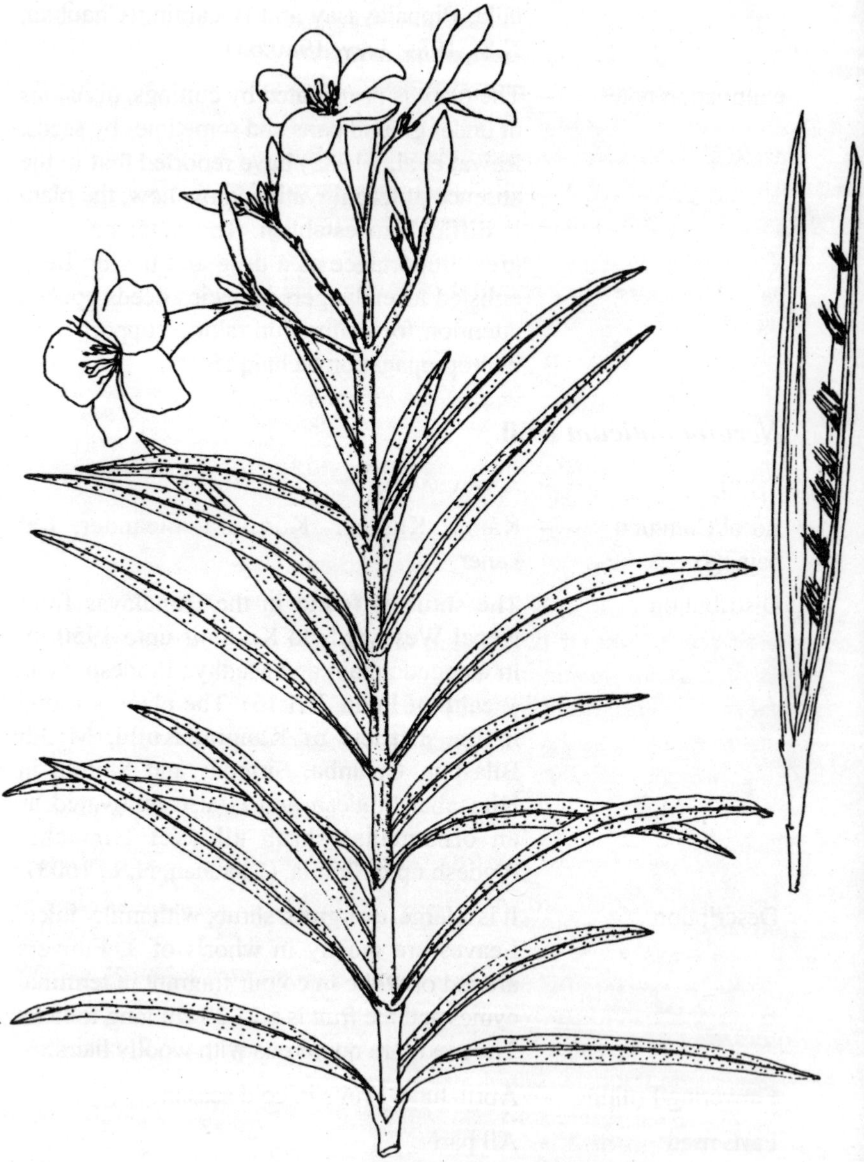

Nerium indicum Mill. (Kaner, Karveer)

Main ingredients	– Root, bark and seeds contain cardioactive glycosides formerly designated as neriodorin, nerioderin, and karabin. Several glycosides with digitalis like action have been isolated from the bark. The bark also contains scopoletin and scopoline. Occurrence of small quantity of tannin, deep red colouring matter, an essential oil, a crystalline base, phlobaphene and a yellow fixed oil is reported in bark. The leaves yield oleandrin, neriodin, ursolic acid, adynerin, a compound similar to rutin; β-D diginoside, β-D-digitaloside and digitoxigenin. Root bark yields α-amyrin, β-sitosterol, kaempferol and odoroside-B (CCRAS 1990).
Uses	– It is regarded poisonous; is reported as anthelmintic, cardiac, carminative, diaphoretic, febrifuge, ophthalmic, powerful heart poison and repellant. Its local application is useful in swelling, leprosy, skin diseases and ulcers. Alcoholic extracts of root, bark, leaf and flower show antibacterial activity. The wood can be used for making hooka tubes. The bark and wood are used in the manufacture of rat poisons. (The Wealth of India, VII: 16 and Dastur, 123).
Preparations	– Karaviradhya-talia and Leghu-vis-garbha-talia.
Cultivation notes	– It is successfully propagated by cuttings or layering.

Nyctanthes arbor-tristis Linn.

Family	– OLEACEAE
Local/Common name(s)	– Harsinghar, Kuri, Shephali, Shephalika, Regmalpatta, Parijata.

Distribution	– The plant is a native of India, occurring wild in the Sub-Himalayan region from Chenab to Nepal upto 1500 m and in Chhota Nagpur, Rajasthan, Madhya Pradesh and Southwards to Godavari. Also cultivated in gardens for its flowers. It is found in the districts of Chamba and Kangra in Himachal Pradesh. The plant has been collected from Kallanwala Forest, Jwar forest, Palah-da-jungle, Nalagarh etc. and planted at Jogindernagar and UHF botanic garden at Nauni, Solan. (Chauhan, N.S., 225, 1355 and 1877).
Description	– It is a hardy, large shrub or a small tree upto 10 m high, with grey or greenish white rough bark and quadrangular branches. Leaves are opposite, ovate, acute, coriaceous, covered with stiff white hairs. Flowers are small in size, sweet scented, sessile, 3-7 together in pedunculate heads, arranged in trichotomous cymes. Flowers tubular, orange red with white limb. Anthers 2, sub-sessile, inserted near the mouth of corolla tube; capsule upto 1 cm long, orbicular, compressed.
Flowering/Fruiting	– July-October/cold season.
Parts used	– Flowers, bark and leaves.
Current market rate	– Rs. 300/- per kg. (Fresh flowers).
Main ingredients	– Essential oil myctanthin (identical with α-crocetin from saffron), tannin, d-manital and glucose are present in the flowers. The seeds yield a pale yellow fixed oil, which consists of the glycerides of linoleic, oleic, lingoceric, stearic and other acids, including the tannic acid. Leaves contain β-amyrin, β-sitosterol, glycosides, benzoic acid, glucose and fructose. A glycoside and two alkaloid have also been reported from the bark.

Nyctanthes arbor-tristis Linn.

Uses – The bark of the tree may be used as a tanning material and leaves are sometimes, used for polishing wood and ivory. Leaves are antibilious, laxative, draphoretic, diuretic, cholagogue, mild bitter tonic and expectorant and are useful in fevers and rheumatism. A decoction of the leaves is given in sciatica. Leaf juice mixed with little sugar is given to children as a remedy for intestinal worms. Bark is expectorant. Powdered seeds are used for curing the scurvy of scalps. (The Wealth of India, VII, 70); Chopra et al., 1956:177; CCRAS, 1990:164-65).

Preparations – Not available.

Cultivation notes – The plant grows gregariously and covers dry steep hillsides and rocky grounds. It tolerates moderate shade and is often found as an under growth in dry, deciduous forests. It is easily propagated by seeds or cuttings and coppices readily and is not browsed by goats. It is leafless during April-May.

Ocimum sanctum Linn.

Family – LABIATAE/LAMIACEAE

Local/Common name(s) – Ajaka, Tulasi, Tulsi, Baranda, Sacred Basil, Holy basil.

Distribution – The plant is found throughout India, ascending upto 1800 m in the Himalayas and in the Andaman and Nicobar Islands. A well known sacred plant of the 'Hindus', grown in houses, gardens and temples all over India. Cultivated in herbal garden Jogindernagar in bulk to meet the requirement of the state run Ayurvedic Pharmacy.

Description – It is a much-branched, aromatic and erect herb, hairy throughout. Stem is woody, perennial, giving out numerous branches. Leaves are opposite, entire or serrate, toothed and dotted with minute glands. Flowers are purplish or crimson, small, in terminal, closely whorled racemes. Seeds are dark-brown, small, oblong, blunt pointed. The plant is held sacred by Hindus all over India and is

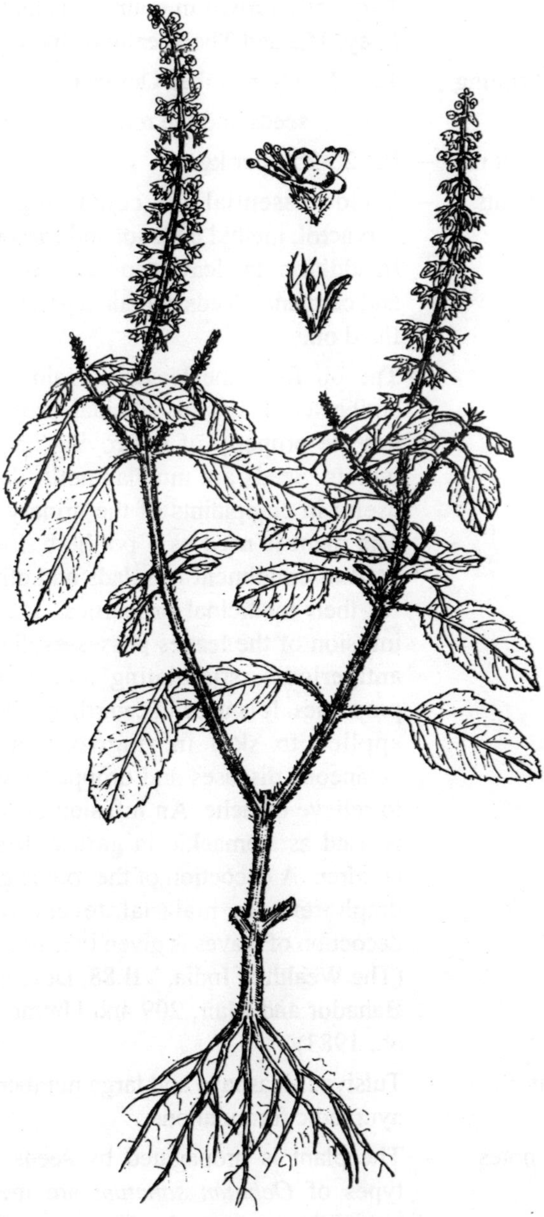

Ocimum sanctum Linn. (Tulsi)

frequently grown in courtyards and temples. (Dey, 152 and The Wealth of India, VII:88).

Flowering/Fruiting – June-July/September-October.

Parts used – Leaves, seeds and the root (entire plant).

Current market rate – Rs. 20-25/- per kg.

Main ingredients – Yields essential oil containing eugenol, carvacrol, methyl eugenol and caryophyllene. In addition, the leaves contains ascorbic acid and carolene. Seeds contain a greenish yellow fixed oil.

Uses – The oil from the leaves is said to possess antibacterial and insecticidal properties i.e. has the property of destroying bacteria and insects. Seeds are mucilaginous, demulcent; useful in complaints of the urinary systems. The plant is used as a pot herb. Leaves are used as condiment in salads and other foods for their medicinal properties. The juice or infusion of the leaves possesses diaphoretic, antiperiodic, stimulating and expectorant properties. It is used in catarrh and bronchitis; applied to skin in ringworm and other cutaneous diseases and dropped into the ear to relieve earache. An infusion of the leaves is used as stomachic in gastric disorders of children. A decoction of the root is given as a diaphoretic in malarial fevers while the decoction of leaves is given in common colds. (The Wealth of India, VII:88; Dey, 1980:152; Bahadur and Mair, 209 and Umrao Singh et al., 1983).

Preparations – Tulsiyadi-Kashaya and large number of other ayurvedic formulations.

Cultivation notes – The plant is propagated by seeds and two types of *Ocimum sanctum* are met within cultivation commonly. These are the green type (Sri-tulsi) and the purple type (Krishna tulsi). Seeds are sown in April transplanted in June and the crop is ready for harvesting in September-October.

Onosma bracteatum Wall.

Family	–	BORAGINACEAE
Local/Common name(s)	–	Ratanjot, Dhamani, Gojihva, Gajavan, Gaojavan, Gule-gaojavan, Shankhahuli.
Distribution	–	The plant is found in the Western Himalayas, from Kashmir to Kumaon between 3600-4500 m elevation. In Himachal Pradesh, it is reported from the districts of Kangra (Chhota & Bara Bhangal), Pangi-Bharmour, Kinnaur, Lahaul and Spiti in drier areas.
Description	–	It is a large, herbaceous perennial with black, woody rootstock, 2.5-5 cm in diameter, terminating in a knotty head from which arise several stems, erect or ascending. Root is purplish red from inside and the stem is simple and rarely branched, thickly studded with calcareous tubercles and armed with bristles. The leaves are entire, thick, petiolate and ovate-acuminate. Flowers are deep blue, later turning purplish in heads, which are thickly covered by white, stiff, bristles.
Flowering/Fruiting	–	August-September/September-October.
Parts used	–	Root, leaves and flowers.
Current market rate	–	Rs. 20-35/- per kg. (Root).
Main ingredients	–	Ratanjot, red dye yielding root, contains mucilage and salts of Sodium, Potassium, Calcium, Iron and Magnesium.
Uses	–	The root is extensively used for colouring purposes. Bruised roots of Ratanjot are applied externally to cutaneous eruptions. The flowers of the plant are used as stimulant and

Onosma bracteatum Wall. (Gojihva)

cardiac tonic. It yields a dye which is used for colouring silk or wool. Plant is used in rheumatism, syphilis and leprosy; useful for relieving of excessive thirst and restlessness; in febrile excitement, in relieving heart palpitation, irritation of the bladder and stomach.

Preparations – Gaojaban Ark and Banafsadi-kwath.

Cultivation notes – Requires well drained light soil in sun with protection against winter. Can be cultivated through seeds and splits. Seeds collected in September-October can be sown during October or March-April after snow has melted. Root slips with stem can also be transplanted.

***Origanum vulgare* Linn.**

Family – LABIATAE/LAMIACEAE

Local/Common name(s) – Sathra, Bantulsi, Wild marjornum, Jakhm-buti, Basloghas.

Distribution – The herb is found in the temperate Himalayas from Kashmir to Sikkim between 1500-3600 m elevation. It has been found at still lower as well as higher levels. It is widely found in districts of Chamba, Kangra, Shimla, Kinnaur, Kullu, Mandi, Lahaul and Spiti, Sirmour and Solan in Himachal Pradesh between 1200-4000 m. The plant is collected from different localities in the state ranging from 1200 m-4200 m elevation. (Chauhan, N.S., 1098, 1153, 1300, 1509, 1975, 2088, 2110, 2210, 2265, 2354, 2647 and 2689).

Description – It is an aromatic, branched, perennial herb, 30-90 cm high. Leaves are broadly ovate and the flowers are in terminal corymbose cymes. Flowers have pale white colour, yet the plants

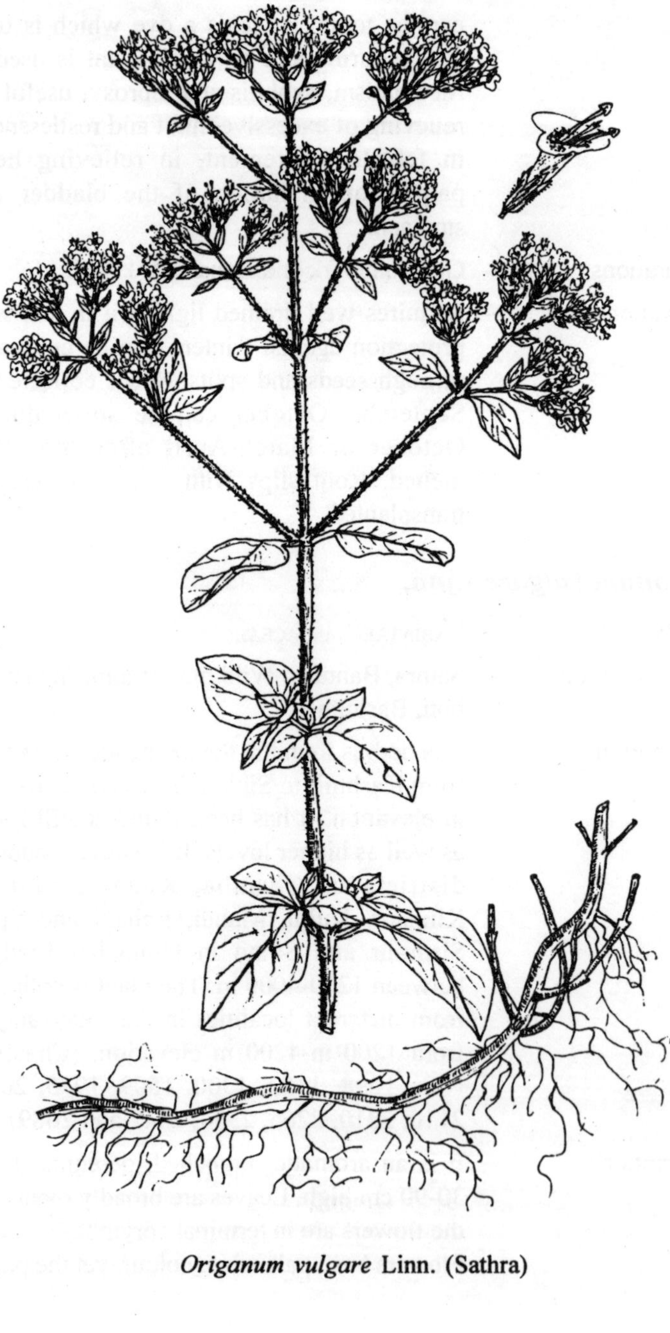

Origanum vulgare Linn. (Sathra)

having flowers with purplish or pinkish tinge are also found. The nutlets are brown in colour. The plant possesses aromatic thyme like flavour.

Flowering/Fruiting – June-December.

Parts use – Leaves, top shoot or entire plant.

Current market rate – Not available.

Main ingredients – The herb contains a volatile oil containing 50 percent thymol. It has also been reported to yield tannin and a bitter principle.

Uses – The leaves and tops, cut prior to blooming are used to flavour foods. The plant is also used as a pot-herb and is used as a vegetable in Lahaul. The origanum oil possesses carminative, stomachic, diuretic, diaphoretic and emmenagogue properties. It is also given as a stimulant and tonic in diarrhoea, colic and is applied in chronic rheumatism; toothache and earache. Also used in cough and bronchitis. In homoeopathy, it is used, for hysteric conditions. The oil is used in baths, healing wounds and stimulates the growth of hair. and has been extensively used in the cosmetic and soap industry. Locally, the tablet made from the leaves is administered orally in bone fractures. The paste of the leaves is applied in fireburns, eczematous skin, boils, cuts and wounds.

Preparations – Not available.

Cultivation notes – The herb is propagated by seeds, cuttings, layers and root divisions. It can be sown during October in the plains and during March-April in the hills. The underground part is perennial and gives out numerous stems during spring.

Oroxylum indicum Vent.

Family	– BIGNONIACEAE
Local/Common name(s)	– Ullu, Arlu, Saona, Shyonaka, Sonapatta, Tatpalanga, Kinnauri phool (Seeds).
Distribution	– The tree is found throughout the greater part of India, upto an altitude of 1200 m, mainly in the ravines and moist places in the forests. In Himachal Pradesh, it is found in the districts of Kullu, Kangra, Una, Hamirpur, Bilaspur, Mandi, Chamba, Solan and Sirmour (Nahan division) upto an elevation of 1200 m above m.s.l. (Chauhan, N.S., 1430, 1731).
Description	– It is a small to medium sized tree and is deciduous in winter. Bark is light greyish brown in colour. Leaves are large opposite, 3-pinnate near the base, 2-pinnate about the middle and simple pinnate above. Flowers are large, purplish, fleshy in lax, terminal racemes as having 60-90 cm long peduncles. Capsule (fruit), septically 2-valved; 45-75 cm long and 5.5-8.5 cm broad. Seeds are flat, surrounded by a broad, transparent, white, papery wing which is usually 5-6.5 cm across.
Flowering/Fruiting	– June-July/November-January.
Parts used	– Root, bark and tender fruit.
Current market rate	– Rs. 18—20/- per kg. (Bark).
Main ingredients	– From the leaves, an anthraquinone derivative aloe-emodin was reported for the first time. Chrysin, baicalein, luteolin; 5, 7, dihydroxy-6-methoxyllavone, glycosides of baicalein and aloe-emodin were isolated from root bark and the leaves. (CCRAS, 1990).
Uses	– Bark is a well known drug in Ayurvedic system and is a tonic and astringent useful in diarrhoea and dysentery. The bark is

Oroxylum indicum Vent. (Arlu)

diaphoretic and is used in rheumatism. The bark, boiled in sesamum oil, is recommended for otorrhoea. Tender fruits are refreshing and stomachic and the seeds are purgative. Leaves are used externally for enlarged spleen, headache and ulcers. The bark is used externally and the seeds internally in veterinary medicine, whereas the stem is useful in scorpion sting. (The Wealth of India, VII:108 and Chopra et al., 1956:182). In Kinnaur, the seeds are bunched in groups with wire and form a floral type ornament worn on caps.

Preparations – Dashmool-ghrita, Dashmul-Taila, Dashmularishta, Chyavanprasha (Ahuja, 1965:55).

Cultivation notes – The tree reproduces naturally by seeds, which germinate in the beginning of the rainy season. Moderate shade is necessary in the early stages. Artificial reproduction may be done by sowing the seeds in the nursery during March-April and transplanting the seedlings in the first or second half of rainy season. The tree can be propagated by transplanting root suckers which are produced in great profusion, often forming a dense growth round the parent stem. The rate of growth of the tree is reported to be fast, with a mean annual girth increment of 4-6.4 cm (The Wealth of India, VII:108).

Paeonia emodi Wall.

Family – RANUNCULACEAE (Paeoniaceae)

Local/Common name(s) – Udsalap, Udsalib, Himalayan Paeony.

Distribution – North-west Himalayas from Kashmir to Kumaon between 1600-3000 m elevation.

Also planted in Herbal garden at Nauni (Chauhan, N.S. 5938, 6648).

Description – A glabrous, perennial herb. Stem 30-60 cm, leafy, erect. Leaves alternate, 15-30 cm long; leaflets 3; usually 3-parted, segments lanceolate, pointed, entire. Flowers few, showy, 7.5-10 cm across, long-stalked, usually solitary in the axils of the upper leaves. Buds globose, sepals 5, orbicular, concave, green, persistent; the outer ones ending in a leaf like point. Petals 5-10, broadly ovate, concave, red or white. Stamens many ovaries 1-3, densely hairy, many ovuled, style short. Follicle ovoid, 2.5 cm approx. Seeds few, large in size.

Flowering/Fruiting – May-June.

Parts used – Tubers/rhizomes

Current Market rate – Rs. 450/- per kg.

Chemical contents – Starch (9.5%), sucrose (5.4%), Malic acid (0.47%), Oxalic acid (0.36%), tartaric acid (0.34%) and benzoic acid are present in roots. Root oil yields n=alkanes, beta-amyrin, butyropermol, cycloartenol, lupeol, linoleic acid etc.

Properties and uses – Tuberous roots are used for intestinal diseases, colic, bilious obstructions, dropsy, epilepsy, convolutions and hysteria; given to children as blood purifier. Seeds are emetic. Plant is CNS depressant, hypothadermic, diuretic, anti-inflammatory and useful in gout.

Preparations – Mostly used in Unani drugs.

Cultivation – It can be cultivated through rhizome cuttings. Suitable sized rhizome pieces with 2-3 clear buds are selected and sown in furrows or pits mixed with FYM during February-March.

Parmelia perlata **(Huds.) Ach.**

Family	– PARMELIACEAE
Local/Common name(s)	– Chharila, Shaileya, Shinkhrao, Mehndi, Janpapro, Stone flower.
Distribution	– Commonly found growing on rocks and on the trunks and branches of oak trees between 1600-3200 m elevation (Chauhan, N.S. 761, 2745).
Description	– A thallus of dirty white or grey colour, 5-10 cm long and made up of numerous fibres spreading from common stalk upon the trees or rock, often forming a rosette like flat mass. Fibres are flat, generally dichotomously, rarely trichotomously branched; marked here and there with irregular depressions, having fruitifications like punched tubercles on the fibre surface and aromatic. Taste is mucilaginous, bitterish rather saline.
Flowering/Fruiting	– Being a lichen, no true flowers/fruits are borne. However, cup shaped fructifications appear on mature thallus on upper surfaces.
Parts used	– The plant (thallus)
Current market rate	– Rs. 35-55/- per kg.
Chemical contents	– Contains lecanoric acid and atranorin, a yellow crystalline matter called chrysophanic acid, identical with the colouring matter of rhubarb. Lichenin, cellulose and sugars have also been isolated from it.
Properties and uses	– Demulcent, tonic, febrifuge and diuretic. It is given in amenorrhoea, chlorosis and leucorrhoea. As a soporific and sedative, it is given to stop nocturnal emissions. Powder is used as a snuff to remove headache. A poultice placed over the lumber and renal region acts as a good diuretic and emollient.

As a liniment, it is applied in headache. It is much used as an incense, as a masala in washing the hair; useful in stomach disorders, dyspepsia, vomiting, pain in the liver or womb. Stated to be useful in scorpion sting and snake bites. Lecanoric acid is an excellent material for the manufacture of litmus (Chopra et al., 1958: 186; Khory and Katrak: 652).

Preparations – Used in Ayurvedic medicines. Also as condiment and a dye.

Cultivation – It grows naturally as a thallus covering rock surfaces, dead, most, bark and branches of old trees especially high level oak i.e. *Quercus semecarpifolia*.

Physochlaina praealta (G. Don.) Miers.

Family – SOLANACEAE

Local/Common name(s) – Sholar, Bajar-bang, Nandru, Laltang.

Distribution – The herb is found in Lahaul Sub-division of Lahaul & Spiti district, Ladakh, North Kashmir and Western Tibet at altitudes of 3300-4650 m. Collected from Pattan valley, Darcha and Keylong area where it grows frequently as a roadside weed often forming patches running to several square metres. Found sporadic in Spiti region around Khab, Youngthang and Mane. (Chauhan, N.S., 2807).

Description – An erect, nearly glabrous, perennial herb, 150-160 cm high. Leaves are irregular, ovate-oblong, wavy; flowers greenish, campanulate, in terminal corymbose cymes; capsules are 2-celled. Seeds are many, sub orbiculate-reticulate.

Physochlaina praealta (G. Don.) Miers.

Flowering/Fruiting – June-September.

Parts used – Leaves, often entire plant.

Current market rate – Not available.

Main ingredients – One of the source of tropane group of alkaloids. Leaves contain hyoscyamine, which is further converted into atropine and minor contents of hyoscine; fairly large amounts of potassium salts like potassium nitrate, potassium chloride and potassium sulphate. Roots contain hyoscyamine and sucrose.

Uses – The leaves of the herb are narcotic and possess mydriatic properties, causing dilation of the pupil of the eye, like belladona. The leaves are applied to the boils. Seeds are used by local people as vermifuge to expel round worms and as an emetic in bilious attack.

Preparations – Used in modern pharmaceutical industry for its hyoscyamine, atropine and hyoscine contents and enters into preparations requiring such contents.

Cultivation notes – The herb grows wild and is abundant in Ladakh and Lahaul-Spiti. It occurs commonly as a weed in Lahaul sub-division on the boundaries of the fields and roadsides. The plant remains dormant under the snow and sprouts out next spring. Being a perennial, the leaves can be gathered, more than once in a year. Leaves should be collected, when plants are in full bloom i.e. towards the end of June and spread open in thin layers for drying in the sun. It is estimated that nearly 15 tonnes of dried leaves can be gathered annually from Lahaul, if systematically collected. The herb can be successfully grown by transplanting seedlings raised from seeds sown in March. Some plants develop black fungal growth on the stem and are said to be highly poisonous to all types of animals.

Phytolacca acinosa Roxb.

Family	– PHYTOLACCACEAE
Local/Common name(s)	– Jharka, Jalga, Sweet belladona, Matazor.
Distribution	– Distributed in temperate Himalaya from Kashmir to Bhutan and in Assam. In Himachal Pradesh, it is found in forest shades in Chamba, Kullu, Mandi, Shimla and Sirmour districts between 2300-2800 m. (Chauhan, N.S. 902, 1091 & 2477).
Description	– A succulent, robust, glabrous, erect herb upto 1.5 m tall. Leaves alternate, broadly lanceolate, 15-25 cm, entire, long pointed, narrowed into a short stalk. Flowers pale green, aggregated in leaf opposed, robust, cylindrical racemes; 5-20 cm long and 2.5 cm to over 5 cm in diameter. Fruits dark purple, succulent, crowded in on erect raceme.
Parts used	– Entire plant (chiefly leaves, roots and seeds).
Current market rate	– Not available.
Properties and uses	– The tender leaves and twigs are cooked as vegetables. The herb is believed to have narcotic effect, which is destroyed on boiling. The fruit is occasionally eaten and used for flavouring purpose. There have been cases that local people mistook *Atropa acuminata* (Jalgi) for *P. acinosa* and suffered serious food poisoning and went into euphoria for a couple of days as a result of this mistake (Chauhan, 1984). For the same resemblance, the roots and leaves of *P. acinosa* have been found mixed in belladona consignments

exported from India (cf. Wealth of India, VIII: 42).

Preparations – No standard formulations are known. As an adulterant of belladona it finds its way into such preparations, where belladona is used.

Cultivation – Can be cultivated through seeds and root splits. In far flung areas, tall, perennial plants are planted in the courtyards of village house. Its tender leaves are used as a vegetable sag.

Picrorhiza kurooa Royle ex Benth.

Family – SCROPHULARIACEAE

Local/Common name(s) – Kutki, Kaur, Karru, Tikta, Karvi, Picrorhiza.

Distribution – The plant is found in the alpine Himalayas from Kashmir to Sikkim, at the elevation of 3000 to 4000 m above mean sea level.

In Himachal Pradesh, kutki is found in Thamsar and Dainasar areas of Bara and Chhota Bhangal respectively in Kangra district; Pangi-Bharmour (Jutadhar and Mani Mahesh) of Chamba district, Lahaul, Kinnaur. Rohtang Pass in Kullu and Shimla districts above 3200 m elevation especially in rocky slopes. (Chauhan, N.S., 22, 1290, 1314, 2338, 3072).

Description – The plant is a trailing herb with jointed and zig-zag underground stems (rhizomes), giving off shoots at joints and also rooting at the nodes. Leaves are spathulate and dentate. Flowers are bluish white, showy and are arranged in dense, terminal and spicate racemes. Capsules are ovoid and conical (Chauhan, N.S., 1984).

Flowering/Fruiting – June-August-September (Capsules ripen and shed seeds).

Picrorhiza kurooa Benth. (Kutki, Karvi)

Parts used – Rhizomes and stolons. Collection is done during September-October.

Current market rate – Rs. 150-165/- per kg (dried).

Main ingredients – A brown resinous glucoside, picrorhizin and its aglycone, picrorhizetin have been reported from the rhizome by earlier workers (Rastogi et al., 1949). Rastogi and Sen (1953) isolated a glycosidal bitter principle kutkin, a non-bitter substance kurrin, vanillic acid, kutkiol occurring as acetate and kutkisterol from the drug. Kutkin, apocynin, alkanol, alkane and β-sitosterol have also been reported from the rhizomes (Anonymous, 1990).

Uses – It is considered to be a valuable bitter tonic like Gentian, stomachic, laxative in small doses and cathartic in large doses. It is reputed to have beneficial action in dropsy and is considered to be anthelmintic, appetiser, blood purifier, blood pressure reducer, car-diac, expectorant, ferbifuge and is useful in asthma, cold and cough, bile trouble, jaundice, leprosy, constipation and stomach troubles. Clinical trials on this drug have shown its therapeutic efficiency in a number of condi-tions of immunological disorders like asthma, arthritis, urticaria etc. (Anonymous, 1980).

Preparations – Arogyavardhini-vati, Katukadya-lauha, Tiktadi-kwath, Tiktadya-ghrita, Amritastak-kwath, Mahasudarshan-churna, Aragbadh-yadi-kwath, Raktashodhak-vati, Sudarsan-churna, Sarva-jwar-hara-lauha, Amritarista, Phalatrikadi-kwath, Punarnava-mandur, Yograj-guggulu, Kumaryasava, Sarivadya-rishta.

Cultivation notes – Vegetative propagation from rhizomes, stolons and offsets is quite easy. Prefers porous soil layer with sufficient sun to facilitate horizontal spreading of the rhizomes underneath, which ultimately gives up sprouts from the nodes. It has been successfully domesticated in the herbal garden at Rahla, Distt. Kullu.

Pinus roxburghii Sarg.

Family	–	PINACEAE
Local/Common name(s)	–	Chir pine, Chirh, Saral, Biroja (Resin).
Distribution	–	Vast stretches of forests occur in Himachal Pradesh in Shivalik belt, lower parts of many districts between 450-2200 m elevation. Highest limit of Chirpine is found in Parvati valley (Malana Khad-Jari area) in Kullu district.
Description	–	A moderately tall tree with a spreading crown; branches more or less whorled, bark dark grey, often reddish, deeply fissured and rough. Leaves are long, 20-30 cm acicular, triquetrous, light green in clusters of 3. Male flowers about 1.5 cm long, arranged in the form of cones. Female cones are solitary, often 2-5 together, ovoid, 10 to 20 cm × 7.5 to 13 cm when ripe, brown and woody. Seeds are provided with long, membranous wings.
Flowering/Fruiting	–	April-June.
Parts used	–	Oleo-resin (Terpentine oil etc.)
Current market rate	–	Resin Rs. 30-35/- per kg.
Chemical contents	–	Oleoresin yields turpentine oil and rosin. Turpentine oil contains alpha-pinene, 20-30 per cent; beta-pinene, 5-10 per cent; -Δ carene, 55-65 per cent. Longifolene and longicyclene has also been reported. Needles yield pine oil.

Properties an uses – The turpentine oil is used in pharmaceutical preparations, perfumery industry, synthetic pine oils, disinfectants, insecticides and denaturants. The oil is valued in medicines and is included in the Indian Pharmacopoeia. It is feebly antiseptic and as expectorant is useful in chronic bronchitis. It is especially recommended in the treatment of gangrene of the lungs and has been found beneficial as a carminative in flatulent colic, also arrests minor haemorrages in tooth sockets and nose. As enema, the oils is useful in obstinate constipation, tympanitis and seatworm infestations. Externally, it is used as rubefacient in various thematic affections, such as lumbago, arthritis and neuralgia (I.P.C., 192-93). The oil from chir pine needles is reported to have anti-bacterial activity against a number of organisms including *Staphyllococcus aureus* (Wealth of India, VIII, 76-77).

Preparations – Rosin and turpentine are used for a variety of products. Camphor obtained during processing of resin can also be used in pharmaceutical industries.

Cultivation – Seedlings are raised through seeds in polybags and transplanted in pits in the plantation areas. Being a hardy species, the survival rate is very high. Several forests have been raised by plantation in Himachal Pradesh.

Pinus wallichiana A.B. Jackson

Family – PINACEAE

Local/Common name(s) – Kail, Chil, Blue pine.

Distribution – Forms considerably good forests either pure or mixed with Deodar, fir, spruce and broad leaved trees of *Acer, Aesculus, Prunus* and *Quercus* between 2000-3500 m in the districts of Sirmour, Shimla, Kullu, Kangra, Kinnaur and Chamba. It has been noticed to go upto the last limit of tree life even higher than high level silver fir in Holi Range in Bharmour, distt. Chamba.

Description – A tall, evergreen tree with blue-green foliage. Leaves (needles) 12-20 cm long in clusters of 5, drooping, except when quite young. Male cones ovoid, short. Female cones on maturity are cylindric 15-25 cm long, erect when young, pedulous afterwards. Tip of the scales slightly thickened with obtuse beak. Seed winged.

Flowering/Fruiting – May-October (cones ripen in the autumn of the year after flowering)

Parts used – Resin

Current market rate – Not available.

Chemical contents – Resin obtained from the tree is much more superior as compared to the resin obtained from *P. roxburghii* having high pinene content.

Properties and uses – The yield of oleoresin and turpentine oil is about half of chir pine, but the oil is of superior quality and has high pinene content and is suitable for all purposes, where American turpentine oil can be used. Needle oil also shows antibacterial property (cf. Wealth of India, VIII, 80-81). The resin from the young sapling is used as liniment for healing of the cuts and wounds. The bark in circular cylinders, taken out of pole size plants is used as a bandage on the dislocated/ fractured bones as a plaster for humans, cattle, sheep as well goats. The resin and circular bark tied on the organ keeps the bones in place and facilitates calcification and unification of fractural pieces (Chauhan, N.S. 1984).

Preparations – Not available.

Cultivation – Seedlings are raised from seeds and planted in forest lands. It also regenerates naturally and has been observed to come up as a coloniser in newly exposed sites, land slips etc.

Pistacia khinjuk Stew. ex Brandis

Syn: *P. integerrima* Stewart.

Family – ANACARDIACEAE

Local/Common name(s) – Kakrasingi, Karkatsringi, Kakar, Kakru.

Distribution – The tree is found in the Himalayas from Indus to Kumaon at altitudes of 350-2400 m.

It is found in the districts of Chamba, Kangra, Una, Hamirpur, Bilaspur, Kullu, Mandi, Kinnaur, Shimla, Sirmour and Solan. (Chauhan, N.S., 112, 907, 997, 1651 and 2455). It has been observed that the gall formation is more prevalent in lower warmer areas and almost absent at the high altitude.

Description – A moderate sized deciduous tree with rough, grey bark. Leaves are odd or even pinnate 15-23 cm long and the leaflets are nearly opposite, lanceolate, shining, entire. Flowers are uni-sexual and dioecious. Drupe is dry and glabrous. The gall formed on the leafstalk by some insects has different shapes and sizes; sometimes twisted, hornlike or circular having purplish brown colour. When broken, the hollow of the hornlike gall is filled with white, greyish green dust and often small insects.

Flowering/Fruiting – March-May/June-October.

Pistacia khinjuk Stew. ex Brandis

Parts used – Galls are formed on the leaves especially the petioles by the insects and are called as 'kakarsingi'.

Current market rate – Rs. 150-325/- per kg.

Main ingredients – The galls contain tannins, a resin and an essential oil, two isomeric triterpenic acids designated as pistacienoic acid A and pistacienoic acid B; a triterpene alcohol and a waxy compound. Two crystalline compounds provisionally named as pistacin and pistacinin have been isolated from the drupes (CCRAS, 1990).

The galls yield 1.3 percent of essential oil containing α-pinene, camphene dl-limonene, cineole, α-terpeneol, aromadendrone and gum mastic, a resinous substance and two crystalline acids. (Mukerji, 207).

Uses – Galls are used for tanning and dyeing, besides being used in asthma, phthisis and other diseases of the respiratory tract and in dysentery. The oil, found in the galls, is used as a carminative. Leaves are used as fodder. The wood is used for construction and beautification purposes.

Preparations – Shringyadi-Churna, Karkatadi-Churna, Balchaturbhadra-Churna, Dashmoolarishta, Chyavanprasha, Balachaturbhadra churna etc.

Cultivation notes – The trees grows on hot, dry slopes with shallow soil or on open rocky ground and limestone soils. Natural seedlings spring up under bushes and force their way out. The rate of growth is moderate with annual girth increment of 1.8-2.0 cm. (The Wealth of India, VIII, 120). For seed propagation, the mature seeds can be collected in June and sown after shade drying during June-July.

Germination takes 2-4 weeks' time. Softwood cuttings can also be used for propagation by the application of 2000 ppm I.B.A.

Plumbago zeylanica Linn.

Family – PLUMBAGINACEAE

Local/Common name(s) – Chita, Chitrak, Chitra, Chitrakmula, Leadwort.

Distribution – The species is widespread and is possibly indigenous to South-East Asia. It is found wild in peninsular India and West Bengal and cultivated in gardens throughout India.

In Himachal Pradesh, it is found growing sporadically near field borders, often in shrubberies in Mandi, Hamirpur, Kangra, Chamba, Bilaspur, Solan, Sirmour upto 1400 m elevation and has been collected from Sitla-Mandir, Shivbari, Dada Siba, Hamirpur, Nalagarh, Jogindernagar, Solan-Nauni. (Chauhan, N.S., 228, 411, 1486, 1810).

Description – A rhambling, subscandent perennial herb or under shrub with woody, striate stem. Leaves are ovate, glabrous and the flowers are white in colour in elongated spikes, capsules are oblong, pointed, covered with mucilage and are contained in viscid, glandular, persistent calyx tube. The roots are 6-20 mm in diameter and seldom branched. Externally, these are dark, reddish brown, somewhat shrivelled and marked here and there by small warty projections. Internally, it is brown and striated. It tastes acrid and biting; considered poisonous.

Flowering/Fruiting – Nearly all the year around except in winters.

Parts used – Roots (dried).

Current market rate – Rs. 30/- per kg.

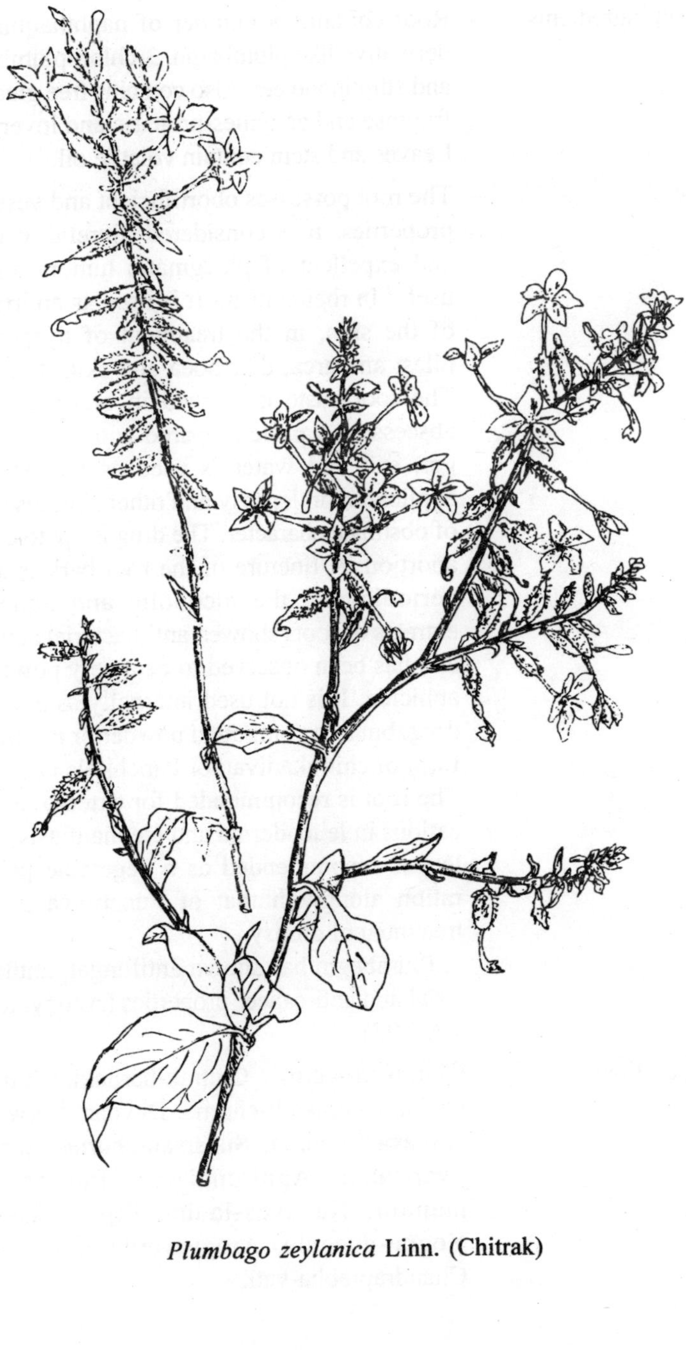

Plumbago zeylanica Linn. (Chitrak)

Main ingredients – Root contains a number of naphthaquinone derivative like plumbagin, 3-chloroplumbagin and slliptinone etc. Also contains free glucose, fructose and enzymes-protease and invertase. Leaves and stem contain volatile oil.

Uses – The root possesses obortifacient and vesicant properties. It is considered diuretic, caustic and expellent of phlegmatic tumors and is useful in rheumatism. It is used as an irritant of the skin, in the treatment of dyspepsia, piles, anasarca, diarrhoea and skin diseases. The root paste is applied in order to open abscesses. A paste prepared with milk, vinegar, salt and water is used as an external application in leprosy and other skin diseases of obstinate character. The drug is apt to cause abortion. A tincture of the root bark is antiperiodic and the alcoholic and aqueous extracts of roots showed anti-bacterial activity and has been observed to be a very powerful aphicide. It is not used internally as a single drug, but as a compound powder or pill in the form of chitrakadivati or Panchkola choorna. The root is recommended for external applications in leucoderma and elephantiasis. The leaf is recommended as a vegetable preparation alongwith that of Punarnava in the treatment of dropsy.

Plumbagin has shown antifungal, antibacterial and anti-cancer properties (Anonymous, 1990:97).

Preparations – Chitrakadi-guttika, Chitrak-haritaki, Chitrak-ghrita, Chitrakadi-churna, Devdaradi-kwath, Amlayadi-churna, Sudarsan-churna, Sarva-jwar-lauha, Agnitundi-vati, Punarnava-mandur, Navayas-lauha, Pippalyasava, Yograj-guggulu, Arogyavardhini-vati and Chandraprabha-vati.

Cultivation notes	–	Through seeds, cuttings and splits. The plant throws up several suckers, which can be separated with root part and transplanted. It thrives well near field borders, debris heaps and loose soils. Easily propagated through seeds and branch cuttings in the university herbal garden.

Podophyllum hexandrum **Royle**

Syn: *P. emodi* Wall.

Family	–	PODOPHYLLACEAE (Berberidaceae)
Local/Common name(s)	–	Bankakri, Kanda-ri-mokri, Giriparpat, Rikhpat, Papra, Indian podophyllum.
Distribution	–	Temperate Himalaya between 2600-4500 m. In Himachal Pradesh, it is found sporadically in Rohru, Kullu, Kangra, Chamba, Nichar and Lahaul & Spiti forest divisions (Chauhan, N.S. 1547, 4255).
Description	–	An erect, glabrous, succulent herb; 35-60 cm tall with creeping, perennial, knotted rhizomes bearing numerous roots. Leaves usually 2; long, stout, stalked, orbicular-reniform, palmate, peltate with lobed segments. Flower-solitary, in the axil of the two leaves; creamy white, cup shaped. Fruit is an oblong, elliptic berry, 2-5 cm to 5.0 cm in diameter, red on ripening and containing many seeds embedded in the red pulp. Fruit is edible and relished by shepherds and gujjars.
Flowering/Fruiting	–	April-June.
Parts used	–	The rhizome and the roots. The rhizomes is irregular, knotty, tortuous, about 2-5 cm long, 1-2 cm thick and dorsiventrally flattened. External colour is yellowish brown to earthy brown . Transversely cut surface is smooth and irregularly circular in outline. Adventi-

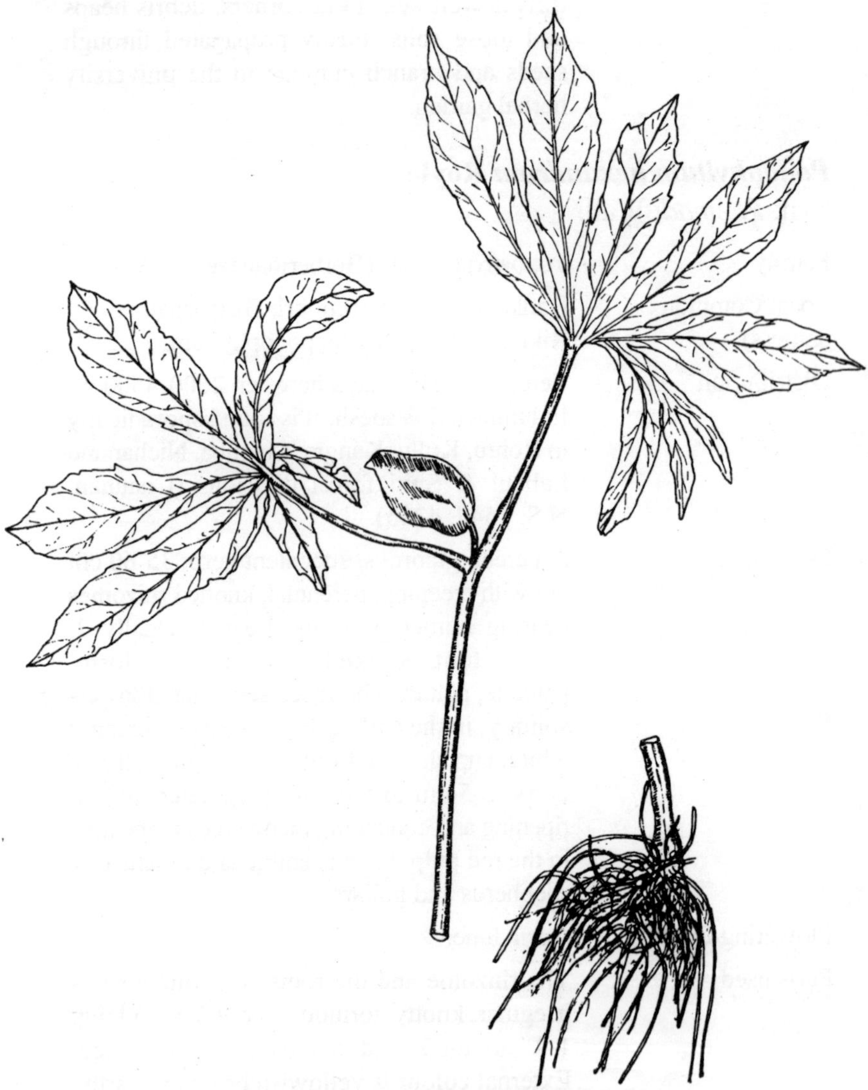

Podophyllum hexandrum Royle

tious roots are upto 7 cm long. Taste is disagreeably bitter and acrid (I.P.: 488).

Current market rate – Rs. 70-90 per kg.

Chemical contents – The Indian Podophyllum yields resin in the range of 7-15%; whereas the American plant, *P. peltatum* yields 4-8% resin. It yields podophyllin and podophyllotoxin as the active principles and also podophyllol, quercetin, kaempferol, astragalin, an essential oil, wax and mineral salts. The roots are richer in resin than the rhizome. Fresh and spring (May) collected roots are rich in resin. It is lost considerably after long storage (Chopra et al., 1958: 288; Wealth of India, IX: 171).

Properties and uses – The rhizomes and roots are considered hepatic, stimulant, cholagogue, purgative, alterative and bitter tonic. The resin or podophyllin is a drastic purgative and produces copious watery stools. It is thus, given with *Belladona* and *Hyscyamus* to avoid gripping. Podophyllin has been tried on various diseases like warty lesions of the skin and neoplasms of the body parts accessible to the tropical therapy and is considered the drug of choice in 'Condyloma acuminatum'. It is an effective vermifuge, first stimulating, then paralysing and finally killing ascaris. During recent years, podophyllin has acquired special importance for its possible use for some forms of cancer. The limiting factor has been the severe gastro-intestinal discomfort followed by its use in high doses. The cytotoxic effect causes mitotic arrest, nuclear fragmentations and other cell damage. As a karyoplastic, it affects both dividing and non-dividing cells. Podophyllotoxin and peltatins have high potency in reducing damage in sarcoma, carcinoma and melanoma (Cancerous compounds). The tumor damaging potency of the podophyllin compounds is said to be closely associated with their stereo-chemistry (Wealth of India, IX: 174).

Preparations – None in Indian system of medicine, but used in anti-cancer formulations in modern medicines.

Cultivation – It can be cultivated through seeds and rhizome splits. Seed generated plants take 3-4 years to develop; whereas those grown through rhizomes pick up fast growth and development.

Polygonum aviculare Linn.

Family – POLYGONACEAE

Local/Common name(s) – Nismoli, Machoti, Bannalia, Hunraj, Anjubar, Hunter's tea, Knotgrass.

Distribution – The plant is distributed from Kashmir to Kumaon at altitudes of 1800-3600 m. Found sporadic in Himachal Pradesh in temperate and sub-alpine areas. Common in billing Nalla slopes (Aswal, 10444).

Description – A highly variable annual herb, with prostrate stems and ascending branches upto 60 cm long having partly purple tinted colour. Roots are spindle shaped at the top and branched at the bottom, producing several, many branched, jointed stems that usually spread in every directions. Root is very tough, often woody. The leaves are elliptic or lanceolate, set at the nodes and right upto the tip of branches. Flowers are small, green, tipped with red or white and are in axillary clusters. Nuts are triangular, blackish purple and minutely wrinkled.

Polygonum aviculare Linn. (Anjubar, Knotgrass)

Flowering/Fruiting – Late spring to late autumn. (June-September)

Parts used – Herb (at full flowering stage).

Current market rate – Rs. 30/- per kg.

Main ingredients – A source of Vitamin E, the herb has been reported to contain tannins, mucilage, sugars, gallic, caffeic, oxalic, silicic, chlorogenic and p-coumaric acids; d-catechol, leucoanthocyanins, oxymethylanthraquinone, guercitrin, avicularin and several other flavonoids and polyphenols, which include derivatives of quercetin, kaempferol, esculetin and scopoletin.

Uses – The herb is known as Hunter's Tea or Homerian Tea. It is said to possess astringent, tonic, antipyretic, anti-septic, diuretic, haemostatic and vermifuge properties. It is used in the treatment of diabetes, rheumatism, fever, external and internal ulcers. A decoction of the herb is given in dysentery, diarrhoea, bronchitis and bleeding piles and to check profuse menses. Intravenous injections of an acqueous or alcoholic extract of leaves lowers blood pressure in cats, rabbits and dogs. The extract is thought to be useful as a haemostatic agent. The seeds are aromatic and are powerfully emetic and cathartic. The herb is reported to have been successfully tried for feeding silkworms in mulberry industry. (The Wealth of India, VIII, 1969:195). It is used as a diuretic for gall bladder and kidney problems and for rheumatic complaints. It is also used as a cure for tuberculosis of the lungs (Dorfier & Rosett, 1989: 226).

Preparations – Enters into various Ayurvedic and Unani preparations.

Cultivation notes – Can easily be cultivated through seeds, cuttings and splits.

Pongamia pinnata **(Linn.) Merr.**

Syn: *P. glabra* Vent.

Family	– LEGUMINOSAE/FABACEAE
Local/Common name(s)	– Karanja, Puti-Karanj, Karanjpatra, Pongam, Indian Beech.
Distribution	– The plant is found practically throughout India, very common as a sea-coast tree in the Andamans and Nicobar Islands. (Dey, 76). Found sporadic in Shivalik range in Himachal Pradesh and also planted in gardens and avenues. (Chauhan, N.S., 1000, 1020).
Description	– A handsome tree with drooping branches; leaves unequally pinnated with 2-3 pairs of ovate, acuminate shining green leaves laden with rose tinted white flowers. Flowers arise in axillary racemes, about half the length of the leaves. Pods are with a short decurved point, turgid, woody, glabrous with brownish-green colour and are one seeded, rarely two seeded. Seeds are white, marbled with brownish lines.
Flowering/Fruiting	– April-June.
Parts used	– Bark, leaves, flowers and seeds.
Current market rate	– Rs. 35/- per kg.
Main ingredients	– Seeds contain 27-36.4 percent of a bitter fatty oil and traces of an essential oil and three crystalline substances—karanjin, pongamol and glabrin. Roots also contain karanjin, the principle responsible for curative properties.
Uses	– The poultice of leaves is applied to worm infested ulcers. Oil is antiseptic, useful in skin diseases. Bark is astringent, carminative, cooling, demulent, parasiticide and stimulant. The plant is indicated in abdominal en-

		largements, diabetes, diarrhoea, dyspepsia, epilepsy, flatulence, insanity, leprosy, skin diseases-scabies, herpes and other cutaneous diseases; ulcers, fistulous sores, bleeding piles (fresh bark), and whooping cough. Seeds and roots are also used as fish poison.
Preparations	–	Karanjadi-churna, Karanjadi-ghrita, Karanjadi-taila, Karanjadi-yoga and Vidangadi-lep.
Cultivation notes	–	The tree can grow in a variety of soils and climate. It is grown as an avenue tree all over India. It is drought resistant, moderately frost hardy and tolerant of salinity. It is a shade bearer and is advocated for plantation in pastures as grass grows well in its shade. Propagation is through seeds, root suckers and root-shoot cuttings. It is fast growing and coppices well. The tree starts bearing flowers and fruits at the age of 4-7 years.

Potentilla fulgens Wall.

Family	–	ROSACEAE
Local/Common name(s)	–	Bajra-danti
Distribution	–	Distributed in temperate Himalaya between 2000-3000 m elevation. In Himachal Pradesh, the plant has been collected from Chhajpur (2400 m), Khabal-Dhamiala (2800 m) and Dhamkhari-Chansil Pass (3200 m) (Chauhan, N.S. 2236 and 2262).
Description	–	An erect, perennial, stout, ascending herb, 15-75 cm tall with a thick rootstock. Leaves are pinnate, clothed with silvery hairs beneath. Flowers are orange yellow in corymbose panicles.
Flowering/Fruiting	–	August-October.
Parts used	–	The roots

Current market rate – Not available.

Chemical contents – Not available.

Properties and uses – The root is used to strengthen gums and teeth (Chopra *et al.*, 1969:83).

Preparations – Roots are used in proprietary preparations like Vicco-Bajradanti tooth powders and tooth pastes.

Cultivation – Can be cultivated through seeds and root slits.

Prinsepia utilis Royle

Family – ROSACEAE

Local/Common name(s) – Bhekhal, Bhekhla, Bhenkal.

Distribution – Distributed in China, India, Bangladesh and Farmosa. In India, it is distributed throughout Himalayan range from West to East between 1600-3000 m elevation. In Himachal Pradesh, it is frequently found in Shimla, Sirmour, Kinnaur, Mandi, Kullu, Kangra and Chamba between 1400-2800 m. (Chauhan, N.S. 606, 1534 & 2014).

Description – A thorny shrub upto 3.5 m in height. Leaves are lanceolate or elliptic-lanceolate. Flowers yellowish-white in axillary racemes. Drupe-oblique, ellipsoid 1.3 to 1.8 cm × 1.0 cm; deep purple to almost black; fleshy on ripening. Stone (putamen) are ellipsoid, smooth, enclosing a single brown seed, having thick oily cotyledons.

Flowering/Fruiting – Flowers during winter starting from December to March and fruits ripen in summers i.e. May-July.

Parts used – Oil (from kernels collected in June-July).

Current market rate – Rs. 35-40/- per kg. approx.

Prinsepia utilis Royle

Chemical contents – Seeds yield 35-40% semidrying, pale yellow fatty oil. Fatty acid composition of the oil consists of 1.8% myristic, 15.2% palmitic, 4.5% stearic, 0.9% lignoceric, 32.6% oleic and 43.6% linoleic acids. Resin acid is about 1.4%. The unsaponifiable matter contains sitosterol.

Properties and uses – The oil is used for cooking purposes and as in illuminant in the hills. Also reported suitable for hydrogenation and soap making. The oil possesses rufefacient properties and is applied externally in rheumatism and pains resulting from over fatigue.

Preparations – Consumed locally. The physico-chemical properties of Bhekhla oil resemble closely to those of olive oil.

Cultivation – Can be cultivated easily through seed. Before sowing, seeds should be immersed in water for 24-36 hours to soften the hard seedcoat.

***Prunella vulgaris* Linn.**

Family – LABIATAE/LAMIACEAE

Local/Common name(s) – Austakhaddus, Ustakhadus, All-heal, Self-heal.

Distribution – The plant is found from Afghanistan to Bhutan in the North temperate zone at an altitude of 1500-3600 m.

In Himachal Pradesh, it is found in different parts of Chamba, Kinnaur, Mandi, Kangra, Sirmour, Kullu, Lahaul & Spiti and Shimla districts in Himachal Pradesh, in moist shady localities between 2000 m-3800 m. (Chauhan, N.S., 1303, 1346, 1688, 2083, 2174 and 2581).

Description – It is an annual herb with many, spreading or ascending stems, 10-30 cm in height. Leaves

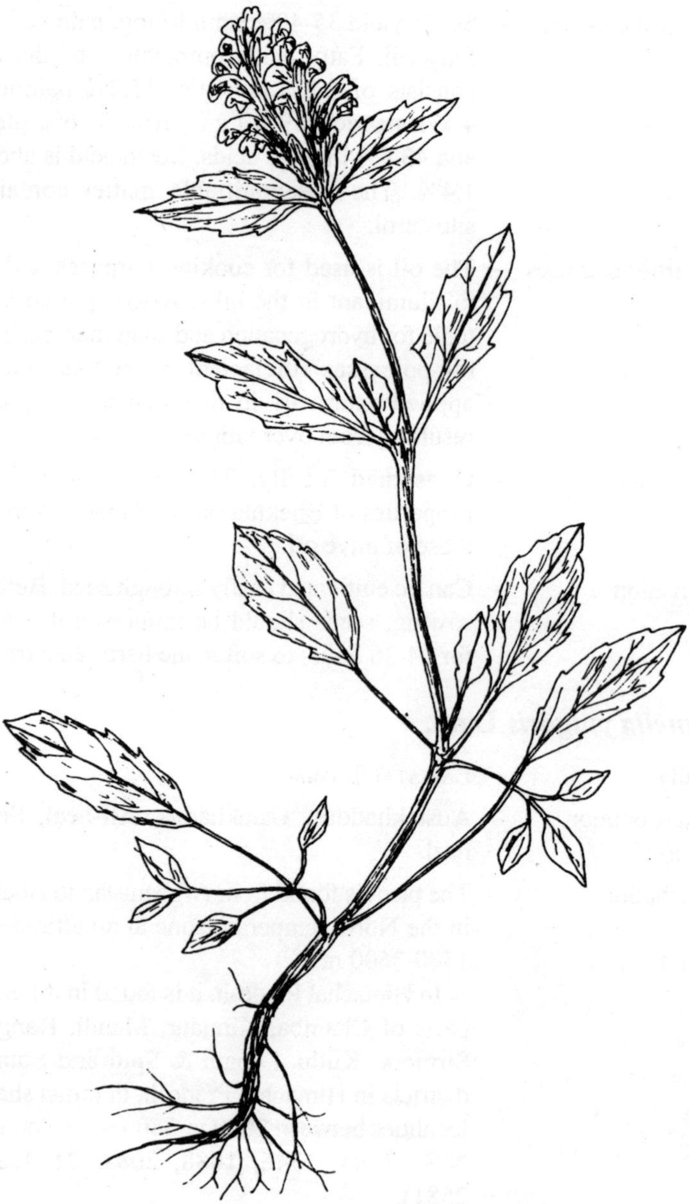

Prunella vulgaris **Linn. (Austakhaddus)**

are opposite, ovate, obtuse and hairy, arising immediately below the bract and spreading beyond the inflorescence. Flowers are bluish, pink, purplish and in terminal spikes.

Flowering/Fruiting – June-October.

Parts used – The dried herb.

Current market rate – Rs. 150/- per kg.

Main ingredients – The herb yields a bitter principles and essential oil. The oil mainly consists of d-camphor and d-fenchone and traces of penchyl alcohol. (Chauhan, N.S., 1984).

Uses – The herb is considered antiseptic, expectorant, anti-rheumatic, alterative, tonic, astringent, carminative, anti-spasmodic, stimulant and is used in fever and coughs. It is also used in haemorrhages, diarrhoea, relaxed throat and ulcerated mouth in the form of an infusion and is useful for the treatment of internal bleeding piles. The green leaves coated with castor oil, warmed over fire are applied to the rectum to cure the painful piles. The infusion is used as the mouth wash also. It can be used as a syrup for the internal wounds and for curing the headaches. An old German saying is popular about the plant that, "He needs neither physician, nor surgeon, that hath self-heal to help himself."

Preparations – It is an important ingredient of Unani medicines.

Cultivation notes – Propagation is through seeds.

Prunus cerasoides **D. Don.**

Syn: P. *puddum* Roxb.ex.Wall.

Family – ROSACEAE

Local/Common name(s) – Padmakh, Pajja, Paddam, Padmakashta.

Distribution – The tree is indigenous to the temperate Himalayan region, extending from Kashmir to Bhutan and in Akai and Khasi hills in Assam and in Manipur at altitudes of 900-2300 m. It is found in the districts of Chamba, Kangra, Hamirpur, Una, Bilaspur, Kullu, Sirmour and Shimla upto an elevation of 1800 m; rarely occurring at higher elevations in exposed sites. (Chauhan, N.S., 1728 and 1042).

Description – It is a small or a middle sized attractive tree and the only tree to bear attractive pinkish-white flowers in winters/autumn. Bark is smooth and brown in colour. Leaves are variable in length and breadth; 7.5-12.5 cm × 2.5-5 cm; ovate-lanceolate, dentate. Flowers are in fascicles or umbels and rose-red to white in colour. Fruit is ovoid, obtuse, yellow or red on ripening. Seed is rounded and furrowed. The species is reported to have many geographical races.

Flowering/Fruiting – November-May.

Parts used – Bark, sticks of branches.

Current market rate – Rs. 12/- per kg. (Bark and wood pieces).

Main ingredients – The stem bark contains flavonone sakuranetin, flavone genkwanin, isoflavone prunetin and isoflavonone padmakastin, along with smaller quantities of the glycosides sauranin and padmakastin. (The Wealth of India, VIII, 266). Also contains amygdalin, prunacetin, sakuranetin, puddumetin etc. (Dey, 1980: 113).

Uses – It is considered as an anti-abortifacient, analgesic, carminative, conceptive, expectorant, febrifuge and tonic; useful in stomach trouble, burning sensation of the body, cold and cough, seminal weakness and in pregnancy. Bark is

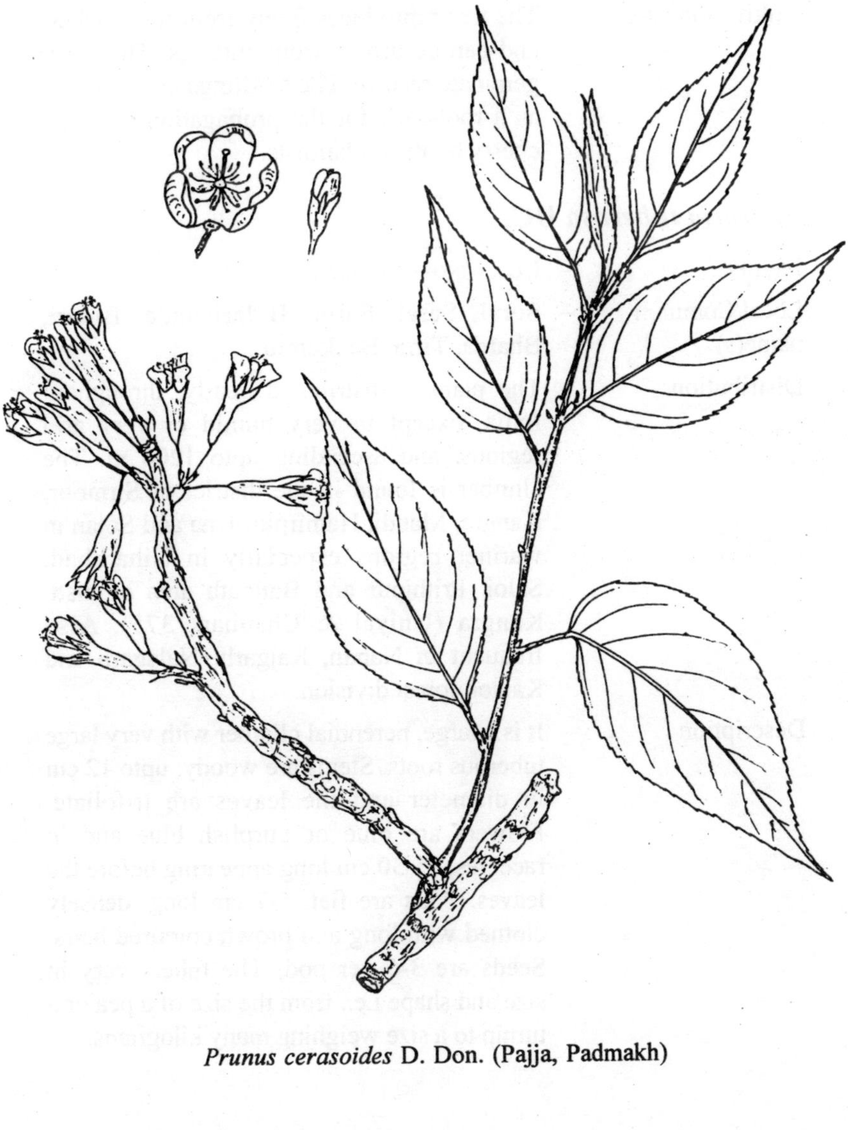

Prunus cerasoides D. Don. (Pajja, Padmakh)

mainly used for tannin. The smaller branches are crushed and soaked in water and taken internally to stop abortion. Seeds are used in stone and gravel. (Chopra et al. 365; Dey, 1980, 112).

Preparations – Sudarshan-churna, Guduchyadi-kwatha etc.

Cultivation notes – The tree reproduces freely from root suckers and can be grown from cuttings. The seeds geminate readily. The seedlings are also used as a rootstock for the propagation of sweet cherry by the orchardists.

Pueraria tuberosa DC.

Family – LEGUMINOSAE/FABACEAE

Local/Common name(s) – Sural, Saral, Sarur, Bidarikanda, Bidari, Bharda, Tirra, Bankumra.

Distribution – The plant is distributed nearly, throughout India, except in very humid or very arid regions, and ascending upto 1200 m. The climber is found in the district of Sirmour, Kangra, Mandi, Hamirpur, Una and Solan in warmer regions especially in Niharkhad, Saloi, Prithipur and Baijnath area of distt. Kangra (Uniyal & Chauhan, 375). Also frequent in Nahan, Rajgarh, Nalagarh and Karsog forest division.

Description – It is a large, perennial climber with very large tuberous roots. Stems are woody, upto 12 cm in diameter and the leaves are trifoliate. Flowers are blue or purplish blue and in racemes, 15-30 cm long appearing before the leaves. Pods are flat, 5-7 cm long, densely clothed with long and brown coloured hairs. Seeds are 3-6 per pod. The tubers vary in size and shape i.e., from the size of a pea or a turnip to a size weighing many kilograms.

Flowering/Fruiting – May-July

Parts used – Tubers.

Current market rate – Rs. 12/- to 16/- per kg.

Main ingredients – Tubers contain vitamins B_1, sucrose, glucose and fructose. (Dey, 1980:161).

Uses – Tubers are eaten raw or boiled and may also be used for the extraction of starch. The roots are demulcent and refrigerant and are used as cataplasm on swollen joints and as a lactagogue. (Ambasta, 1980:504). The tubers are aphrodisiac, cardiac, emetic, expectorant and febrifuge and are used in heart weakness and rheumatism. (Dey, 162). Leaves are used as fodder for horses and cattle. The tubers on keeping become sweet and are eaten raw by the natives in scorching summers.

Preparations – Chavyanprash.

Cultivation notes – The plant is quite useful for soil binding, soil cover and restoration of soil fertility. It can be easily propagated from seeds and crowns. The production of the seeds is prolific. The tubers are collected during or after the rainy season.

Punica granatum Linn.

Family – PUNICACEAE

Local/Common name(s) – Daru, Darmu, Daran, Dadim, Darim and Pomegranate.

Distribution – The tree is considered to be a native of Iran, Afghanistan and Baluchistan. It is found growing wild in the warm valleys and outer hills of the Himalayas between 900-1800 m and is cultivated throughout India. Forms vast patches in open dry slopes in Mandi, Kangra, Sirmour, Bilaspur, Chamba, Kullu and Shimla and Solan in Himachal Pradesh. Vast patches of the plant are found in Guma, Seri and Drang areas of Jogindernagar and Solan—Giripul in Solan district. (Chauhan, N.S., 957, 879, 1577).

Description – It is a large deciduous shrub or a small tree and is often armed. Leaves are opposite, 2.5-6.3 cm long and the flowers are 3.8-5.0 cm long. Calyx is greenish red and the flowers are bright red, tubular, having numerous stamens. Fruit is 3.8-7.5 cm in diameter and the seeds are numerous and angular. (Mukerji, 115 & Dymock et al., 44).

Flowering/Fruiting – May-September.

Parts used – Pomegranate rind, which consist of the dried pericarp separated from the fruit of *P. granatum* Linn. and seeds as Anardana.

Current market rate – Rs. 55-60/- per kg, even upto Rs. 100/- per kg. (Anardana)

Main ingredients – The rind contains about 28 per cent of gallo-tannic acid together with a yellow colouring matter. Flowers yield sitosterol, ursolic acid, maslinic acid, elegic and gallic acid. Edible portion of the fruit contains glucose, fructose, maltose, starch, carotene, thiamine, riboflavin, nicotinic acid, vitamin C, pectin, amino acid etc. (CCRAS, 1990:126).

Uses – In action, it is anthelmintic, astringent, cardiac, cooling, expectorant; useful in brain affections, coughs and colds, diarrhoea and dysentery, heart tonic; increases seminal flow and stops bleeding from the nose. Fruit is a good source of sugar and vitamin C. Seed juice is a favourite drink and blends well with other fruit juices and may also be used for

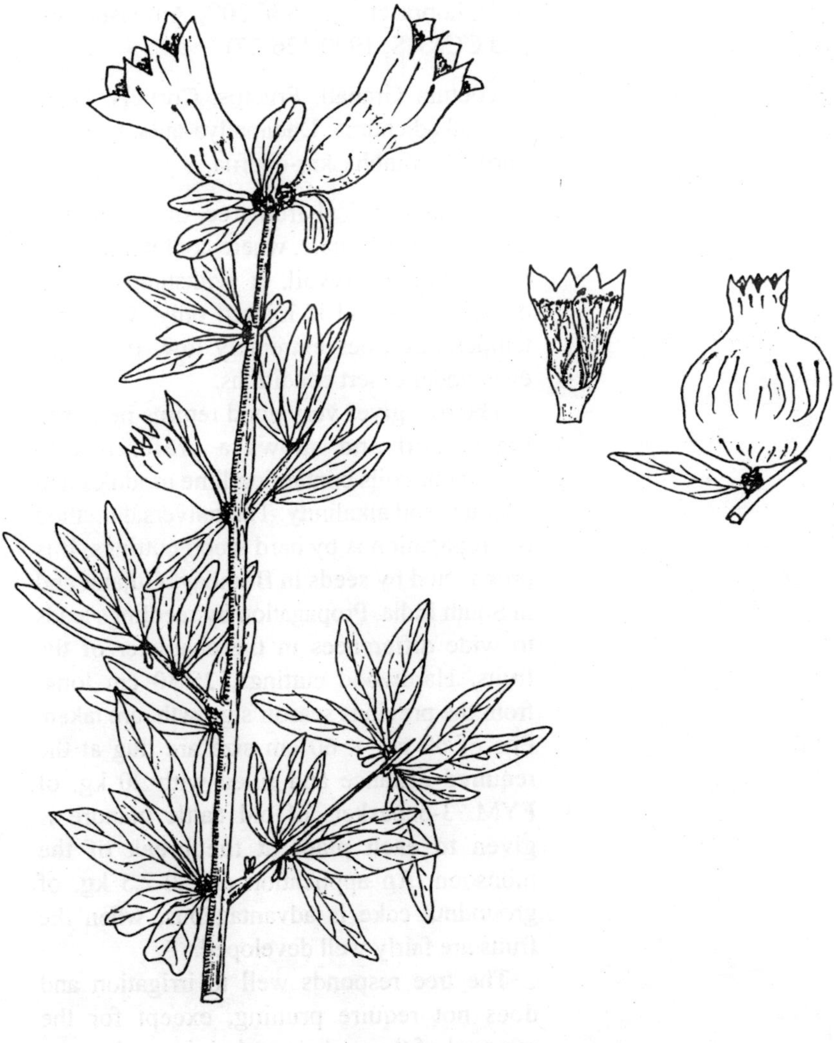

Punica granatum Linn. (Darim, Daru)

making wine. Fruit rind is used as a tanning material. Flowers yield a red dye. Bark is used to expel tapeworms. Rind is used as an astringent in diarrhoea and dysentery. Flower-buds are used in bronchitis, (Dey, 1980:39-39, Chopra et al., 1956:207; Ambasta, 505 and CCRAS; 1990:126-27).

Preparations – Decoctum Granati, Fructus, Corticis. Dadimashtakadichurna, Dadimadya-taila, Dadimaghrita, Lavanbhaskar-churna.

Cultivation notes – It is a sub-tropical fruit tree, grows best in the semi-arid climate, where cool winters and hot summers prevail. It withstands considerable frost and is injured only where the temperature goes extremely low. It thrives, even under desert conditions.

The tree gives very good returns in a deep loamy soil and shows a preference to calcareous soils containing lime modules and tolerates soil alkalinity. The universal method of propagation is by hard wood cuttings. It is propagated by seeds in Bombay, Gujarat and in South India. Propagation by seedlings leads to wide differences in the character of the fruits. Hardwood cuttings, 25-30 cm long, from the previous season's growth, are taken. Pits of 60 cm × 60 cm size are dug at the required distance and filled with 20 kg. of FYM. 3-6 baskets of old cattle manure is given to each plant at the break of the monsoon. An application of 2.0-2.5 kg. of groundnut cake is advantageous, when the fruits are fairly well developed.

The tree responds well to irrigation and does not require pruning, except for the removal of the suckers and giving a shape to the tree.

***Putranjiva roxburghii* Wall.**

Family	–	EUPHORBIACEAE
Local/Common name(s)	–	Putranjiva, Poutijia, Juti, Jiaputa.
Distribution	–	Found wild or cultivated almost in all parts of India, ascending upto an altitude of 750 m. (The Wealth of India, VIII, 1969:325). In Himachal Pradesh, the tall trees are found in Dehragopipur, Dadasiba; planted at Jogindernagar and Nauni (1200 m). (Chauhan, N.S., 240).
Description	–	Mostly a dioecious, evergreen tree, attaining a height of upto 18 m and a girth upto 2 m having grey bark. Leaves are upto 5-10 cm long, elliptic-oblong to ovate-lanceolate, unequal sided at the base, dark green in colour and shining in appearance. Flowers are small, male in dense rounded clusters, yellowish in colour; female flowers solitary or 2-3 together, green in colour. Drupe is 15-18 mm long, ellipsoid or globose and white-tomentose. Stones are pointed, rugose, very hard and normally having a single seed. The seed is with copious albumen.
Flowering/Fruiting	–	March-April/July-September.
Parts used	–	Leaves and Stones.
Current market rate	–	Rs. 12/- per kg.
Main ingredients	–	Glucoputranjivin, Glucocochlearin, Glucojiaputin, Gluco-cleomin, Mannitol, Saponin glucoside and fatty oil.
Uses	–	Leaves and stones are given in decoction for cold and fever and are also used in rheumatism. In Bihar, crushed leaves are reported to be applied to swollen throats of the cattle.

The nuts are made into rosaries and necklaces. It is reported to have the properties to effect conception in sterile women and attributed with the birth of a male child, hence the name 'Putranjiva or Jiyapota'

Preparations – With shivlingi, Asoka, halfmoon of peacock feather, cannabis seeds, jaggery and milk of cow having a male calf, its preparation is reported to effect the birth of a male child.

Cultivation notes – The tree usually grows on alluvial soil along the rivers or in swamps or evergreen forests and is sometimes gregarious. It is occasionally met with in rather stunted form in drier situations. Natural reproduction takes place through seeds during the rainy season and is quite abundant. Dissemination of seeds takes place through deer and bats which like the fruit. Artificial reproduction may be done through transplantation of one or two years old, nursery raised seedlings. Transplantation is best carried out in the rainy season. Young plants stand moderate shade and are sensitive to drought and frost. The tree is often planted in gardens, for hedges and on roadsides, chiefly for shade. Several insect pests have been recorded on twigs, leaves and fruit.

***Pyracantha crenulata* (D. Don) M. Roemer**

Syn: *Crataegus crenulata* Roxb.

Family – ROSACEAE

Local/Common name(s) – Hawthorn, Choota seb.

Distribution – Sutlej valley to Bhutan between 1000-2600 m. In Himachal Pradesh, it has been found in Solan, Mashobra (Shimla) and even in Lahaul-Spiti. *Crataegus oxycantha* is found in Udaipur range of Lahaul (Chauhan, N.S. 1923, 2094 & 2228).

Description – A spiny, evergreen shrub with usually crowded, narrow-oblong, shining, leathery leaves and numerous small white flowers in axillary clusters arranged along the branches. Leaves usually 1.5-3 cm; with obscure usually rounded teeth, spines stout, terminating lateral branches. Flowers 9 mm across, petals rounded; calyx fused to the ovary, with 5, blunt lobes. Fruit globular, approx. 6 mm; orange-red; nuts protruding from the persistent calyx.

Flowering/Fruiting – April-June.

Parts used – Fruits

Current market rate – Fruits Rs. 20-25/- per kg.

Chemical contents – Leaves contain leucocyanidins flavonoids principle-vitexin-4-rhamnoside, vitexin glycoside, oligomeric leucoantho- cyanidin and polymeric flavone (apigenin, luteolin); rutin, tannins and saponins. Fruits contain flavonols-kaempferol, quercetin; flavone derivatives, rutin, hyperoside, glycosides of vitexin and orientin; procyanidins, tannins, saponins and cyanogenetic glycosides.

Properties and uses – Fruits are credited with cardiotonic, coronary vasodilator and hypotensive properties. It has been used for cardiac failure, myocardial weakness, paroxysmal tachycardia, hypertension, arteriosclerosis and Burgor's disease. Wood is used for walking sticks.

Preparations – Infusion of dried fruits, liquid extracts and tinctures.

Cultivation – Can be cultivated through seeds.

Randia dumetorum (Retz.) Poir

Family	– RUBIACEAE
Local/Common name(s)	– Madanphal, Mainphal, Menphal, Mindha, Rada.
Distribution	– Found throughout India upto an elevation of 1350 m in the hills. It is found chiefly in warm dry regions of Himachal Pradesh, especially in Nahan-Paonta range, Sarkaghat Hamirpur, Una-Kangra and Nalagarh areas (Chauhan, N.S., 937, 1380).
Description	– It is a large deciduous, thorny shrub or a small tree, with grey bark, often long and stout axillary spines. Leaves 2.5-5 cm long, obovate, obtuse or acute, narrowed into the short, marginate stipules. Flowers are greenish or whitish-yellow in colour, fragrant, solitary or 2-3 together at the ends of short, lateral branchlets. Fruits are ovoid, of the size of a lemon, yellow on ripening. Seeds are many and flat in appearance.
Flowering/Fruiting	– December-January.
Parts used	– Bark, Root and Fruits.
Current market rate	– Not available.
Main ingredients	– Saponin, Valerianic acids, vax, resin, colouring matter, d-mannitol and essential oil. (Nandkarni, 1047). A new glycoside and lupeol have been isolated from the ripe fruit. Bark gave randialic acid-A and randialic acid-B; whereas root contains scopolatin and d-mannitol. Pulp contains citric and tartaric acid, tannins, pectin, mucilage, saponins, dumetaronins and free sugars (CCRAS, 1990:215).
Uses	– Fruits are edible and are credited with a number of medicinal properties. The pulp of

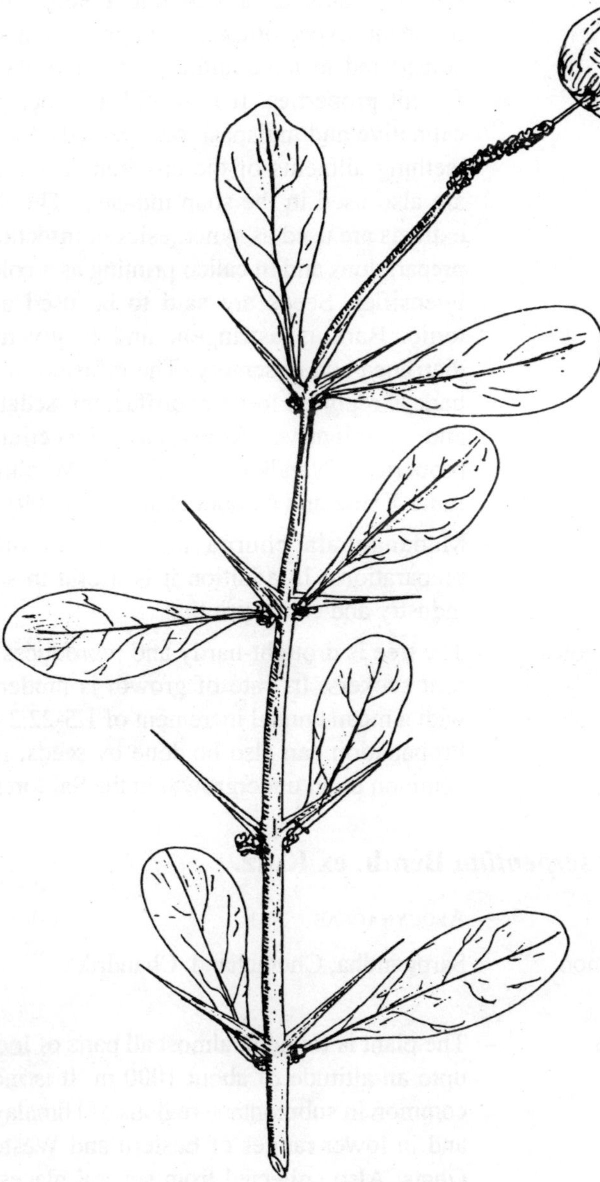

Randia dumetorum Lamk (Madanphal)

the fruit acts as a valuable emetic. It is nauseant, expectorant and diaphoretic and is considered to have anthelmintic and abortifacient properties. It is useful as a nervine, calmative and antispasmodic remedy for the teething ailments of the children. The fruits are also used in the soap industry. The fruit extracts are used as synergesics in insecticide preparations and in calico printing as a colour intensifier. Seeds are said to be used as a tonic. Bark is astringent and is given in diarrhoea and dysentery. The infusion of the bark is reported to be abortifacient, sedative and carminative. Roots have insecticidal properties. (Nandkarni, 1047; The Wealth of India X:362 and Chopra et al., 1956:209).

Preparations – Madanaphala churna and various other preparations. In addition it is useful in soap industry and calico printing.

Cultivation notes – The tree is drought-hardy and reproduces by root suckers. Its rate of growth is moderate with a mean annual increment of 1.5-22.2 cm Propagation can also be done by seeds. It is common as an undergrowth in the Sal forests.

Rauvolfia serpentina Benth. ex Kutz.

Family – APOCYNACEAE

Local/Common name(s) – Sarpgandha, Chotachand, Chandrika.

Distribution – The plant is found in almost all parts of India, upto an altitude of about 1000 m. It is more common in submontane regions of Himalayas and in lower ranges of Eastern and Western Ghats. Also collected from several places in the plains of Bengal, Bihar, U.P. and Maharashtra. In Himachal Pradesh, this plant is found in the Paonta Sahib range of Sirmour district. It has now become an endangered species. Collected from Dhaulakuan, Khara forests and IRISM, Jogindernagar. (Chauhan, N.S., 402, 1388).

Description – An erect, glaborous undershrub, 30-75 cm high. Leaves are whorled, 8-20 cm long, gradually tapering into a short petiole. Flowers are white or pinkish, about 1.5 cm long in small clusters. Fruits are small, rounded, dark purple or blackish when ripe.

Flowering/Fruiting – May-September.

Parts used – Roots.

Current market rate – Rs. 140/- per kg. (Roots).

Main ingredients – Rauvolfia roots yield 0.8-1.3 per cent total alkaloids, consisting of serpentine, reserpine, ajmaline, ajmalinine and ajmalicine. Besides these alkaloids, the root contains an oleoresin and a steroid, serposterol. (Chopra et al., 1956:2). As many as 48 alkaloids have since been isolated from the plant.

Uses – The chief use of the drug is as a sedative and hypnotic and for reducing blood pressure. The drug is now largely used in insanity and high blood pressure. The alkaloids stimulate the central nervous system. It is more suitable for cases of mild anxiety or patients of chronic mental illness. The drug has tranquillising effect. (Mukerji, 1953:218). Root is also considered as anthelmintic and an antidote to snake venom. Its decoction is given during labour pains to increase uterine contraction. Juice of the leaves is used for the cure of corneal opacity of the eyes.

Preparations – Sarpagandhadi-churna, Sarpagandha-yoga,

Rauvolfia serpentina Benth. (Sarpgandha)

Sarpagandha-vati, and various patent medicines as serpina tablets etc.

Cultivation notes – The plant can be cultivated almost anywhere in plains, in evergreen forests and sub-Himalayan tracts. It flourishes well in hot, humid conditions. It grows in a wide variety of soils from sandy alluvial loam to red laterite loam or stiff dark loam. The plants are best raised from root cuttings, and are also raised from seed. The roots are collected during the rainy season. Seedlings are raised in the nursery beds during April-May and transplanted in the field during July-August at a spacing of 45 × 30 cm. It requires 90 kg N and 45 kg P_2O_5/ha. Half dose of nitrogen (45 kg. N) and full dose P_2O_5 is applied as a basal dose and the remaining half dose of nitrogen is applied after 75 days of transplanting and in the succeeding spring i.e. February-March. The crop should be irrigated frequently, initially for better establishment of the seedlings and kept free of weeds for better root development and good quality roots. The crop is harvested (roots are dug out) after 15-18 months of transplanting (preferably in November-December). Roots should be dug out thoroughly from the soil and dried well to reduce the moisture to 8 per cent before storing. Thus, an-average yield of about 10 q/ha of dried roots can be obtained bringing an expected gross income upto Rs. 20,000/- (Verma et al., 1988:113). Germination of seeds poses problems. Seeds are given hot water treatment, acid treatment and scarification as its seed coat is very hard.

Rheum australe D. Don.

Syn: *Rheum emodi* Wall. Ex. Meissn.

Family – POLYGONACEAE

Local/Common name(s) – Revandchini, Revatchini, Gandhini, Chukri, Archa, Chuchi, Tukshu, Motininai, Rhubarb, Ushare-revand.

Distribution – The herb is distributed in the Himalayas from Kashmir to Sikkim at altitudes of 3300-5200 m. (The Wealth of India, IX; 1972:4). In Himachal Pradesh, it is found growing in Chhota Bhangal & Bara Bhangal of Kangra; Pangi Bharmour of Chamba, Parvati valley of Kullu, Dodra kawar, Khashadhar and Rohru ranges of Shimla district, Kinnaur, Lahaul & Spiti districts (Uniyal & Chauhan, 159; Chauhan, 729, 2788, 4237, 4276).

Description – A robust perennial herb with thick, long roots; 1-15 m tall and grows on rocky lands. Roots are stout and the stem is hollow. Lower leaves are 20-50 cm in diameter, orbicular and are broadly ovate. Flowers are small and dark purple in colour. Fruits are purple and angled.

Flowering/Fruiting – June-July/August-September.

Parts used – Dried rhizomes and roots. Collection is done during September-November. The roots are cut into circular pieces dried and stored.

Current market rate – Root Rs. 55/- per kg. Its extract, Ushare-Revand sells Rs. 750/- per kg.

Main ingredients – Chrysophanic acid, emodin, starch, rhein, anthraquinone derivatives, essential oil, calcium oxalate and a resinous substance have been obtained from the roots. Leaves contain oxalic acid.

Uses – Rhizomes/roots are purgative, astringent, tonic, stomachic and aperient. The petioles

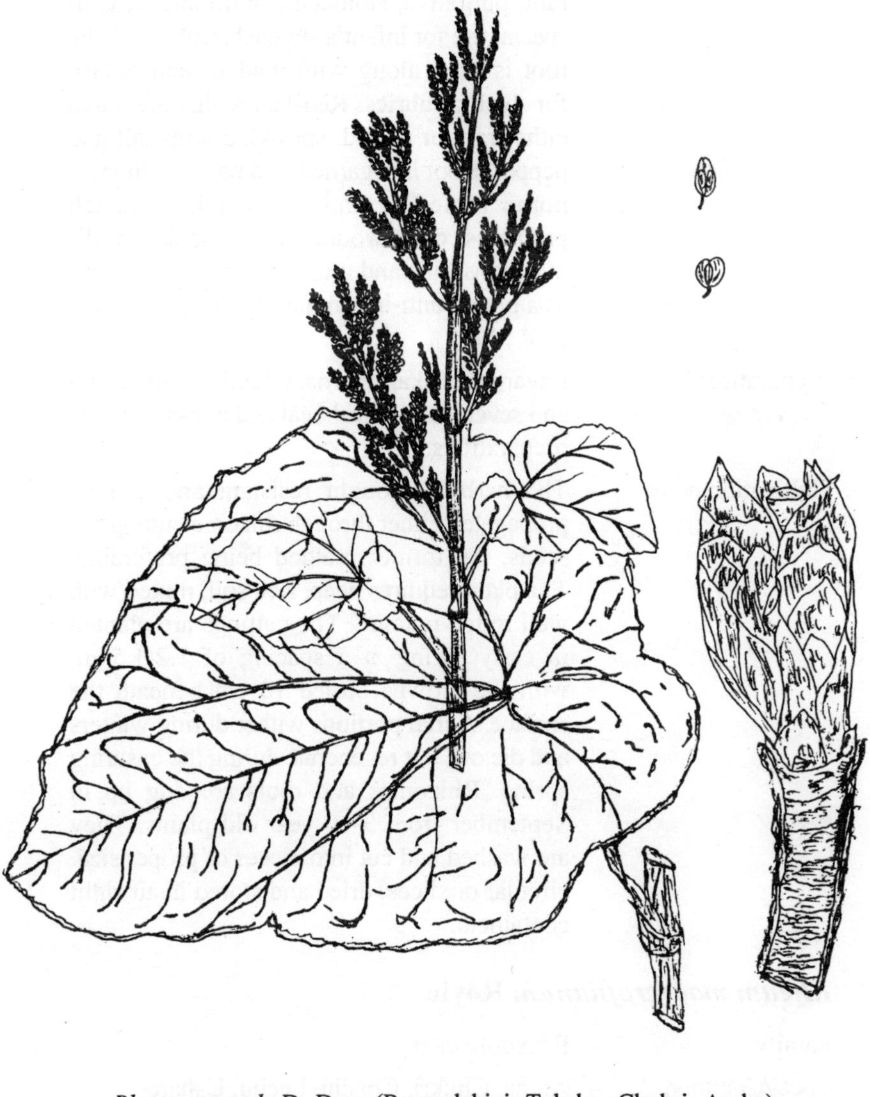

Rheum australe D. Don. (Revandchini, Tukshu, Chukri, Archa)

are pickled. Powdered roots are used for cleaning teeth and are sprinkled over ulcers for quick healing. In action, it is aperient, astringent, diuretic, emmenagogue, expectorant, purgative, stomachic and tonic. It is of special use for infant's stomach problems. The root is used along with madder and potash for dyeing fabrics. Red-leaf stalks are eaten either raw or boiled, sprinkled with salt and pepper. Root is regarded as a panacea in local home remedies and is used in stomach problems, cuts, wound and muscular swellings, tonsillitis and mumps. It has been found as a potent anti-inflammatory drug (Chauhan et al., 1992).

Preparations – Lavangabhaskar-churna, Ghuttis, Gripe water and several anti-diarrhoeal and anti-dysenteric preparations.

Cultivation notes – The herb is drought resistant and can be propagated either through rhizome cuttings or seeds, the former method being preferable. The plant requires deep, rich soil, mixed with well rotten manure. The cuttings are planted in early spring at a spacing of 1.2-1.5 m, with the crowns buried 10 cm beneath the surface. Aerial portions wither during winters and die off, but regenerate during the ensuring spring. Rhizomes and roots are dug up in September from 3-10 year old plants. They are washed and cut into pieces of proper size, circular or slices, dried and stored in air right containers.

Rheum moorcroftianum **Royle**

Family – POLYGONACEAE

Local/Common name(s) – Archa, Chukri, Chuchi, Lechu, Usharerevand, Revandchini, Rhubarb.

Rheum moorcroftianum Royle (Revandchini, Lechu)

Distribution – *R. moorcroftianum* is found at altitudes of 3000 to 5200 m in the Himalayas, chiefly in Garhwal and Kumaon. In Himachal Pradesh, the plant is found chiefly in Lahaul-spiti, Kinnaur, Kangra and Bharmour areas in sub-alpine and alpine pastures (Chauhan, N.S., 2137, 2176, 2424, 2512, 5333, 5427 & 5432).

Description	–	A stout, perennial herb, variable in size from 0.3-1.8 m high, stem lacking with radical leaves 10 to 60 cm in diameter, thick, coriaceous, margins undulate, usually cordate often puberulous beneath. Flowers are pale-yellowish arising in stout, glabrous racemes from the middle of the large, stout radical leaves. Fruit is broadly oblong or orbicular in appearance and pink-purple in colour.
Flowering/Fruiting	–	June-September.
Parts used	–	Roots
Current market rate	–	Dried roots Rs. 50-55/- per kg. Extract: 750/- per kg.
Main ingredients	–	Rhubarb contains as its chief constituents, certain anthraquinone derivatives, glucoside rhaponticin, chrysophanic acid, emodin, eugenol, a terpene alcohol and a product believed to be methyl heptyl ketone. Leaves contain oxalic acid.
Uses	–	The drug is used as purgative. Since it also contains tannins, after purgation it creates astringent effect as well. It is not suitable in cases of chronic constipation, but must be used in mild cases. Due to its astringent action, the drug is also given in certain types of diarrhoea. In villages, the lotion/paste made in water is used in cuts, wound swellings, rashes etc. It is similar in effect to that of iodine.
Preparations	–	Baby's tonic and anti-diarrhoeal and anti-dysenteric medicines like Gripe-Water and ghuttis, Lavamgabhaskar-churna.
Cultivation notes	–	Same as for *Rheum australe*.

Rhododendron anthopogon D. Don.

Family	– ERICACEAE
Local/Common name(s)	– Dhoop, Nichni, Rattanbat, Nera, Kaizaban, Morua, Talisa, Talisri, Talsar, Talispatra.
Distribution	– The plant is found in the Himalaya from Kashmir to Bhutan at altitudes of 3000-5200 m. It Himachal Pradesh, the bush is found in Chansal peak rocks, Chansal-dhamkhari, Chounti slopes, Buraskandi, Mani-Mahesh, Rohtang peak, Thamsar and Dainasar areas between 3200-4500 m elevation. (Chauhan, N.S., 1121, 2382, 2499, 2536).
Description	– The plant is a small shrub or a scrub with an aromatic odour. Branchlets are scabrous and scaly. Leaves are 2.5-4 cm long, elliptic or broad-oblong, shining above and brown tomentose beneath having a short petiole. Flowers are yellow and in dense, terminal corymbs; calyx lobes are oblong, elliptic, membranous or ciliate; corolla is tubular with a dilated mouth. Tube is .2 cm long. Lobes are obovate, entire and spreading. Stamens are 6-8; ovary is scaly and the capsules are .5 cm long and ovoid. Seeds are oblong and sub-acute.
Flowering/Fruiting	– June-July.
Parts used	– Leaves.
Current market rate	– Rs. 20-30/- per kg. (leaves).
Main ingredients	– Chemical screening needs to be done on this alpine shrub.
Uses	– Leaves possess stimulant properties. These are aromatic and are administered as an errhine to produce sneezing.
Preparations	– Talisadi-churna, Talisadi-vati, Lavanbhasker churna etc.

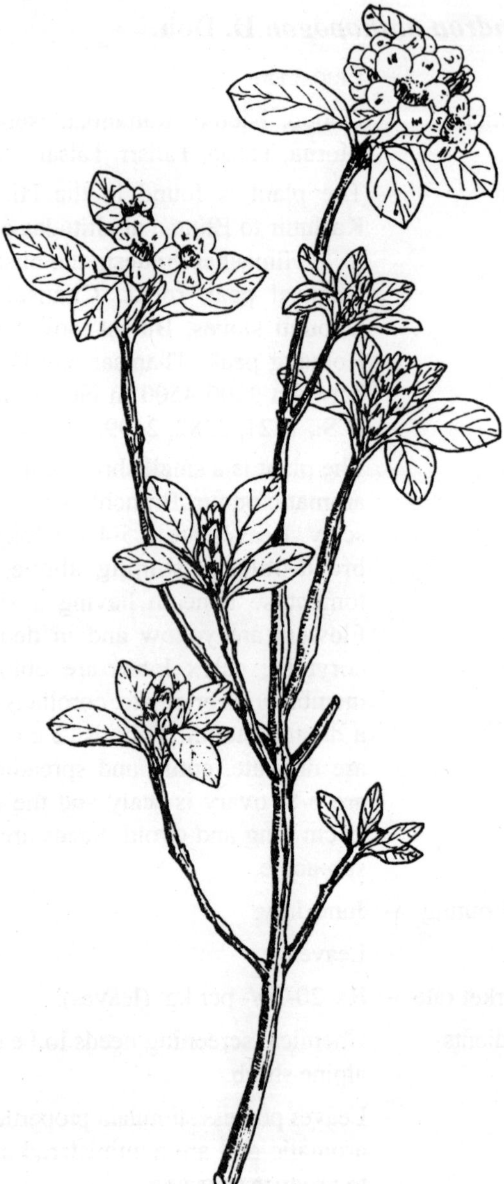

Rhododendron anthopogon **D. Don. (Talispatra)**

Cultivation notes	–	No work is reported on its cultivation. As per habit of the scrub, it is possible to propagate the plant through seeds; divisions of suckers and layering. Harvesting of leafy, top shoots is done during September-October.

Rhododendron arboreum Sm.

Family	–	ERICACEAE
Local/Common name(s)	–	Bruans, Bras, Buras, Barah-ke-phool, Rhodo-flowers.
Distribution	–	The plant is found in the Himalayas from Kashmir to Bhutan and in the hills of Assam and Manipur at altitudes of 1200-4000 m. In Himachal Pradesh, it is found in Chamba, Kangra, Kullu, Shimla, Mandi, Kinnaur and Sirmour districts in sporadic or scattered form. It is quite common in Shimla catchment, Gallu forest in Jogindernagar and Kalaghat in Solan. (Chauhan, N.S., 1445, 1563, 1636, 912).
Description	–	An evergreen, much branched tree, upto 14 m in height and 2.4 m in girth. Bark is reddish-brown, soft, rough, exfoliates in thin flakes. Leaves are lanceolate or oblong, crowded towards the ends of the branches. Flowers are showy, red, in dense, globose cymes. Corolla tube spotted. Capsules are oblong, curved. Seeds are minute, dark brown, compressed, oblong.
Flowering/Fruiting	–	March-April/June-September.
Parts used	–	Flowers, Leaves.
Current market rate	–	Not available. (Flowers are collected and processed into juice which has gained market popularity as Rhodojuice).
Main ingredients	–	Green leaves are reported to contain a glucoside, ericolin (arbutin), ursolic and α-amyrin, epifriedelinol, a new triterpenoid, named

Rhododendron arboreum Smith (Buras, Bras)

companulin, quercetin and hyperoside. Bark contains leuco-perlargonidin. Flowers also contain ursolic acid, quercitrin and resin (The Wealth of India, X, 14-15).

Uses – Young leaves are poisonous; applied to forehead for headache. Extracts of leaves, stem and bark cause hypotension in rats and inhibit intestinal movements in rabbits. The acetone and chloroform extracts and a resinous fraction from the alcoholic extracts of leaves have been reported to depress respiration. Flowers are used in diarrhoea and dysentery (The Useful Plants of India, 521). The fresh petals are used in chutneys. The flowers and leaves are fitted in long ropes made of munja grass and tied around the houses including temples as decorations during Baisakhi (Bishoo) festival. (Chauhan, N.S., 1984).

Preparations – A sub-acidic jelly or preserve is made from the petals. Rhododendron juice/sharbat has become a famous market commodity.

Cultivation notes – Propagated through seeds in March-April. Cuttings are also used for propagation. *R. arboreum* thrives best on moist loam without lime, but can grow on rocky ground, provided there is sufficient soil-moisture and humus.

***Rhododendron campanulatum* D. Don.**

Family – ERICACEAE

Local/Common name(s) – Cherailu, Shargar, Simrung, Gaggar, Simrath, Kashmiri-patha.

Distribution – The plant is commonly found throughout the Himalayas at altitudes of 2400-5200 m. In Himachal Pradesh, it forms vast rhambling patches in Chansil, Thamsar base, Dainasar pastures, Mani-Mahesh, Rorag thatch in Parvati valley; Muraldanda; Shikari Devi in Mandi and Churdhar in Sirmour district between 2800-4200 m. (Chauhan, N.S., 1230, 1517, 2061).

Description – The plant is an evergreen, rhambling shrub or a small tree, 4-6 m in height and is up to 90 cm in girth. Bark is thin, cinnamon-coloured or grey, smooth and peeling off in thin papery flakes. Leaves are elliptic or ovate, leathery 7-15 cm × 3-6 cm and under

Rhododendron campanulatum D. Don. (Kashmiri-patha)

		surface with rusty brown tomentose (Cinnamon colour). Flowers are white, in various shades of purple or pink, spotted within in corymbose heads and the capsules are cylindric and more or less curved. Seeds are linear oblong and compressed.
Flowering/Fruiting	–	May-July/Oct.-December.
Parts used	–	Leaves.
Current market rate	–	Rs. 35/- per kg.
Main ingredients	–	Contains a tonic substance, andromedotoxin, besides, ericolin, ursolic acid, α-amyrin, friedelin, epifriedelinol, companulin, quercetin, a bitter, yellowish-brown resin and four unidentified compounds.
Uses	–	Leaves are said to be used in chronic rheumatism, syphilis and sciatica. They are mixed with tobacco and used as snuff to cure hemicrania and colds. However, the leaves are poisonous to livestock.
Preparations	–	In preparations concerning skin diseases like scabies for external applications, hemicrania and snuffs.
Cultivation notes	–	Regenerates naturally through seeds/suckers and forms compact patches of its own.

Ricinius communis Linn.

Family	–	EUPHORBIACEAE
Local/Common name(s)	–	Arandi, Eranda, Arand, Eran, Erand, Castor-bean.
Distribution	–	The plant is cultivated throughout India and also found wild in the fields, gardens, village thickets, along roadsides, near habitations and on waste lands in the hills. It is cultivated in most tropical and warm temperate countries upto an elevation of 2000 m. In H.P., the

plant is found in Kangra, Hamirpur Una, Bilaspur, Joginder Nagar, Oachghat, Solan; Shaklal-Simoli Kiar areas of Rohru, foot hills of Kullu and Sirmour. Grows profusely in dumping grounds (Chauhan, N.S., 71, 201, 704, 1052, 1210 and 1580).

Description – It is a tall shrub, sometime becoming tree like. Leaves are very large, broad, roundish in outline, but partly divided into 5-7 (sometimes) palmate lobes with toothed margins. Flowers are large in terminal branches. Fruit is a prickly capsule rather marked into three parts. Seeds are oblong having crustaceous seed-coat.

Flowering/Fruiting – May-August.

Parts used – Seeds.

Current market rate – Seeds Rs. 10/- per kg.

Main ingredients – Seed contain about 45-50 per cent of fixed oil known as castor oil. The cake left after expression of the oil contains a crystalline principle ricinine, the poisonous taxalbumin ricin, and a very active lipase and other enzymes. Defatted castor meal contains 1.8 per cent poly-saccharidic protein. (Mukerji, 1953:222).

The other constituents are the albuminoid matters, 2.2 per cent of sugar and mucilage and about 18 per cent of glucose (Bentley and Trimen, IV, 237).

Uses – Castor seed is poisonous and two or three seeds have been known to prove fatal. Castor oil is used as purgative. It is best administered with milk or fruit juice. It is also used in ointments as a soothing agent and as an oil vehicle in eye-drops. The use of castor oil for facilitating child birth is doubtful, rather it

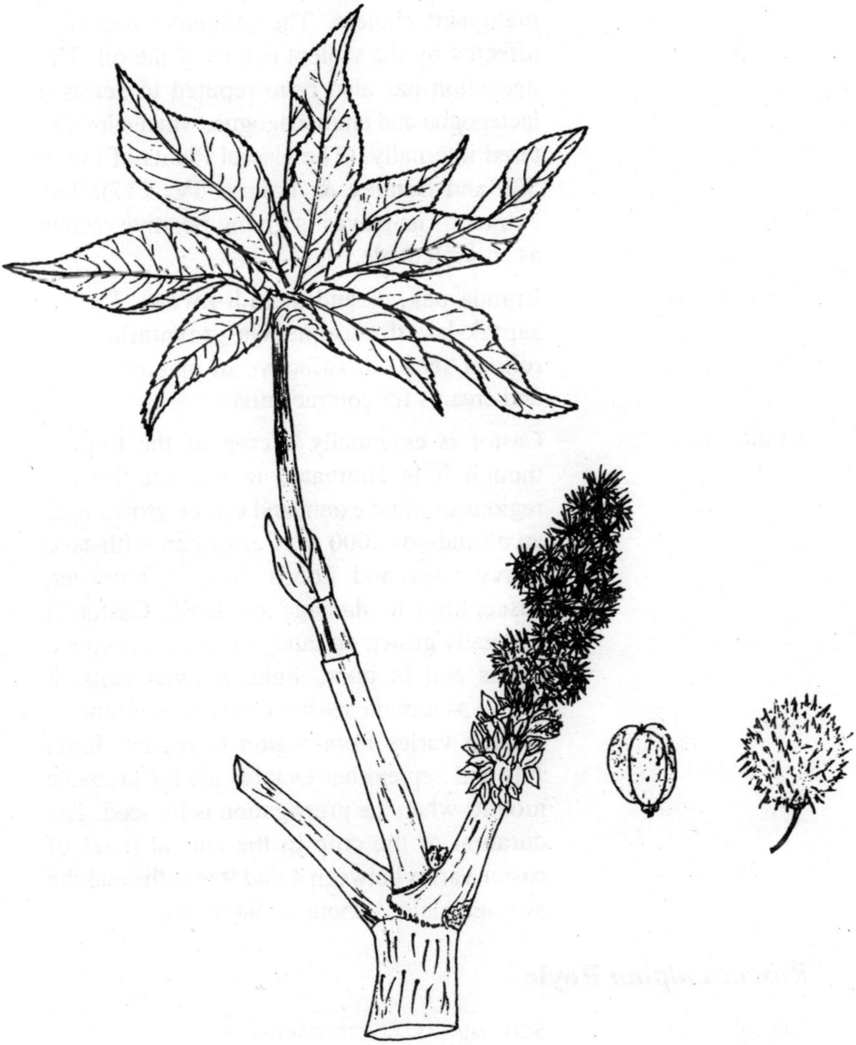

Ricinius communis Linn. (Eranda, Castor plant)

should be sparingly used even as purgative for pregnant women and during menses. It is used in making contraceptive jellies. It has also been strongly recommended as an illuminant in lamps. It is also used for curing malignant cholera. The purgative action is affected by the violent nature of the oil. The decoction has also been reputed to act as a lactagogue and emmenagogue, when administered internally. (The Useful Plants of India, 526 and Bentley & Trimen, IV, 237). The extract of its leaves is giving positive results as an insecticide.

Preparations – Eranda-pak, Erandimuladi-kwath, Rasnasaptak-kwath, Laghu-vis-gravataila, and refined seed oil. Effective in making jellies and creams for contraceptives.

Cultivation notes – Castor is essentially a crop of the tropics, though it is cultivated in the sub-tropical regions to some extent and can be grown upto an altitude of 2000 m. Castor can withstand heavy rains and floods; but, is, however, susceptible to damage by frost. Castor is generally grown on sandy or clayey, deep red loams and in good, light, alluvial soils. It grows as a main season crop and the time of sowing varies from region to region. June-July and Septem-ber-October are the probable months when the propagation is by seed. The duration of the crop in the annual types of castor varies between 4 and 9 months and the average yield is about 20-90 qt./hac.

Roscoea alpina Royle

Family – SCITAMINEAE/ZINGIBERACEAE

Local/Common name(s) – Safed-musli, Meenachari, Kakoli.

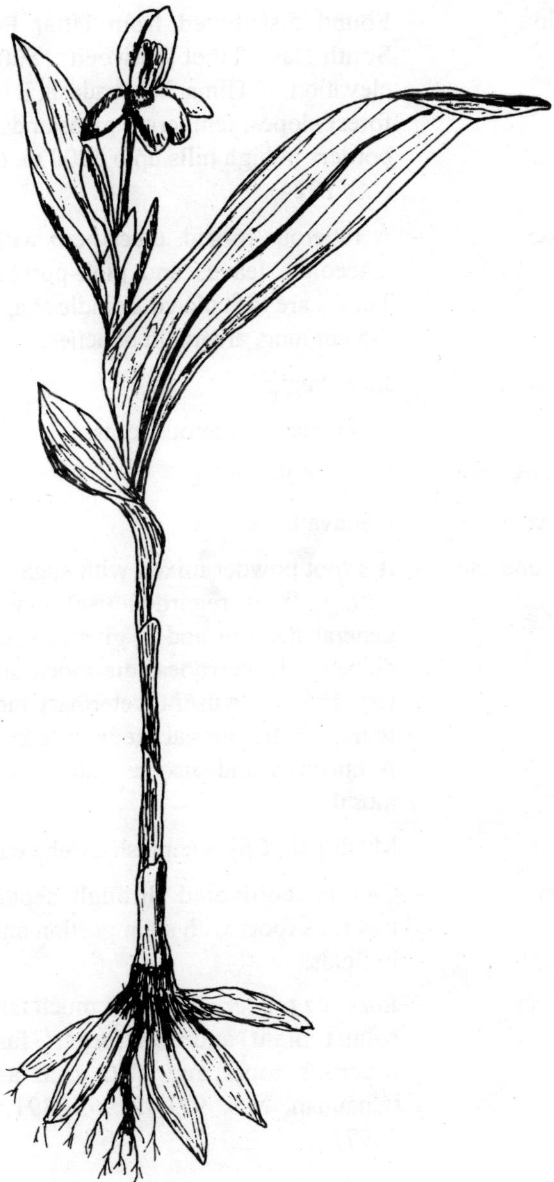

Roscoea alpina Royle

Distribution – Found distributed from Uttar Pradesh to South-East Tibet between 1500-3000 m elevation. In Himachal Pradesh, it is found in forest slopes, temperate grasslands and field borders in high hills upto 3200 m. (Chauhan, N.S. 2324).

Description – A tuberous rooted, erect herb with oblong-lanceolate leaves and pink-purple flowers. Tubers are fasiculated, spindle shaped, white, 2-5 cm long, arising in bunches.

Flowering/Fruiting – June-August.

Parts used – Fasciculated tuberous roots.

Current market rate – Rs. 50-60/- per kg.

Chemical contents – Not available.

Properties and uses – It's root powder mixed with sugar and taken with milk is regarded useful as tonic in general debility and is given in impotency, diabetes, leucorrhoea, diarrhoea and dysentery. It is also a useful veterinary medicine. It is used in traditional medicines as Kakoli in Ashtavarga and also as substitute of safed musali.

Preparations – Musli-pak, Chyavanprash-avleha etc.

Cultivation – Can be cultivated through separation of tuberous roots with stem portion and planted in fields.

Other species – *Roscoea procera* Smith. is much taller, more robust plant and has larger; fasiculated tuberous roots and also used as Kakoli (Chauhan, N.S. 971, 2196, 2391, 2440 & 2497).

Rubia manjith **Roxb. ex Fleming**

Syn: *R. cordifolia* Linn.

Family	– RUBIACEAE
Local/Common name(s)	– Manjistha, Bala-meshika, Manjit, Manjith, Tushusho, Indian madder.
Distribution	– The plant is common throughout India, ascending upto an altitude of 3750 m. In Himachal Pradesh, it is widely distributed in almost all the climatic zones and has been collected from Mariarha-Bhujjling (2400 m). Jogindernagar (1300 m), Bagsiad forest (2000 m). Thujji-Barot (2000 m), Shikari Devi (3200 m) and University campus, Nauni (1250 m) (Chauhan, N.S., 452, 811, 1015, 1366, 1429, 1639, 1719, 1974, 2079, 2285, 2320, 2614 and 2722).
Description	– The plant is a very variable, a prickly creeper or climbing herb upto 10 m long. Rootstocks are perennial and the roots are long, cylindric, flexuous with thin, red bark. Stems are 4 angled. Leaves are very variable, cordate-ovate to ovate-lanceolate, 2-8 in a whorl, normally 4 and sometimes having one larger pair. Flowers are small, white or greenish, or in shades of red and yellow; sweet scented in many or few flowered terminal panicles or cymes. Fruits are globose, or slightly 2-lobed, dark purplish or black in colour and fleshy with 2 small seeds, sweetish in taste, edible.
Flowering/Fruiting	– June-September/October-November.
Parts used	– Roots.
Current market rate	– Rs. 70/- to 80/- per kg.
Morphology of the official parts used	– The part of commerce consists of short rootstocks with numerous cylindrical, smooth and straight roots, about the size of a quill. These are covered with a thin, brownish cork,

Rubia manjith Roxb. ex Fleming (Manjistha)

which peels off in flakes, exposing a red brown bark, marked by longitudinal furrows. It is sweetish in taste first, acrid afterwards and finally bitter.

Main ingredients – Roots contain purpurin, munjistin, xanthopurpurein or purpuroxanthin, pseudo purpurin, alizarin and anthraquinones (Chopra et al., 1956:216; CCRAS, 1990:210).

Uses – Roots are credited with tonic, astringent, antidysenteric, antiseptic and deobstruent properties. They are used in rheumatism and form an ingredient of several ayurvedic preparations. Roots are said to be active against *Staphylococcus aureus* and are made into a paste for application into ulcers, inflammations and skin troubles. It is reported that after oral administration of the root decoction, the urine and bowels of the patient show a red tinge. Roots are also used for colouring medicinal oils. A decoction of the leaves and stems is used as a vermifuge. The extract enters into the drug, Septilin which is employed in the treatment of rhinosinal infections. Ethanolic extract of the aerial parts of the plant showed hypo-glycaemic activity in albino rats and the maximum tolerated dose is 1.0g/kg of body weight (The Wealth of India IX, 1972:83). The root is greatly in demand by the beauty parlours for improving facial beauty and affording pink-red colour to creams and lotions.

Preparations – Manjisthadi-kwath, manjisthadiardak, Karanjadi, Laghu-vis-garva-taila and Pinda-taila.

Cultivation notes – Cultivation through seeds has given satisfactory results. Propagation can be done through divisions. Since the roots go too deep in the soil, deep, porous, well-aerated soils are most suitable for the crop. Field borders bunds, newly exposed sites and shrubberies provide ideal sites for its growth.

Salvia moorcroftiana Wall. ex Benth.

Family – LABIATAE/LAMIACEAE

Local/Common name(s) – Kalijari, Shobri, Thuth, Halu, Papra.

Distribution – The plant is found in the temperate Himalayas from Kashmir to Kumaon. In Himachal Pradesh, it has chiefly been found growing in patches in Kullu-Bajaura, Kiarighat, Shoghi, Deothi-Sabathu and Military area Solan upto 1800 m elevation (Uniyal & Chauhan 1008, Kullu and Parbati valley). Also raised in Herbal garden at UHF, Nauni.

Description – It is a tall, robust, perennial 45-90 cm tall herb. Leaves are thick, long stalked, ovate or oblong, sinuately and irregularly lobed, hairy; lower surface white tomentose. Flowers are pale blue or lilac or nearly white with rose coloured bracts arranged in distant whorls on a long spike. Nutlets are sub-globose, pale green in colour and with dark reticulations.

Flowering/Fruiting – March-April/May-June.

Parts used – Roots, Leaves and Seeds.

Current market rate – Not available.

Main ingredients – Essential oil and mucilage.

Uses – Roots are used in colds and coughs. Leaves are used for guinea-worm and are applied as poultice for boils, wounds and chronic affections of the skin. Seeds are emetic and are used in haemorrhoids, colic and dysentery. These are edible after peeling. Locally the root is used in stomach pains.

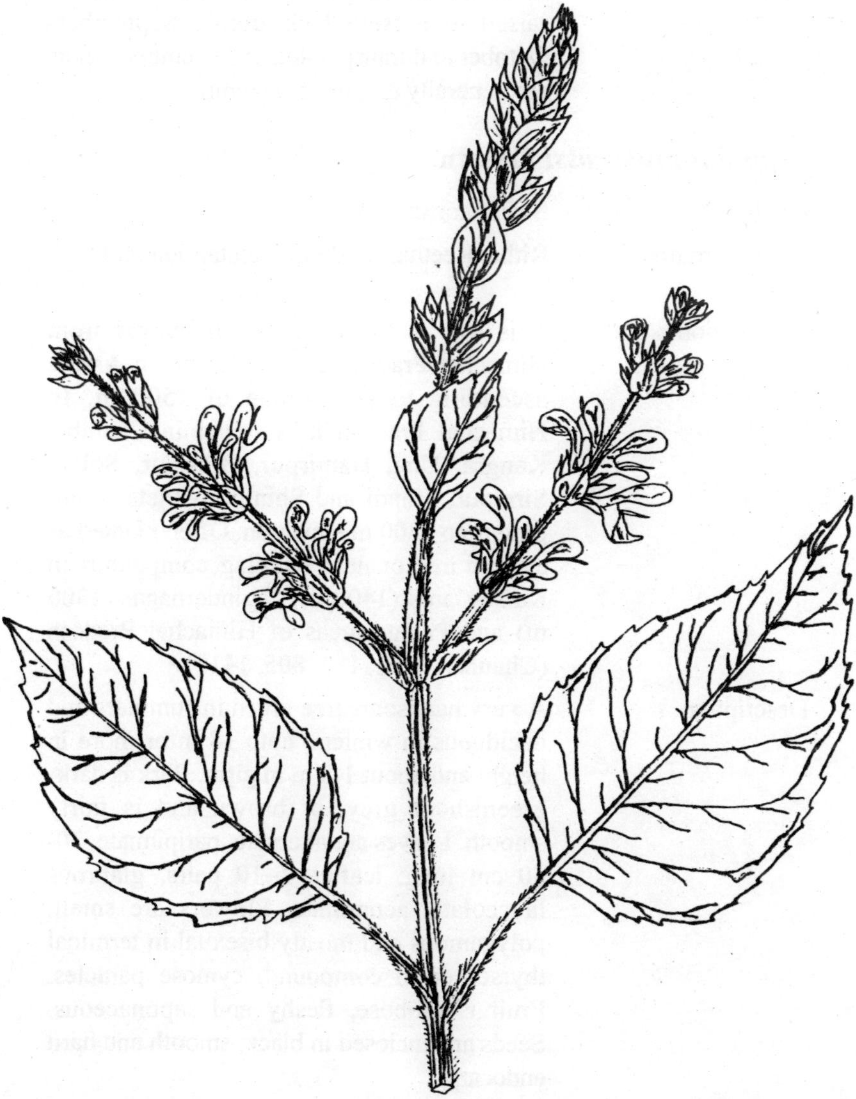

Salvia moorcroftiana Wall. (Thuth)

Cultivation notes – Propagation can be best done through seeds collected during May-June. Seedlings can be raised in nursery beds during September-October and transplanted in December. Roots are generally dug out in autumn.

Sapindus mukorossi Gaertn.

Family – SAPINDACEAE

Local/Common name(s) – Ritha, Reetha, Arishta, Phenita, Soapnut.

Distribution – This tree is found in the Himalayas from Himachal Pradesh eastwards and in Assam ascending to an altitude of 1500 m. In Himachal Pradesh it is found in Chamba, Kangra, Una, Hamirpur, Bilaspur, Solan, Sirmour, Mandi and Shimla districts extending upto 1300 m elevation. Often planted as avenue tree or near building compounds in Karsog area (1400 m) Jogindernagar (1300 m) and Paonta areas of Himachal Pradesh (Chauhan, N.S., 152, 805, 1419).

Description – A very handsome tree green in summers and deciduous in winters, upto 18 m or more in height and about 1-8 m in girth. Bark is dark-greenish or grey, or brown and is fairly smooth. Leaves are alternate, paripinnate, 30-50 cm long; leaflets 5-10 pairs, glabrous lanceolate, acuminate. Flowers are small, polygamous and mostly bisexual in terminal thyrses or in compound, cymose panicles. Fruit is globose, fleshy and saponaceous. Seeds are enclosed in black, smooth and hard endocarp.

Flowering/Fruiting – May-June/October-January.

Parts used – Fruits.

Current market rate – Rs. 12-14/- per kg.

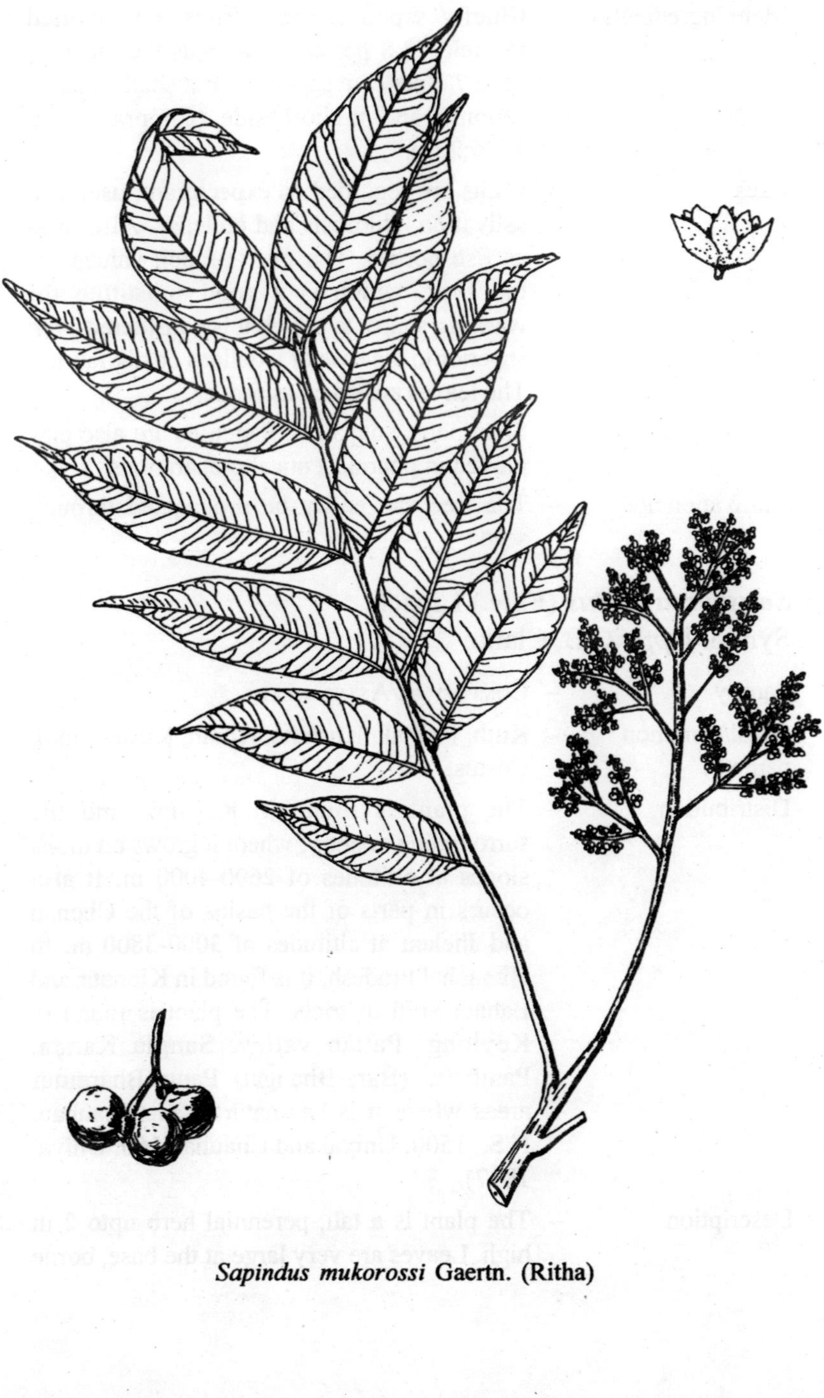

Sapindus mukorossi Gaertn. (Ritha)

Main ingredients – Chiefly saponin. Fresh fruits are reported to yield 13.8 per cent, whereas the old nuts gave 7.2 per cent saponin. Fruit shell contains saponin and mukorosside (Chopra et al., 1856:221).

Uses – Fruits are regarded as expectorant, useful in salivation, chlorosis and epilepsy. Also used as fish poison. The tree is much valued for the fruit, which is a good substitute for washing soaps, to which it is considered even superior in the case of woollens or silk fabrics. The leaves are given as fodder.

Preparations – Shampoos, soaps etc. Saponins are also employed in pharmaceutical preparations.

Cultivation notes – The tree can easily be propagated through seeds.

Saussurea costus **(Falc.) Lipsch.**

Syn: *S. lappa* C.B. Clarke

Family – COMPOSITAE/ASTERACEAE

Local/Common name(s) – Kuth, Pachak, Kushtha, Kooth, Kuste-Hindi, Costus.

Distribution – The plant is found in Kashmir and the surrounding country, where it grows on moist slopes at altitudes of 2600-4000 m. It also occurs in parts of the basins of the Chenab and Jhelam at altitudes of 3000-3800 m. In Himachal Pradesh, it is found in Kinnaur and Lahaul-Spiti districts. The plant is found in Keylong, Pattan valley, Sangla Kanda, Panihartu (Bara-Bhangal) Pangi-Bharmour areas where it is has naturalised (Chauhan, N.S., 1500, Uniyal and Chauhan 160, Uniyal 1817).

Description – The plant is a tall, perennial herb upto 2 m high. Leaves are very large at the base, borne

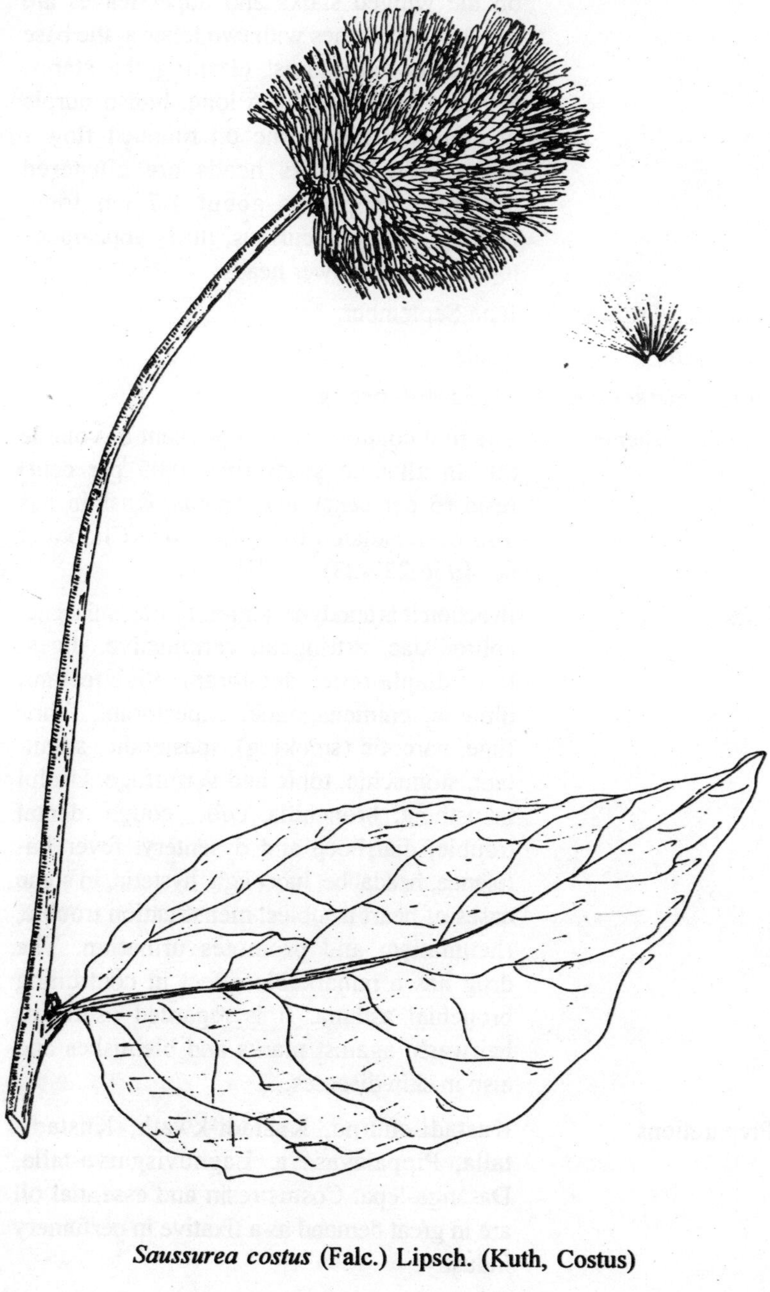

Saussurea costus (Falc.) Lipsch. (Kuth, Costus)

on the winged stalks and upper leaves are smaller, sometimes with two lobes at the base of the leaves, almost clasping the stems. Flowers are about 2 cm long, bluish purple or almost black, borne on rounded flower heads; few flowers heads are clustered together. Pappus is about 1.7 cm long, feathery, giving a curious, fluffy appearance to the fruiting flower heads.

Flowering/Fruiting – June-September.

Parts used – Roots.

Current market rate – Rs. 35-40/- per kg.

Main ingredients – The root contains 1.5-2.5 per cent of volatile oil, an alkaloid saussurine (0.05 per cent) resin (6 per cent) and tannins. Kushtin has also been isolated from the roots (Chopra et al., 1956:222-23).

Uses – In action it is anodyne, anti-arthritic, antiseptic, aphrodisiac, astringent, carminative, digestive, diaphoretic, deodorant, disinfectant, diuretic, emmenagogue, expectorant, febrifuge, narcotic (smoking), spasmodic, stimulant, stomachic, tonic and vermifuge. Useful in asthma, bronchitis, colic, cough, dental trouble, diarrhoea and dysentery, fever flatulence, headache, hiccough, hysteria, in some cases of heart troubles, menstruation trouble, rheumatism and promotes urination. The drug has a remarkable effect in controlling bronchial asthma. It is regarded useful as hairwash, against marks and blemishes and also in skin diseases.

Preparations – Kustadi-churna, Kustadi-kwath, Kustadi-taila, Pippalayasava, Laghuvisgarva-taila, Dasanga-lepa. Costus resin and essential oil are in great demand as a fixative in perfumery industry.

Cultivation note	–	Kuth requires a cool and humid climate, but the areas, where the blast of the monsoon is heavy, should be avoided. A deep, rich, porous soil is preferred and plants growing on such soils develop long and thick roots. Kuth can be propagated either by root cuttings or by seed. Roots are harvested during October. A yield of 2.5-3.5 tonnes of dry roots per hectares can be obtained. (The Wealth of India, X, 242-243 and Rawat & Purohit, 1988:179).

Selinum vaginatum C.B. Clarke

Family	–	UMBELLIFERAE/APIACEAE
Local/Common name(s)	–	Bhutkeshi, Mathosla.
Distribution	–	Distributed in Western Himalaya between 1800-3800 m. Found in temperate and alpine moist pastures in Chhota-Bara Bhangal area of Kangra, Holi Range of Chamba, Parvati Valley and Rohtang area in Kullu, Rampur and Rohru division of Shimla district (Chauhan, N.S. 746, 1257, 1311 & 2734).
Description	–	A glabrous, stout herb, 1.0 to 1.5 m tall with robust, aromatic rootstock beset with thick tuft of grey fibres, left over by the dying leaves. Roots, 12 to 18 cm long, abruptly tapering from the middle. Stem is hollow and finely grooved. Leaves, 1 to 2 pinnate; segments lanceolate, serrate, lobed or pinnatifid. Flowers-white, in long-stalked, compound umbels. Fruit-glabrous, flattened, ovate-oblong, 4 to 8 mm long with winged ridges. The roots are stouter than that of *S. wallichianum*.
Flowering/Fruiting	–	July-September.
Parts used	–	The roots.

Current market rate – Rs. 30-35/- per kg.

Chemical contents – Roots contain coumarins like selinidin, vaginidin, angelicin, oroselol and lomatin. A flavone derivative selinone has also been reported. Dried roots on steam distillation yield an aromatic oil (1.54%) containing 45.5% alpha-pinene, 25.3% linonene, 5.7% camphene, 5.2% beta-phellanetrine, 1.2% alphathujene, 3.2% fenchyl alcohol, 3.8% terpeneol and 2.6% an unidentified ketone.

Properties and uses – The roots possess a sweet, musky odour with a bitter taste and are employed as incense. They are also used as nervine sedatives. The roots of *S. vaginatum* and *S. wallichianum* are often offered as a cheap substitute to *jatamansi* and also adulterated with the latter. The oil obtained from the roots, has hypotensive, sedative and analgesic properties. Large scale collection from the area is possible and proper time is October.

Preparations – Roots are consumed by Ayurvedic Pharmacies for drugs. Also used in the manufacture of dhoop and incense and locally used in local beverage called, 'Dheli'.

Cultivation – Like *S. wallichianaum*, it is easy to cultivate the plant through seeds.

Selinum wallichianum **(DC.) Raizada and Saxena ex Raizada**

Syn: *S. tenuifolium* Wall.

Family – UMBELLIFERAE/APIACEAE

Local/Common name(s) – Muramansi, Mathosal, Nesrawlo.

Distribution – Temperate Himalaya between 2000-4000 m. In Himachal Pradesh, it is usually found in temperate pasture lands, forest shades and alpine pastures in Kangra, Kullu, Chamba, Kinnaur, Sirmour and Shimla district between 2400-4200 m. (Chauhan, N.S. 1067, 1256, 2053, 2367 & 2569).

Description – An erect herb, 0.6 to 2.4 m tall with a perennial rootstock. Leaves-pinnately compound, deeply cut into narrow, linear-lanceolate, acute segments. Flowers white in compound umbels. Fruit 0.5 cm, much dorsally compressed, lateral ridges are the broadest.

Flowering/Fruiting – July-September.

Parts used – The rootstock. It is long, greyish with a thick tuft of fibres on the upper portion left over by the withering leaves. This fibrous portion is used as official part and is quite aromatic.

Current market rate – Rs. 25-30/- per kg.

Chemical contents – Three furanocoumarins-bergapten, heraclenin and heraclenol alongwith another coumarin ST-1 have been isolated from the roots. Angelicin and xanthotoxol are the two other coumarins and a mixture of long chain saturated fatty oils has also been detected. Seeds yield essential oil. The volatile oil from the roots contains an acetylenic compound-3, 5-nonadiyne.

Properties and uses – Root is regarded as a nervine sedative and used as a substitute for Jatamansi. Frequently used as incense and in local brew, called as 'Dheli'. The oil from the root has hypotensive, sedative and analgesic properties. Leaves are regarded as a nutritious fodder for sheep and goats.

Preparations – Dhoop incense, Dheli (locally) and Ayurvedic medicines.

Cultivation – Easily cultivated through seeds. The crop is harvested in the third year during October.

Sesamum indicum Linn.

Family	– PEDALIACEAE
Local/Common name(s)	– Til, Sesame, Sinsim, Gingilly.
Distribution	– Cultivated as a Kharif crop in most parts of Himachal Pradesh either as a single crop or mostly as intercrop with horsegram, buck-wheat and rainfed paddy.
Description .	– An erect, much branched, strongly, smelling, hairy annual herb cultivated during the rainy season. Leaves-large, thin sparsely hairy, upto 12 cm long, lower ones often lobed, toothed. Flowers-axillary, solitary, whitish-pink having an offensive smell. Capsules-oblong, erect 2.5 cm long, four sided, splitting open from above to half way down. Seeds-small in size and varying in colour from white through brown to black (1000 seeds weigh 2.0 to 3.5 gm).
Flowering/Fruiting	– August-October.
Parts used	– Seeds
Current market rate	– Rs. 30-40/- per kg.
Chemical contents	– Fatty oil of commerce named as til or sesame oils much used as an article of diet and confectionery with sugar. Oil forms basis of most scented and medicated oils.
Properties and uses	– The leaves are highly mucilaginous and are used in the form of infusion as demulcent in infantile diarrhoea, cholera, dysentery, catarrh, bladder troubles, acute cystitis, strangury and respiratory troubles. Externally, the leaves are used as emollient poultice and the decoction of leaves and roots is useful as a hair wash. The seeds are emollient, nourishing, tonic, lactagogue, laxative, aphrodisiac, demulcent, diuretic, expectorant, styptic and

suppurative. Useful in burns, bleeding piles constipation, cough, as dentrifice, in dysentery, as hair growth promoter, in urinary complaints, scalds and rheumatism.

The oil in large doses acts as abortifacient. Oil with salt is taken twice a day in chest complaints. Externally the oil is rubbed over the head for the treatment of migraine and vertigo and over the abdomen for promoting abortion. The oil forms a basis for medicated oils and is used as a substitute for olive oil in the preparation of liniments, plasters, ointments and oil dressings for ulcers, suppurating-wounds etc. A mixture containing equal parts of the oil and lime water is a household remedy for burns and scalds.

Preparations – Tiladi-gutika, Tiladilepa, Laghuvishgarbha-taila. Various sweets for use in winter season and scented oils for hair and ointments are made of it.

Cultivation – Cultivated through seeds as a kharif crop/ rainy season crop.

Sida cordifolia Linn.

Family – MALVACEAE

Local/Common name(s) – Bala, Kharenti, Bariyar, Barela, Country mallow, Sida.

Distribution – The plant is a common weed distributed throughout tropical and sub-tropical India and Ceylon in waste places and scrub jungles upto an altitude of 1050 m. It is found in Chamba, Kinnaur and Shimla districts in Himachal Pradesh (Chowdhery & Wadhwa, 117). Common in Ramshahar-Nalagarh and Paonta Sahib areas of Himachal Pradesh (Chauhan, N.S., 1415, 1895).

Sida cordifolia Linn. (Bala, Barela)

Description – The plant is an undershrub, branched, softly hairy and with stellate hairs, nearly in all the parts of the plant. Leaves are 2.5-5 cm long, cordate, ovate-oblong, crenate, obtuse or subacute and not acuminate and the petioles are 1.2-3.8 cm long. Pedicels are solitary and few together, short, some upto 1.2-2 cm long bearing yellow flowers. Calyx is 6-8 mm long with ovate and acute lobes. Corolla exceeds the calyx slightly. Fruit is 6-8 mm in diameter; strongly reticulated with two spiny projections on the top.

A transverse section of the root is circular in outline with a large central woody region and thin outer bark (Ahuja, 1965:69).

Flowering/Fruiting – August-October.

Parts used – Entire plant.

Current market rate – Not available.

Main ingredients – Sida contains an alkaloid, a fatty oil, phytosterol mucin, resin, resin acids and potassium nitrate. The leaves, stem and root yield 0.085 per cent, and the seed about 0.3 per cent alkaloid. The main portion of the alkaloid is ephedrine. (Mukerji, 227). Mehra, P.N. (1986) has reported 0.4 per cent rutin in the plant.

Uses – The plant is reputed for its tonic and aphrodisiac (seeds) properties. The plant parts are used for the fever, fits, leucorrhoea, micturition, gonorrhoea, colic, nervous disorders, general debility and heart irregularity. The root juice is used for the healing of the wound and the juice of the whole plant is given for spermatorrhoea. The bark of the plant is effective in curing facial paralysis and sciatica and the leaves are used for curing the bloody flux. Useful in ophthalmia, rheumatism;

		shivering fits and improves sexual strength (The Useful Plants of India, 1986:573, Dey, 1980:22).
Preparations	–	Baladi-kwath, Baladya-ghrita, Baladyarista, Chandanbala-lakshaditaila, Sudarshan-churna and Kukuvadi churna.
Cultivation notes	–	Cultivation of the plant is done through seeds. The species grows well under dry conditions. The plants are collected during the rainy season. (Ahuja, 1965:69 and The Wealth of India X, 324).

Sida rhombifolia Linn.

Family	–	MALVACEAE
Local/Common name(s)	–	Svetbarela, Mahabala, Pitakharenti, Pit-barela.
Distribution	–	The plant is a common weed distributed throughout tropical and sub-tropical India and Ceylon in waste places and scrub jungles upto an altitude of about 1800 m. The plant is also found in Paonta-Majra, Nalagarh, Jogindernagar, Kunihar, Bilaspur, Nahan, lower parts of Shimla and other areas of Himachal Pradesh upto 1300 m elevation (Chauhan, N.S., 853, 1494, 1845).
Description	–	It is an erect, undershrub, very variable; annual/perennial upto 1.5 m high. Leaves are highly variable, rhomboid-lanceolate to lanceolate; sub-glabrous above, grey-pubescent hoary beneath. Flowers are yellow or white, axillary, solitary or in pairs. The branch bearing inflorescence, often appears like a panicle due to the reduction of the leaves.
Flowering/Fruiting	–	July-September.
Parts used	–	Whole plant.
Current market rate	–	Not available.

Main ingredients	– Leaves contain large amount of mucilage.
Uses	– The plant is regarded useful in tuberculosis and rheumatism. In some parts, the leaves are used as the substitutes for tea. Stem abounds in mucilage and is employed as demulcent and emollient. It is also used in the treatment of rheumatism and leucorrhoea. Ethanolic extract of the plant depresses the activity of the smooth muscles of the ileum of pig. (The Useful Plants of India, 574). The pounded leaves are applied on swellings (Chopra et al., 1956:227).
Preparations	– Baladi-kwath, Balaya-ghrita, Baladyarista, Chandan-bala-lakshadi-taila, Sudershan-churna and Kukuvadi churna.
Cultivation notes	– Propagation is through seeds and cuttings.

Solanum khasianum C.B. Clarke

Family	– SOLANACEAE
Local/Common name(s)	– Ban-bhindi, Kandiari, Kantkari, Bari kandiari.
Distribution	– Found in the Khasi, Jaintia and Naga hills of Assam and Manipur, Nefa, Sikkim, West Bengal, Orissa, the upper Gangetic plains upto an altitude of 1600 m (Wealth of India, IX, 1972:382). Common in Himachal Pradesh in the waste places, ravines and near habitations upto an elevation of 1500 m (Chauhan, N.S., 5, 8, 205, 865, 2607 & 2677).
Description	– The plant is a stout, much-branched, under shrub, upto 1.5 m tall, with almost straight prickles. Leaves are ovate, lobed and the lobes are lanceolate or triangular, hirsute and prickly on both the surfaces. Flowers are white in colour and are in lateral, 1-4 flowered racemes. Berries are yellowish or greenish,

		globose and 2.5 cm in diameter. Seeds are smooth, brown and compressed.
Flowering/Fruiting	–	May-October.
Parts used	–	Fruits in phytochemical industry and the entire plant in indigenous system of medicines often as a substitute for *Solanum indicum* (Brihati).
Current market rate	–	Rs. 10/- to 12/- per kg. (dried herb) Seeds Rs. 200/- per kg.
Morphology of the official parts used	–	Entire plant is used in Ayurvedic medicines and berries, which are globose, about 2.5 cm in diameter, turning yellow on ripening are used in Pharmaceutical industry for its high solasodine content, a precursor of 16-DPA.
Main ingredients	–	The berries contain solasodine, solakhasianin (containing galactose and rhamnose) and diosgenin. Its berries are richest source of Solasodine (upto 4 per cent in berries).
Uses	–	Alcoholic extracts of the plant have been tested for the biological activity. They affect the contraction of the isolated ileum of guinea pig and also influence the central nervous system (Wealth of India IX, 1972:382 and Ambasta, 1986). It is much valued for its solasodine used in various formulations including birth control pills. In Ayurvedic medicines it is used as a substitute for *Solanum indicum* (Brihati).
Preparation	–	Kantkarikshara, various steroidal/hormonal preparations in allopathic system.
Cultivation notes	–	Propagation is through seeds. The seeds are sown in February-March and the transplanting should be done in May-June. Harvesting is done in the months of August-September. The cultivation can be done by using the hedge row planting or single row planting technique. Spineless varieties have, also been developed.

Application of NPK at the rate of 90:90:45 kg./ha respectively as basal dose is recommended for the crop.

Solanum nigrum Linn.

Family	– SOLANACEAE
Local/Common name(s)	– Makoi, Kakamachi, Bandakh.
Distribution	– The plant is found throughout India in dry parts, upto an elevation of 2100 m. In Himachal Pradesh, the plant is found all over the state upto 2600 m, especially as weed near villages, footpaths and fields. Collected from Nauni, UHF Garden, Chhajpur, Ravingarh, Bagi, Bijahi, Kanda (Mandi), Sundernagar, Nalagarh, Bag-Baniya, Jwar forest, Unchagaon, Kunihar, Bhujjling, Chamba and Devi Bardhar (Chhota Bhangal area) (Chauhan, N.S., 821, 1618, 1757, 1825, 1846, 1908, 1916, 1958, 2075, 2100, 2191 & 2602).
Description	– An erect or rhambling, sparingly or often much branched, usually glabrous herb; 30-45 cm tall. Leaves are ovate or oblong, sinuate-toothed or lobed, narrowed at both ends. Flowers are white in colour and the berries are red, yellow or black (3 different types of plants have been observed in this species). Seeds are smooth and yellowish in colour.
Flowering/Fruiting	– June-October.
Parts used	– Entire plant.
Current market rate	– Herb Rs. 42/- per kg. Seeds Rs. 200/- per kg.
Main ingredients	– Leaf is a rich source of riboflavin, nicotinic acid and vitamin 'C'. Besides these, β-carotene and citric acid are present. Fruits

Solanum nigrum Linn. (Kakamachi, Makoi)

contain glucose and fructose, vitamin C and β-carotene. Seeds contain, about, 17.5 per cent protein on dry weight basis and yield an oil. The immature green fruit of the plant contain four steroidal glyco-alkaloids viz. solamargine solasonine and α and β-solanigrine and all of them yield solasodine, as the aglycone. Diosgenin is also present in leaves, stem and fruits.

Uses – The herb has antiseptic and antidysenteric properties and is given, internally for neuralgia and gripe. An infusion of the plant is used as an enema to infants, having abdominal upsets. It is a household remedy for anthrax pustules and is applied locally. The plant is also credited with emollient, diuretic and laxative properties and its decoction is regarded as an antispasmodic and narcotic. Freshly prepared extract of the plant is effective in the treatment of cirrhosis of the liver and also serves as an antidote to opium poisoning. An alcoholic extract of the leaves is active against *Staphylococcus aureus* and *Escherichia coli*. Decoctions of the plant may be used for the treatment of ascites. Leaves are applied to wounds and sores and are used in the treatment of psoriasis, eczema, piles and syphilis; as a diuretic in gout, rheumatism dropsy, gonorrhoea, cough, splenic and hepatic enlargement. The heated leaves are applied to painful and swollen testicles. Leaves and the tender shoots are edible. Fruits make a delightful jam. Berries are considered to possess tonic, diuretic and cathartic properties and are useful in anasarca and heart diseases. They are a domestic remedy for fevers, diarrhoea, ulcers and eye troubles.

Preparations – Enters into various proprietary and standard preparations in Ayurveda, Unani and modern medicines.

Cultivation notes – It is a profuse seed bearer having a high germination rate. It is, thus easy to propagate the plant through seeds. Nursery can be raised during March-April and trans-plantation is done during June.

Solanum surattense Burm. f.

Syn. *S. xanthocarpum* Schrad.

Family – SOLANACEAE

Local/Common name(s) – Kantkari, Kateli, Laghu-kantkari, Wild egg plant.

Distribution – The plant is found throughout India, Ceylon and Pakistan; also in South-East Asia, Malaya and tropical Australia. In Himachal Pradesh, the herb is found in waste land, roadsides, dry ravines and field borders in warmer areas of the state below 1300 m elevation. (Chauhan, N.S., 803, 922, 995).

Description – A prickly, much-branched herb, usually spreading or diffuse; young branches are densely covered with minute star-shaped hair, prickles are yellow, shining, about 1.5 cm long. Leaves are upto 10 cm long, their midribs and other leaves with sharp, yellow prickles. Flowers are purple, about 2 cm long, few together in small branches, opposite to the leaves. Fruits are 1.5-2.0 cm, round, yellow or pale with green veins.

Flowering/Fruiting – April-August.

Current market rate – Herb Rs. 6-8/- per kg.
Seeds Rs. 200/- per kg.

Parts used – Roots, Berries (Usually entire herb is used).

Solanum surattense **Burm. f. (Chhoti-Kateri)**

Main ingredients – Fruits yield carpesteral and gluco-alkaloid, Solasodine, Solasonine and Solanocarpine. Solanine-on hydrolysis yields alkaloid, Solanidine.

Uses – The drug is used in cough, asthma, pain in chest, and certain kinds of fevers. It is diuretic and is considered useful in controlling of stones in bladders. Antibacterial property in fruits and shoots has also been shown experimentally. Drug is also used in bronchitis, muscular pains, enlargement of liver and spleen, blood cancer, vomiting and fever etc. Antifertility properties are also found. (The Useful Plants of India, 1986:582).

Preparations – Nidigdhi-kwath, Bayaghri-haritaki, Kantakari-ghrita, Bayaghri-taila, Kantkari-avleha, Chyavanprash, Dashmularishta, Sudarshan-churna, Sarvajwar-hara-louha, Devadarvadi-kwath, Sringayadi-churna, Kanakasav.

Cultivation notes – The plant is cultivated through seeds and prefers dry, warm aspects and sandy soil.

Spilanthes acmella Linn.

Family – COMPOSITAE/ASTERACEA

Local/Common name(s) – Akarkara, Pokarmul, Akalkada.

Distribution – Throughout India, ascending to 1700 m, often found occurring as a weed in rice fields. Grown successfully in the university herbal garden at Nauni-Solan (Chauhan, N.S., 1402, 4993).

Description – An annual herb, more or less pubescent, sometimes hairy. Srem 30-60 cm, usually decumbent near the base. Leaves opposite, stalked, ovate-lanceolate, about 5 cm × 3.75 cm, toothed or entire. Flower heads discoid

		or radiate, conical, solitary on long stalks. Flowers are white or yellow in colour, mostly bisexual. Achenes flattened, each enclosed in a scale.
Flowering/Fruiting	–	May-October.
Parts used	–	Leaves, flower heads and roots.
Current market rate	–	Rs. 800/0 per kg. (Roots) Rs. 200/- per kg. (Seeds)
Chemical contents	–	Main ingredient is spilanthol obtained from roots, leaves and flowering heads.
Properties and uses	–	Leaves are chewed to relieve toothache and affections of throat and gums. Tincture made from flower heads is reported to be used as a substitute for tincture of Pyrethrum to treat inflammation of Jaw-bones and caries. Plant boiled in water is given in dysentery and an ether extract of fresh flower-tops is effective against anopheles mosquito larvae. It increases the flow of saliva and is useful in fever especially during summers.
Preparations	–	It forms a vehicle of many compound powders, composition of aphrodisiac pills and lehyams.
Cultivation	–	Easily cultivated through seeds. Once cultivated it regenerates profusely during subsequent years as it produces enormous amount of seeds, shattering on the ground on ripening of the heads.

***Swertia chirayita* Karst.**

Syn. *S. Chirata* Buch. Ham.

Family	–	GENTIANACEAE
Local/Common name(s)	–	Chirayita, Kirata-tikta, Bhunimba, Chirata.

Swertia angustifolia Buch.-Ham. ex D. Don.

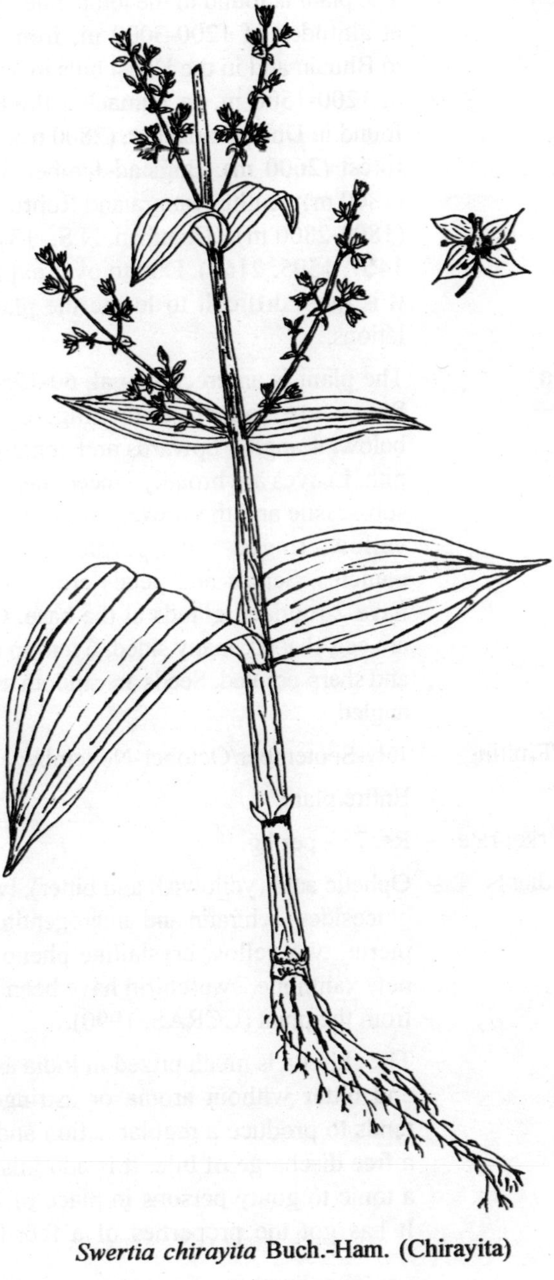

Swertia chirayita Buch.-Ham. (Chirayita)

Distribution – The plant is found in the temperate Himalayas at altitudes of 1200-3000 m, from Kashmir to Bhutan and in the Khasi hills in Meghalaya at 1200-1500 m. In Himachal Pradesh, it is found in Dhaulthanadhar, (2800 m), Luwayin forest (2600 m), Bagsiad-Janjheli in Mandi (1800 m); Kullu, Kangra and Rohru divisions (1800-2800 m). (Chauhan, N.S., 1315, 1358, 1467, 1595, 2164). Due to over exploitation, it is most difficult to locate the plant populations.

Description – The plant is an erect annual, 60-125 cm tall. Stems are robust, branching, cylindrical below, 4 angled upwards and contain a large pith. Leaves are broadly lanceolate, 5 nerved, sub-sessile and the flowers are lurid greenish yellow, tinged with purple and in large panicles. Sepals and petals are 4 and petals have 2 rounded glands at the base. Capsules are egg shaped, many sided, 6 mm in diameter and sharp pointed. Seeds are smooth and many angled.

Flowering/Fruiting – July-September/October-November.

Parts used – Entire plant.

Current market rate – Rs. 75/- per kg.

Main ingredients – Ophelic acid (yellowish and bitter), two bitter glucosides—chiratin and amarogentin, gentiopicrin, two yellow crystalline phenols and a new xanthone, swerchirin have been isolated from the plant (CCRAS, 1990).

Uses – The chiretta is much prized in India as a tonic and bitter without aroma or astringency. It tends to produce a regular action and causes a free discharge of bile. It is administered as a tonic to gouty persons in place of gentian. It has got the properties of a febrifuge. In

Indian medicine, chirreta is prescribed in a variety of forms and combinations in chronic fevers and anaemia. It has got the special reputation as a remedy for bronchial asthma and liver disorders. If taken with the sandalwood paste, it is said to stop internal haemorrhage of the stomach. It is credited with tonic, anthelmintic and anti-diarrhoeal properties in Indian medical systems. Chiretta is prescribed in dyspepsia, in the debility of convalescence and generally in cases in which corroborant measures are indicated. It may be given as a tonic, powder, infusion, tincture or a fluid extract. Chiretta is said to be used for dyeing cotton cloth yellow and is used in the liquor industry as a bitter ingredient.

Preparations – Sudarshan-churna, Kiratadi-kwatha, Panchbhadra-kwatha, Sarwajwarhara-lauha, Phalatrikadi-kwatha and Chandraprabha vati in addition to many modern preparations.

Cultivation notes – The plant spreads quickly from seed which is shed during October-November. The herb can also be cultivated in suitable localities in the temperate Himalayas. The seeds, which are very small, should be sown in a nursery and the seedlings transplanted later in the field.

Annual demand for chiretta in India is reported to be about 37,300 kg. Most of the Indian supplies of chiretta are obtained from Nepal and the East-Himalayan region. Each year, 60,000-100,000 kg of chiretta are imported into India. Some years back India exported chiretta mostly to Italy, Singapore and Afghanistan. Exports have declined in recent years due to the fast decline in its population.

Other species of *Swertia* i.e. *S. angustifolia, S. ciliata* are also used frequently.

***Symplocos paniculata* Wall.**

Syn. *S. crataegoides* Buch-Ham. ex D. Don.

Family – SYMPLOCACEAE (Styraceae)

Local/Common name(s) – Lodh, Lodar, Lodhra, Lojh.

Distribution – The tree is common in the Himalayas from Himachal Pradesh to Assam and in the Khasi hills, occurring at altitudes between 600 to 2400 m. In Himachal Pradesh, it is moderate to sporadic between 1600-2600 m in forests at Kharapathar, Kashdhar and Rohru Ranges; Parvati valley in Kullu; Churdhar forest in Sirmour, Bilang; Barot in Mandi distt., Chhajpur, and Dalhousi-Kalatop forest. Its population is decreasing because the branches are frequently lopped for goat fodder (Chauhan, N.S.; 2709).

Description – A deciduous shrub or a tree, upto 12 m in height. The bark is soft, friable, of light brown colour. The external surface is corky or much fissured transversely; while the internal part is of a lighter colour and of fibrous texture. Taste is faintly balsamic and astringent. The leaves are obovate, serrate, 6.25 cm × 2.5-3.75 cm. Flowers are white, fragrant, in terminal panicles. Fruit is drupaceous, blue on maturity, 1-seeded.

Flowering/Fruiting – May-September.

Parts used – Bark (Lodhra-twak)

Current market rate – Rs. 10-13/- per kg. (Bark)

Main ingredients – The bark contains 2 alkaloids; loturine and colloturine and also a red colouring matter. The alkaloids are chemically related to harmine found in *Peganum harmala*. Seeds contain a fatty oil.

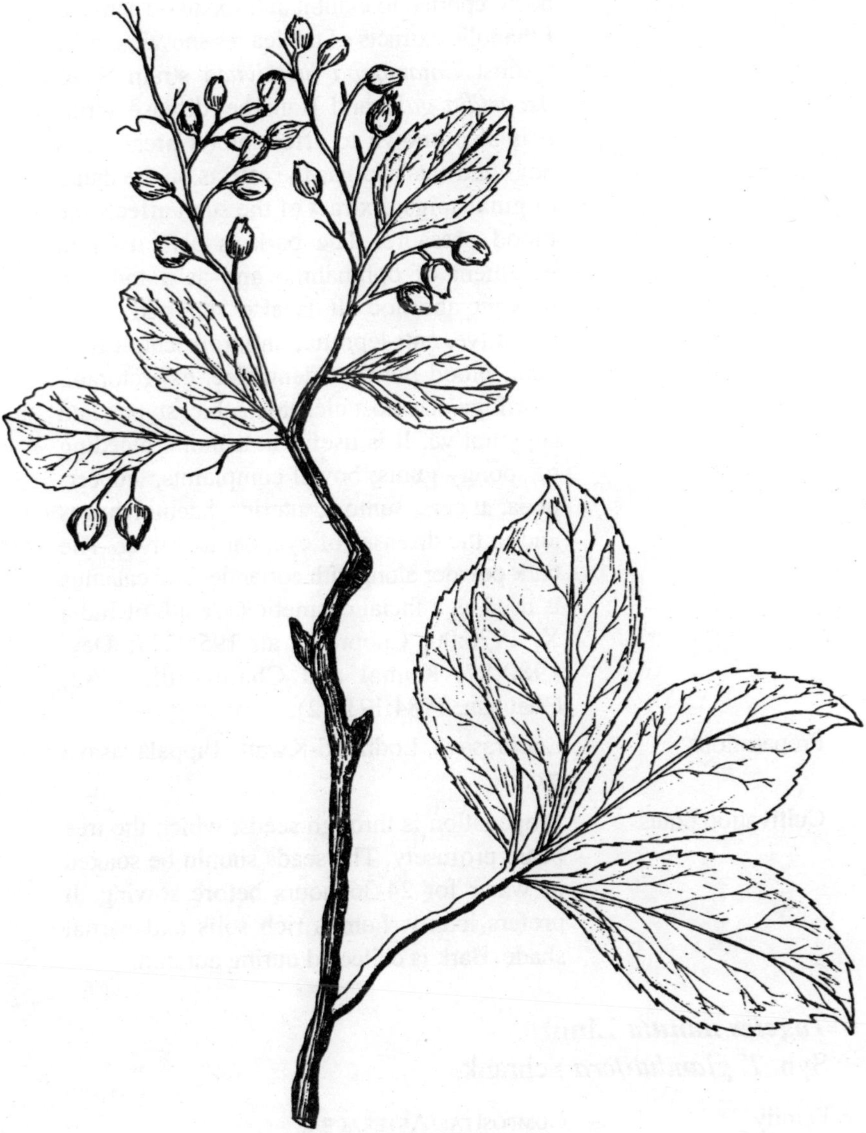

Symplocos paniculata Wall.
Syn. *S. crataegoides* Buch.-Ham. ex D. Don. (Lodhra, Lojh)

Uses – A water soluble fraction from the bark has been reported to exhibit anti-oxytoxic activity. Ethanolic extracts of the leaves shows activity against *Entamoeba histolytica* strain STA, *Ascaridia galli* and Ranikhet-disease virus. It is also known to affect blood pressure in dogs and cats and has the effects on the ileum of guineapigs. Extract of the stem affects the blood pressure. The bark is also used in treatment of ophthalmia and as a tonic to prevent abortion. It is also reported to be alterative, anti-leprotic, antidropsical, astringent, blood purifier, dentrifice, expectorant, febrifuge, haemostatic, stomachic, styptic and suppurative. It is useful in asthma, bleeding of spongy gums; bowel complaints, leucorrhoea, ulcers, tumors, uterine haemorrhages and in the diseases of eye, ear and liver. The bark powder alongwith coriander and calamus is used as a facial cosmetic (Wealth of India X, 1976:90; Chopra et al. 1956:237; Dey, 1980:87; Komal and Chaturvedi, 1982; Chauhan, 1984:171-72).

Preparations – Lodhrasava, Lodhradi-Kwath, Pippalayasava etc.

Cultivation notes – Propagation is through seeds, which the tree bears profusely. The seeds should be soaked in water for 24-36 hours before sowing. It prefers loamy, humus rich soils and partial shade. Bark is collected during autumn.

***Tagetes minuta* Linn.**

Syn. *T. glandulifera* Schrank.

Family – COMPOSITAE/ASTERACEAE

Local/Common name(s) – Gaindri, Jangli-gainda, Gandloghas Tagetes.

Distribution	–	The plant, native to South America, is naturalized near waste places and on dry embankments in the north-west Himalaya, between altitudes of 1250 and 2500 m. In Himachal Pradesh, the plant has been found almost all over the state as a weed of waste places. Robust plant growth has been observed at Rohru, Samoli (1600 m), Chirgaon-Tikkari (2000 m), Dodra kawar (2600 m), Shimla (2400 m) (Chauhan, N.S. 2529).
Description	–	The plant is a highly aromatic annual, 1-2 m tall. Leaves are 7-15 cm long, pinnatisect and the segments are 11-19; 4 cm or more long, linear or lanceolate. Flower-heads are pale-yellow, narrow, tubular, in corymbose clusters and the achenes are black.
Flowering/Fruiting	–	July-November.
Parts used	–	Entire herb/Foliage especially flowering tops.
Current market rate	–	Essential oil Rs. 1400/- per litre.
Main ingredients	–	Aromadendrenem tagetone, phenylethyl alcohol, ocimens, salicylaldehyde, phenylacetaldehyde, eudesmol, linalylacetate, limonene, linalool and an un-identified carbonyl compound have been isolated from the plant.
Uses	–	The oil obtained from the herb, has synergistic activity and shows positive juvenile hormone activity. The ethanolic extracts of the entire herb show anti-viral activity against Ranikhet-disease virus. The roots show nematocidal activity and hence, the plant is grown in the tobacco fields. The plant has strong larvicidal effect and is also a slow repellent. It is suspected of poisoning cows and taints milk and butter. The juice of the herb causes irritation of the eyes and skin. The flowers are used as a stomachic, aperient, diuretic and

diaphoretic. They contain a volatile oil which exhibits tranquilizing, hypotensive, spasmolytic, bronchodilatory and anti-inflammatory properties. (Chandhoke & Ray Ghatak, Indian J. Med. Res., 1969: 57 and the Wealth of India, X, 1976: 111).

Preparations – Perfumery and cosmetics chiefly.

Cultivation notes – Propagation is through seeds. Nursery can be raised in April and transplantation is done during May-June. It grows in any type of soil, but thrives best in loamy, porous, well drained fields. Harvesting of foliage is done at full flowering stage.

Taraxacum officinale Wigg.

Family – COMPOSITAE

Local/Common name(s) – Dugdhapheni, Dulal, Barau, Kanphul, Dudhli, Dandelion.

Distribution – The herb is found throughout the temperate Himalaya, in the Khasi hills of Meghalaya, Mishmi hills of Arunachal Pradesh and in the hills of South India at altitudes of 300-5400 m. (The Wealth of India X, 1976:129). In Himachal Pradesh, the plant has been found almost every where as a weed in fields, roadsides, grasslands, pastures from plains upto alpine grasslands. (Chauhan, N.S., 903, 2069, 2109, 2120, 2282, 2387, 2400 & 2586).

Description – The plant is a small herb, 2-10 cm tall, with milky juice. Leaves are all radical, sessile, variable in shape, irregularly pinnatifid and toothed. Heads are 1-2.5 cm in diameter, solitary on a hollow and leafless stalk, sessile and bright golden yellow in colour. Pappus hairy.

Taraxacum officinale Wigg.

Flowering/Fruiting – March-November (Flowers in early spring at low level and upto September at high altitude pastures in Spiti, Dodra-kawar and Kinnaur).

Parts used – Roots.

Current market rate – Not available.

Main ingredients – The drug contains a bitter crystalline principle, taraxacin and a crystalline substance, taraxacerine; the phytosterols-taraxasterol and homotaraxasterol besides saponin. (Chopra et al. 1956:240). The milky sap contains cerylalcohol, lactuce-roltaraxacin, and choline. The root yields inulin, tannin and some ethereal oil; the leaves contain vitamin C and the flowers, xanthophyll (Dorfter & Rosett, 1989:284). Also contain Potassium and Vitamin A (leaves).

Uses – Root is diuretic, tonic laxative, cholagogue, anti-rheumatic and aperient. The tender leaves are used as 'Salad' when harvested in spring, the plant has a slight inulin content and is used as a bitter for loss of appetite and stomach upsets. The drug promotes the secretion of gall and is a useful remedy for chronic disorders of kidney and liver. It has been in use for gallstones, cholecystitis, jaundice, atonic dyspepsia with constipation, muscular rheumatism and oligurea.

Preparations – Ingredient of liver and gall bladder preparations.

Cultivation notes – The herb is extremely hardy and adaptable to any soil, but prefers rich sandy or light loams. Manures and fertilisers, if added at proper times, hasten the growth and improve the size and succulence of the leaves. The seeds are sown in August, somewhat later in the winter prone areas at a spacing of 25-30 cm in the rows and 35-45 cm in between the rows to be thinned later to 30 × 45 cm. Transplanting can also be done. Irrigation and weeding are highly beneficial. The aerial parts are harvested in the early spring and the rhizome and roots are dug up in the autumns of the second season. The herb should be prevented from becoming a pest. An yield of 1650 kg per hac is expected from a good crop. Carbon tetrachlo-ride is used as a preservative. The rhizomes and roots are washed and dried, either whole or in short pieces; 7-15 cm long and are stored in the containers.

***Taxus wallichiana* Zucc.**

Syn. *Taxus baccata* Hook.f.

Family – TAXACEAE

Local/Common name(s) – Thuno, thuner, Birmi, Rakhyala, Rakhala, Thangi, Himalayan Yew, Sthoneyak, Thoon.

Distribution – The tree is found in the temperate Himalaya at altitudes, between 1800 and 3300 m and in the hills of Meghalaya and Manipur at an altitude of 1500 m. In Himachal Pradesh, it is found mixed in fir and spruce forests between 2600-3200 m in Chamba, Kangra, Kullu, Shimla, Kinnaur, Mandi and Sirmour (Chauhan, N.S.; 709, 740, 1173, 1552).

Description – It is an evergreen tree, usually 6 m or more tall and 1.5-1.8 m in girth. Bark is reddish brown, thin and scaly. Leaves are distichous, linear, with recurved margin, shining above, pale-yellowish brown or rusty red below; flowers are usually dioecious and the male strobili are stalked, globose, arising from the axils of the leaves on the underside of the branchlets of the previous year; each consis-

Taxus wallichiana Zucc. Syn. *T. baccata* Linn. (Talishpatri)

		ting of 6-14 stamens with short filaments. Female strobili are solitary and axillary.
Flowering/Fruiting	–	March-May/September-October.
Parts used	–	Young shoots chiefly needles/leaves.
Current market rate	–	Dry needles as talispatri Rs. 40/- per kg.
Main ingredients	–	All parts of the tree, except the fleshy aril, are poisonous due to the presence of an amorphous basic fraction, named taxine. The leaves are reported to contain, in addition to 'taxine', taxinine, hydrocyanic acid, formic acid, reducing sugars, resins, tannins, ephedrine, a glucoside, taxiphyllin, ferredoxin, acidyrterone and β-sitosterol. Taxol obtained from the needles of Indian yew is being extracted by an Indo-Italian firm and final anticancer drug is manufactured in U.S.A.
Uses	–	A medicinal tincture made from the young shoots has long been used for the treatment of headache, giddiness, feeble and falling pulse, coldness of the extremities, diarrhoea and severe biliousness. The leaves are credited with emmenagogue and antispasmodic properties. They are employed for the treatment of hysteria, epilepsy and nervousness and as a lithic in calculus complaints. It has been reported to be the source of the drug-Zarnab, which is prescribed in the Unani system of medicines. Taxol, obtained from the needles is reported to cure uterine and breast cancers.
Preparations	–	Talisadi-churna, Talisadi-vati, Sudarshan-churna, Lavanbhaskar-churna etc.
Cultivation notes	–	The tree is found in association with the *Quercus samecarpifolia* and *Abies pindrow*; sometimes, but rarely, with *Picea smithiana, Cedrus deodara* and *Q. floribunda*. Propagation is through seeds, which do not

germinate until the second year and from cuttings. The young plants require shelter and deep shade and does not thrive in the areas, from where the forests have been removed. The tree lives to a great age, but the growth is slow; the annual rings are 12-40 per 2.5 cm of radius, giving a mean annual girth increment of 0.4-1.3 cm (The Wealth of India X, 1976:176). Recently, it has successfully been propagated through cuttings and tissue culture techniques have not succeeded. It has caught worldwide attention for its anti-cancerous activity.

Terminalia arjuna W. & A.

Family – COMBRETACEAE

Local/Common name(s) – Arjuna, Arjan, Arjun.

Distribution – The tree is found throughout the greater parts of India in Sub-Himalayan tract, Madhya Pradesh, Bombay and Madras chiefly along water channels, stream banks and moist places. In Himachal Pradesh, it is common in Nurpur and Nahan forest divisions, sporadic else-where and planted in Jacch, Dhaulakuan, Jogindernagar and Nauni (Chauhan, N.S., 6359).

Description – It is a large tree with smooth and thick bark, greenish white in colour and flaking off in large flat pieces. Leaves are usually sub-opposite, oblong or elliptic-oblong, pale-brown beneath. Flowers are sessile, in short, axillary spikes or in terminal panicles. Drupe is 2.5-5 cm long, ovoid or obovoid-oblong, fibrous, woody, glabrous and dark brown in colour with 5 hard projecting wings. The bark

Terminalia arjuna W. & A. (Arjun)

found in the market is in short pieces of pinkish colour which are fibrous, gritty under the teeth and break with a short fracture. Internal surface is of lighter colour, finely striated and have astringent taste (Dey, 1980:10).

Flowering/Fruiting – April-May/Cold Season.

Parts used – Bark and Fruits.

Current market rate – Bark Rs. 8/- per kg.

Morphology of the official parts used – Bark is 1/3 inch thick, smooth, pinkish grey and the sap-wood is reddish white. The heart wood is brown and variegated with dark coloured streaks.

Main Ingredients – The bark contains a crystalline compound arjunin, a lactone-arjunetin, essential oil, tannin, reducing sugars and colouring matter (Chopra et al. 1956:241).

Uses – The preparations from the bark and the fruit are used as tonic, astringent, febrifuge and deobstruent. These are used in bilious affections and for sores. Arjun is a reported cardiac tonic and is useful in blood dysentery, blood pressure, fractures of bone, heart diseases leucorrhoea and earache, and acts as an antidote to poisons.

Preparations – Arjunarishta, Arjun-kwath, Hridayarnava-Rasa. Arjunaghrita and Arjunatwak churna.

Cultivation notes – The tree prefers loose, moist, fertile and alluvial loam. It can be raised by means of seeds, stumps or entire transplant. Fruits are immersed in water for 24 hours or dumped in cowdung. Sown during March-April in raised beds at 20 × 5 cm distance. Germination is reported between 35-70 per cent and the seedlings are ready for transplantation after 1-1½ years. The seedlings are sensitive to drought and frost and the tree is not to be planted in open, dry hill sides.

Terminalia bellirica Roxb.

Family – COMBRETACEAE

Local/Common name(s) – Bahera, Bahira, Vibhitaki, Bibhitak.

Distribution – The tree is found in the deciduous forests throughout the greater part of India, but not in the arid regions. In Himachal Pradesh, it is sporadic in Shivalik zone and is generally seen planted near villages and field borders, often as a single tree here and there. Planted in herbal gardens at Nauni and Jogindernagar (Chauhan, N.S.; 911: 1739; Uniyal & Chauhan, 413).

Description – A large deciduous tree, nearly 20-30 m tall. Leaves are alternate, stalked, broadly ovate with cuneate base, clustered at the end of the branches, and are entire and smooth. Flowers are small, pale-green in colour and have offensive odour. Fruits are 2-3 cm long, ovoid and the stones are hard and contain a sweet oily kernel.

Flowering/Fruiting – April-June/November-February.

Parts used – Fruits.

Current market rate – Dried fruits Rs. 7-10/- per kg.

Main ingredients – The active constituent are the tannins. β-sitosterol, gallic acid, ellagic acid, chebulegic acid, and a number of free sugars have also been reported. Kernels yield oil, which contains palmitic, oleic and linoleic acids.

Uses – Fruits are bitter, astringent, anthelmintic, pungent, laxative and are used in piles,

Terminalia bellirica Roxb. (Bahera)

enlargement of spleen, diarrhoea, eye troubles, leprosy and headache. Kernel is said to be narcotic. Fruits are taken in cough, hoarseness, throat troubles and skin diseases.

Preparation – Bibhitak-taila, Triphala-churna, Talisadi-churna, Lavangadi-vati, Sudarsan-churna, Agnitundi-vati, Sanjivani-vati, Punarnava-mandur, Navayas-louha, Sringadi-churna, Karanjadi-yoga, Laghuvis-garva-taila, Arogyavardhini-vati, Khadirarista, Gokshuradi-guggulu, Chandraprabha-vati, Vidangadi-louha.

Cultivation notes – The tree is a common associate of sal, teak and other important trees, occurring more or less scattered and not gregariously in the moist valleys. The tree is light demander, but can withstand slight shade when young. Young seedling are known to tolerate fairly heavy shade during the first or second year, but heavy shade, afterwards, kills them. It is sensitive to frost and is drought hardy. The tree coppices, fairly well. The fruits ripen during November-February. Freshly fallen fruits are collected from the ground swept clean in advance. The pulp is immediately removed and the seed dried in the sun before storing. The seeds retain viability for about an year. The plant can be raised in the field by direct sowing or by transplanting the nursery raised entire seedlings of one-year growth. Stump plants are not satisfactory as compared to the entire plants. The rate of growth of this species is slow in the early stages, but is moderate to rapid later on. Fruits are collected during December to February.

Terminalia chebula **Retz.**

Family	– COMBERTACEAE
Local/Common name(s)	– Her, Harad, Haritaki, Harde, Chebulic myrobalan.
Distribution	– The tree is found in the outer Himalayas, ascending upto 1600 m from Kangra to Bengal and then throughout the Central and South India. In Himachal Pradesh, it is distributed between 800-1100 meters in, Sirmour, Hamirpur, Mandi, Bilaspur, Kangra and Una districts. Forms a pure patch near Maranda in Palampur and is scattered elsewhere. A tree here and there is found in Kangra, Bilaspur, Hamirpur, Mandi and Nahan regions. Best Harad trees are found in Jukhala area of Bilaspur (Chauhan, N.S., 804, 1535, 6351; Uniyal & Chauhan, 396).
Description	– It is a large deciduous tree, with thick, dark-brown; young parts covered with rusty hairs. Leaves are mostly sub-opposite, ovate or elliptic, 7.5-15 cm long, acuminate Flowers are dull white or yellow, small, strong-scented in terminal, paniculate spikes. Drupe more or less ellipsoid or ovoid, 2.5 cm-5 cm long, with a rough, grooved surface and a hard nut. Based on fruit size and shape, it has as many as seven types.
Flowering/Fruiting	– April-June/January-March.
Parts used	– Fruits.
Current market rate	– Rs. 7/- to 9/- per kg.
Main ingredients	– Contains tannin upto 30 per cent, chebulic acid, 3.5 per cent; chebulinic acid 30 per cent; ellagic acid and ellagotannic acid. Also contains gallic acid, resin and some purgative compounds of the nature of anthraquinone.

Terminalia chebula Retz. (Haritaki, Harad)

Uses – The plant is much used for tanning purposes, for which it has a big foreign market. In action, it is alterative, astringent, dentrifrice, purgative, stomachic and tonic. It is useful in apthae, acsites, asthma, bile trouble, bleeding and ulceration of the gum, blood pressure, burns and scalds, carious teeth, cooling wash to the eyes, cough, dysentery, diarrhoea, piles, vaginal discharge, ulcers, vomiting, worms and is used as a cardiac tonic; in sores and as an antidote to poisons. Harad has been held in high esteem in Hindu medicines. It has been rated above the mother, as she often gets annoyed with her kids, but Harad never does.

Preparations – Abhaya-modak, Amrita-haritaki, Amlakyadi-churna, Danti-haritaki, Sanjivani-vati, Sudarshan-churna, Dasmula-haritaki Triphala, Agastya-haritaki, Chitrak-haritaki, Panchaskar-churna etc.

Cultivation notes – The tree is propagated through seeds. It is cultivated in Sub-Himalayan tracts upto an elevation of 1250 m, preferably in moist regions and clayey as well as on sandy soils.

Thalictrum foliolosum **DC.**

Family – RANUNCULACEAE

Local/Common name(s) – Mamira, Pilijari, Chaitra, Sarphar, Bishkaral.

Distribution – Inner valleys of the temperate Himalayas and West Tibet between 1800-3500 m. In Himachal Pradesh, it is found in Chamba, Kangra, Kinnaur, Kullu, Mandi, Lahaul-spiti, Solan, Shimla and Sirmour districts upto an elevation of 3000 m. (Chauhan, N.S., 963, 987, 1244, 2187, 2222, 2472).

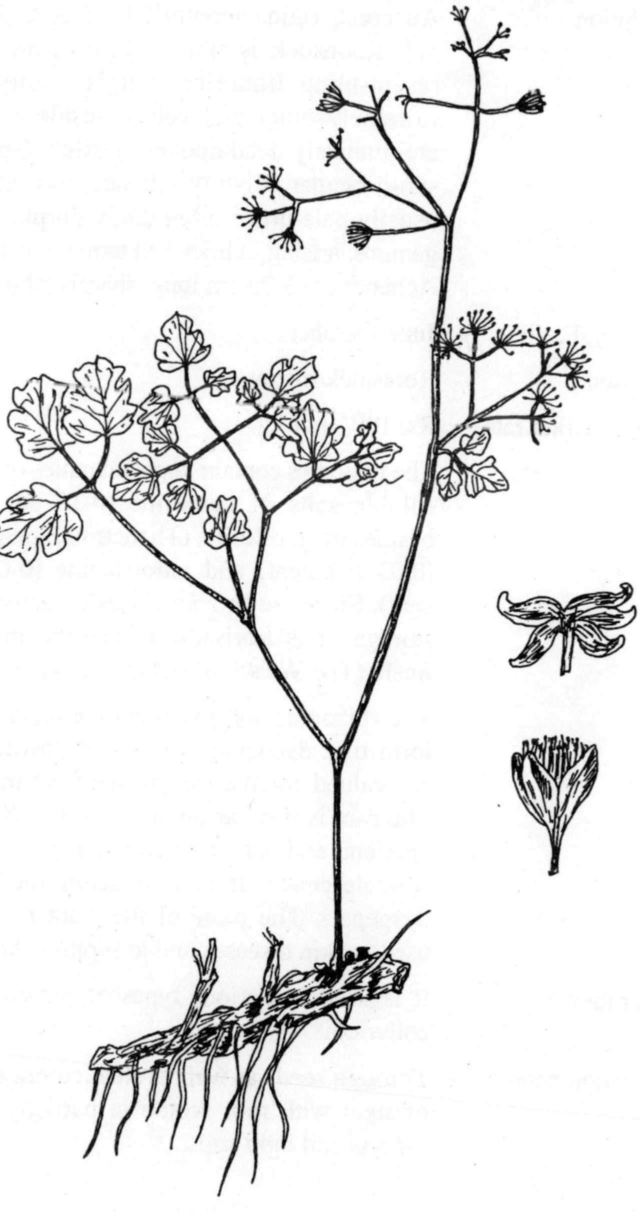

Thalictrum foliolosum DC. (Pilijari, Masuiri)

Description – An erect, rigid, perennial herb upto 2.5 m tall. Rootstock is fibrous yellowish-brown, resem-bling liquorice, but is knotty and extremely bitter and yellow inside. Leaves are pinnately decompound; leaflets 4-6 mm, sub-orbicular, slightly lobulate. Flowers are usually pale-green often dingy-purple, polygamous, arising in branched terminal panicles. Achenes are 3.75 cm long, sharply ribbed.

Flowering/Fruiting – June-October.

Parts used – Rootstock/rhizome.

Current market rate – Rs. 100/- per kg.

Main ingredients – The rhizomes contain large quantities of water soluble salts of berberine (0.35 per cent) besides magnoflorine (Thalictrine), palmitine (0.03 per cent) and jatrorrhizine (0.02 per cent). Since the rhizome loses its activity on storage, it is advisable to use the material fresh. (The Wealth of India, X:204).

Uses – The rhizomes and the roots are used in the form of a decoction, extract or powder and are valued for the treatment of ophthalmia. The root is used as an antiperiodic, diuretic, aperient, and purgative; as bitter tonic during convalescence. It is also given for atonic dyspepsia. The paste of the plant is locally used in skin diseases and to suppress boils.

Preparation – It is used in various types of Surma's and collyriums.

Cultivation notes – Through seeds as well as stem cuttings/splits of stem with root proton in partially shady aspects and light soils.

***Thevetia neriifolia* Juss.**
Syn. *T. peruviana* (Pers) Merrill.

Family	– APOCYNACEAE
Local/Common name(s)	– Pila-kaner, Ashvaha, Ashvaghna, Pit-kaner, Yellow oleander.
Distribution	– It is found throughout India and in Himachal Pradesh, it is found in Una, Daulatpur, Gagret, Mubarakpur, Kangra, Mandi, Nalagarh, Bilaspur, Nahan and Paonta regions in field borders river banks, and wastelands. (Chauhan, N.S. 1418, 1860; Uniyal & Chauhan, 312).
Description	– A large, evergreen shrub, 4.5-6.0 m tall. Leaves are 10-15 cm long, linear and acute. Flowers are bright yellow in terminal cymes, softly scented. Drupes are triangular, fleshy, with 2-4 seeds.
Flowering/Fruiting	– July-September.
Parts used	– Bark, Leaves, Roots, Seeds (whole plant).
Current market rate	– Not available.
Main ingredients	– Glucosides thevetin, thevetoxin, a phytosterolin, abouain, kokilphin are obtained from the kernel of the seed. Nuts yielded a new heteroside neriifolin. Seeds contain glucosides neriifolin, acetylneriifolin and thevetin. (Chopra et al. 1956:244).
Uses	– Kernel of seed used as acro-narcotic poison. Bark is a bitter tonic, cathartic, febrifuge and useful in different kinds of intermittent fevers. All parts of the plant including latex are poisonous. (The Useful Plants of India, 636). The cardiac glycosides obtained from the kernels, bark and flowers are useful for heart diseases.

Preparations – Used in traditional/domestic preparation and its glucosides have been found to have cardiotonic activity.

Cultivation notes – The plant is widely cultivated in gardens in the plains throughout India. It requires little attention except sunshine and well-drained soil with open compound. It is propagated from seeds and cuttings. Cuttings root easily.

Thymus serpyllum Linn.

Family – LABIATAE/LAMIACEAE

Local/Common name(s) – Banajwain, Lepto, Lepte, Masho, Hasha, Pangduma, Padumba, Wild thyme.

Distribution – The plant is found in the Himalayas from Kashmir to Nepal at altitudes of 1500-4500 m and is reported to be grown in the gardens of Western India. In Himachal Pradesh, the plant is frequently found in the open, rocky slopes and ghasnis from 1800 m onwards extending beyond 4500 m in Spiti. (Chauhan, N.S., 918, 1165, 1249, 1405, 1592, 1630, 2185, 2291, 2659, 2686, 5492, 5538).

Description – The plant is a variable, aromatic, hairy, prostrate and evergreen perennial herb; 10-25 cm high and is one of the most effective ground covers, forming dense mats. Rootstock is woody and the leaves are small, elliptic or oblong-ovate or lanceolate, 6-15 mm long, subsessile and glandular. Flowers are polygamous, pink, in globose, terminal whorls. The male flowers are larger in size. The plant is pungent in taste.

Flowering/Fruiting – April-September.

Parts used – The Plant.

Current market rate – Herb Rs. 15-20/- per kg.

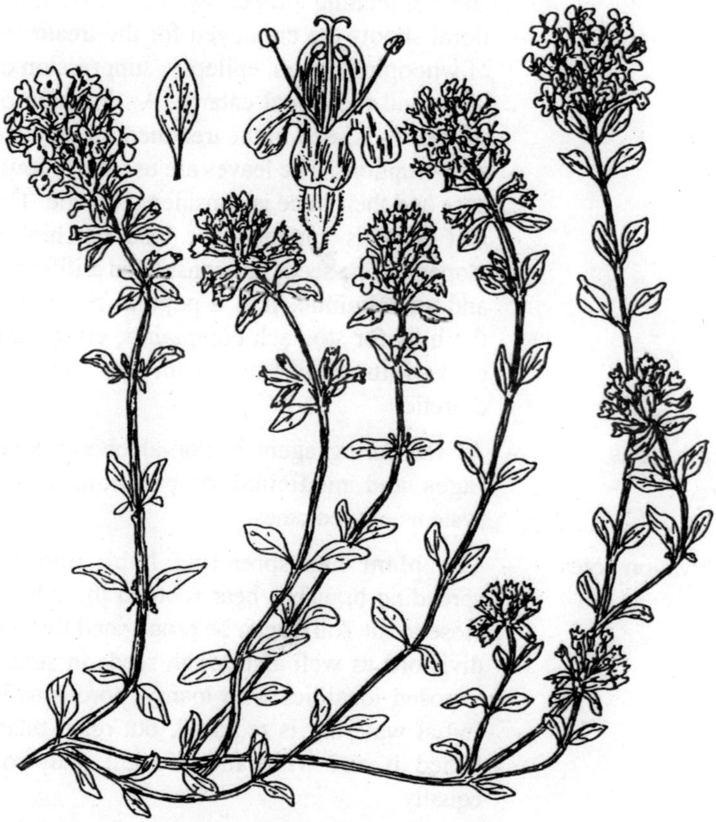

Thymus serpyllum Linn. (Banajwain)

Main ingredients – The leaves and floral tops yield a volatile oil, known as oil of wild thyme 0.27 per cent from fresh and 0.60 per cent from dried plants. Three flavonoides, i.e. scutellarin heteroside, luteolin 7-glucoside, and apigenin, have been isolated from the plant. The floral tops and leaves are reported to yield 0.75 per cent of ursolic acid. Its nectar is rich in ascorbic acid. Oil contains chiefly thymol.

Uses – The herb possesses anti-spasmodic, antiseptic, expectorant, carminative, anthelmintic and stimulating properties. The leaves and floral shoots are employed for the treatment of whooping cough, epilepsy, suppression of urine and menstrual catarrh. An infusion of the leaves is used in the treatment of itch and skin eruptions. The leaves are used in biliousness and their juice is considered emetic. The leaf juice is applied over bald patches of alopecia. The seeds are considered antifebrile and anthelmintic. It is a popular remedy in the hills for stomach complaints, cough and colds. Infusion of leaves with 'gur' acts as a diuretic.

Preparations – As flavouring agent in non-alcoholic beverages and medicinal preparations in all systems of medicines.

Cultivation notes – The plant has spreading habit and the spreading branches bear roots in the portion close to the soil. It can be propagated through divisions as well as through seeds in sunny, exposed localities with loamy, porous soils. Initial watering is required, but once established it can withstand drought and frost equally.

***Tinospora cordifolia* (Willd.) Miers.**

Family – MENISPERMACEAE

Local/Common name(s) – Gulancha, Giloe, Guduchi, Amrita, Gulje.

Distribution – The climber is found throughout the tropical regions of India. In Himachal Pradesh, it is frequented as a robust climber in Una, Paonta, Hamirpur, Kangra, Mandi and Chamba regions, upto 1200 m elevation above m.s.l.

Its associated species i.e. *T. malabarica*, grows upto 1300 m. (Chauhan, N.S., 848, 1449; Uniyal and Chauhan, 380).

Description – It is a glabrous, climbing, succulent shrub with corky grey dotted bark. Leaves are 10-20 cm in diameter, broadly ovate, deeply cordate and shortly acuminate. Flowers are small and greenish-yellow on the old wood in 7.5-15 cm long racemes; slender, usually solitary in the female and clustered in male. Fruits is of the size of a pea and red in colour.

Flowering/Fruiting – March-June/Cold season.

Parts used – Fresh stem/dried stem pieces.

Current market rate – Dried stemp pieces Rs. 10-20/- per kg.

Main ingredients – Tinosporin and a furanoid diterpene identical with columbin have been isolated from the plant. Fresh stem bark yields giloin, giloinin and gilosterol. Berberine and a waxy substance are also present. A new hypoglycaemic agent and a new phenolic lignan have also been isolated (Mukherji, 1954:239; CCRAS, 1990:86).

Uses – It is an alterative, anthelmintic, anti-arthritic, antiperiodic, anti-pyretic, aphrodisiac, bitter tonic, blood purifier, cardiac, carminative digestive, diuretic and expectorant (The Useful Plant of India, 640). Starch from the roots and stems is used in chronic diarrhoea and dysentery. Juice of the fresh plant is diuretic and used in gonorrhoea (Chopra et al. 1956:244). Columbin has shown anti-diabetic properties (CCRAS, 1986).

Preparations – Guduchayadi-churna, Guduchyadi-kwatha, Sudarshan-churna, Amritarista, Kaishore-guggulu, Rasnapanchak-kwath, Yograja-guggulu, Sanjivani-vati etc.

Cultivation notes – The plant is sometimes cultivated for ornamental value and is propagated by cuttings. It is so easy to propagate that even if a twig of it is placed on a branch of a tree, it will establish itself as a giant climber in a couple of years.

Trillidium govanianum (D. Don.) Kunth.

Syn: *Trillium govanianum* Wall. ex D. Don.

Family – LILIACEAE (Trilliaceae)

Local/Common name(s) –

Distribution – Pakistan to Bhutan between 2700-4000 m in forests. In Himachal Pradesh, it is found growing in patches in forest shade rich in humus at Panju in Chhajpur and Rahla-rohtang slopes in Kullu (Chauhan, N.S., 2257; Chauhan & Harish, 6998).

Description – A tuberous rooted herb often upto 30 cm tall, growing frequently in forest shades with three leaves in one whorl at the summit of the stem and a solitary, purple flower in the centre. Leaves are broadly ovate, acute and conspicuously stalked, 3.5-10 cm long. Flowers brown purple with narrow, spreading petals; the outer 3 narrowly lanceolate, the inner 3, linear. Anthers large, yellow; styles 3; long and conspicuous. Fruit is a red, globular berry, 1-2 cm in diameter.

Flowering/Fruiting – May-June.

Parts used – Roots

Current market rate – Not available.

Chemical contents – Roots contain Trillarin, which on hydrolysis yields 2.5% diosgenin—a corticosteroid hormone.

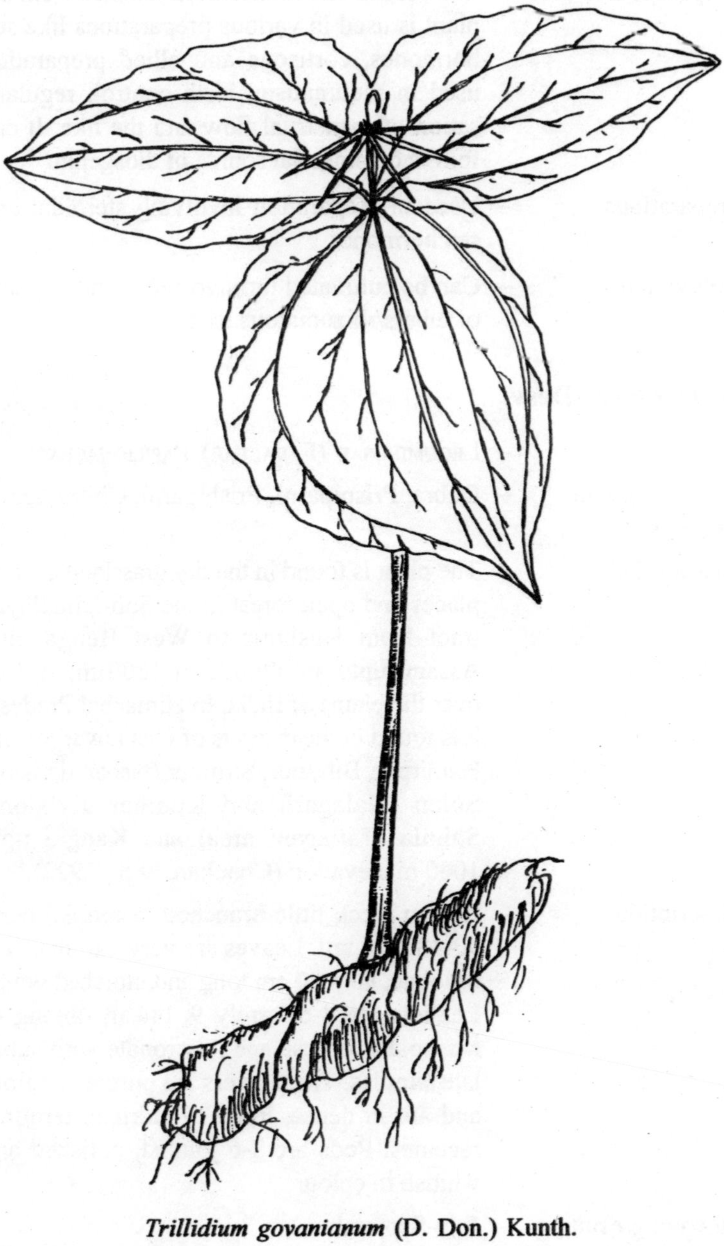

Trillidium govanianum **(D. Don.) Kunth.**

Properties and uses – The cortico-steroid hormone isolated from the plant is used in various preparations like sex hormones, cortisone and allied preparation used in rheumatism, birth control, regularisation of menstrual flow and the like. It can form a commercial source of diosgenin.

Preparations – Used in preparation involving steroidal and sex hormones.

Cultivation – Can be cultivated through seeds and division of tubers/rhizomatous roots.

Uraria picta Desv.

Family – LEGUMINOSAE (FABACEAE) PAPILIONACEAE

Local/Common name(s) – Dabra, Prisniparni, Prishtparni, Chitraparni.

Distribution – The plant is found in the dry grasslands, waste places and open forest in the Sub-Himalayan tract from Kashmir to West Bengal and Assam, upto an altitude of 1800 m, and all over the plains of India. In Himachal Pradesh, it is found in the districts of Una (Jwar forest), Hamirpur, Bilaspur, Sirmour (Nahan division, Solan (Nalagarh and Kunihar divisions, Shimla (Tattapani area) and Kangra upto 1000 m elevation (Chauhan, N.S., 1922).

Description – It is an erect, little-branched, perennial herb, 90-180 cm tall. Leaves are very variable, 1-3 foliolate, upto 30 cm long and blotched white. Leaflets are 4-6, rarely 9, linear, oblong or lanceolate, obtuse and mucronate with subulate stipules. The flowers are purple in colour and are in dense, long, cylindrical, terminal racemes. Pods are 3-6 jointed, polished and whitish in colour.

Flowering/Fruiting – July-September.

Parts used – The herb.

Current market rate – Not available.

Main ingredients – Not available.

Uses – The plant is one of the components of 'Dashmoola' and is credited with fracture healing properties. Its total extract has been found to effect better and quicker healing of fractures in the experimental animals due to early accumulation of phosphorus and more deposition of calcium. In Ghana, the plant is used for treating the heart trouble. The root is credited with aphrodisiac properties and its decoction is prescribed for coughs, chills and fevers. The leaves are considered to be antiseptic and are used in gonorrhoea. The roots and pods are employed in Bihar to treat prolapse of anus in infants and for the treatment of sore-mouth in children.

Preparations – Dashmoolarishta, Dashmoolaghrita, Dash-mool-kashaya etc.

Cultivation notes – Propagation is through seed.

Alternative source – *Uraria lagopodioides* (L.) Desu.

Urginea indica (Roxb.) Kunth.

Syn: *Scilla indica* Roxb.

Family – Liliaceae

Local/Common name(s) – Jangli-piyaz, Ban-piyazi, Ban-palandu, Indian squill, White squill, Sea-onion.

Distribution – Western Himalaya below 1400 m; Bihar, Konkan and Coromandal. In Himachal Pradesh, it is found growing in lower parts of Solan and Sirmour, Bilaspur, Mandi and Hamirpur. Planted in Herbal Garden at Nauni-Solan (HP).

Description – A glabrous herb. Bulb dull white or pale, ovoid, 2.5-5 cm diameter or more. Leaves radicle, flat, narrow pointed. Flowers appearing before the leaves, drooping or spreading, distant in a terminal raceme, 15-50 cm long. Perianth 10.8 cm long, bell shaped. 6-parted; segments white, with 3 green ribs in the center, tapering downwards; ovules several in each cell. Capsule 1.25-2 cm, oblong, 3-valved; seeds many, flat, black.

Flowering/Fruiting – May-September.

Parts used – Bulbs

Current market rate – Not available.

Chemical contents – Fresh squills yield at least two glycosides, scillarin-A, which is crystalline and scillarin-B, which is amorphous.

Properties and uses – The bulb in action is anthelmintic, cardiac, stimulant, deobstruent, digestive, diuretic, emmenagogue, and expectorant. Useful in asthma, ailments of hearts, calculous affections, cough and bronchitis, paralytic affections, promotes urination, in rheumatism and skin diseases. Bulbs are source of raticide, and is used as an expectorant, stimulant, and cardio-tonic in small doses. Squill is used chiefly in chronic bronchitis and asthma. Alcoholic extracts of the bulbs possess anticancer activity against human epidermoid carcinoma. Also employed in dropsy, rheumatism and skin troubles. Externally used for removing corns and warts.

Preparations – Preparations are panale (syrup), and tincture. Various other preparation are also made of the drug. Much used in preparations of silver.

Cultivation – Can be cultivated through daughter bulbs.

Valeriana hardwickii Wall.

Family	– VALERIANACEAE
Local/Common name(s)	– Muskbala, Tagar, Nihani.
Distribution	– Pakistan to South-west China, Burma, South-east Asia, frequent on moist slopes in the Himalayas and Khasi hills in shrubberies and open slopes between 1500-4000 m elevation. In Himachal Pradesh, it has been found in Bharmour division of Chamba, Kanda area of Karsog and Chansil of Rohru forest division.
Description	– Erect, pubescent herbs, 20-60 cm tall. Radical leaves few, ovate-orbiculate, entire or crenulate, pilose, cauline leaves 3-7 pinnate, terminal lobe the largest. Lobes ovate-lanceolate, crenate, serrate. Flowers white in axillary or terminal paniculate cymes or corymbs. Achenes pubescent, longer than the upper bracts.
Flowering/Fruiting	– June-September.
Parts used	– Roots (Rootstock dark brown, aromatic, going vertically deep in the soil).
Current market rate	– Rs. 105/- per kg.
Chemical contents	– Contains a crystalline acid. A new oxide, hydrocarbons a polyoxygenated crystalline solid together with alpha-eudesmol, beta-eudesmol, edemol, angelicin, 4-hydroxy thymol dimethyl ether have been isolated.
Properties and uses	– The roots are used as insecticide. It is also used in the mental disorders. Also used for scenting hair. Locally much used as a hair tonic and in dhoop manufacturing. It is a substitute for the drug valerian obtained from *V. officinalis* L.

Preparations – Preparation made are mansadi-kwath, raksha-ghna-ghrita and Sarvousadhi-snan, Dasanga-lepa, Laghu-vis-garbha-taila, Pippalayasav and Sudarsan-churna.

Cultivation – Can be cultivated through seeds and vegetatively by multiplying rootstock.

Valeriana jatamansi Jones.

Syn. *V. wallichii* DC.

Family – VALERIANACEAE

Local/Common name(s) – Tagar, Sugandhbala, Mushkbala, Naknihani Nihani, Indian Valerian.

Distribution – The plant occurs abundantly in the temperate Himalayas from Kashmir to Bhutan at altitude of 1300-3800 m and in the Khasi hills between 1250-1800 m. In Himachal Pradesh, it is quite common in forest undergrowth all over between 1600-3600 m. Collected more frequently from Chamba, Kangra, Kullu, Sirmour, Mandi, Kinnaur and Shimla districts, (Chauhan, N.S., 63, 730, 925, 1401, 1608, 1630, 2136, 2182 and 2447).

Description – A pubescent, perennial herb, having horizontal, thick root stock with thick, descending fibrous roots; stem 15-45 cm high, usually tufted. Basal leaves are deeply cordate, ovate, usually acute and toothed. Stem leaves are much smaller, few, entire or pinnate. Flowers are white or tinged with pink in terminal corymbs; often unisexual. Fruit hairy or nearly hairless.

Flowering/Fruiting – March-May-June.

Parts used – Roots and Rhizomes.

Current market rate – Rs. 45-50/- per kg.

Main ingredients – Indian valerian roots yield a volatile oil (0.5-2.12 per cent). Fresh roots and rhizomes also

Valeriana jatamansi Jones. (Sugandhbala)

contain a new group of monoterpene derivative called, 'Valepotriates' upto 3.82 per cent (Sood M., 1992).

Uses – The drug is much used in making perfumed powder and cardiac preparations. It is regarded as aphrodisiac, antiseptic, cardiac, stimulant, carminative, diuretic, emmenagogue, expectorant, febrifuge, nervine tonic, ophthalmic, sedative, suppurative and tonic. Useful in hysteria, epilepsy, chorea, shell shock and neurosis. Dried rhizomes are also employed in hair oils and perfumes, as incense and in medicines for hysteria and nervous problems.

Preparations – Sudarshan-churna, Pippalayasava, Dasangalep and various other preparations.

Cultivation notes – The plant is cultivated easily by dividing the old roots, either in the fall or in the spring and setting the divisions at about 30 cm distance in rows 60-90 cm apart. The spacing of 30 × 30 cm has been considered to be the best. The yield from the plant is maximum with the application of the nitrogenous, potassic and phosphatic fertilizers at a dose of 40 kg, 30 kg and 20 kg/ha respectively. (Sood M., 1992).

It may also be propagated through seeds. The plant grows well in all types of ordinary soils, but prefers a rich and rather heavy moist loam.

Vanda tessellata **Lodd. ex Loud.**

Syn: *V. roxburghii* R. Br.

Family – ORCHIDACEAE

Local/Common name(s) – Rasna, Atirasa, Vanda.

Distribution – Found in Bengal, Chhota Nagpur Bihar, Madhya Pradesh, West Peninsula and Travancore. In Himachal Pradesh, it is found in Kangra, Mandi and Sirmour districts growing as an epiphyte on mango trees. (Chauhan, N.S.; 838, 933).

Description – It is an epiphytic orchid, 30-60 cm high, with leafy stem. Leaves are thickly coriaceous, recurved, plicate, obtusely keeled. Flowers are greenish yellow, mottled with brown on the mid lobe of lip with purple caruncles. *Rhynchostylis* is used as adulterant for Rasna.

Flowering/Fruiting – April-June.

Parts used – Whole plant.

Current market rate – Rs. 6-7.50/- per kg. (Leaves).

Main ingredients – Whole plant contains an alkaloid and a glucoside. The herb also contains a bitter principle, tannins, resin, saponin, sitosterols, fatty oil and a colouring matter. (The Wealth of India, Vol. X, 432).

Uses – In action, it is anti-arthritic and has been found useful in rheumatism and allied disorders. Rasna enters into the composition of various medicated oils for external application in diseases of the nervous system and rheumatism. Paste of the leaves is applied to the body in fever. Leaf juice is a remedy for otitis media (Chopra et al. 1956:252).

Preparations – Rasnasaptak-kwath, Rasna-guggal, Rasna-dasmula-kwath, Rasna-panchak-kwath, Maharasnadi-kwatha, Mahamash-taila etc.

Cultivation notes – Vanda can easily be cultivated by cuttings/divisions. Vandas require more light than other orchids, but light shade is required during the very bright weather. The plants need high temperature and abundance of moisture during their growth and a compatible host for shelter and support. Mango is found to be its favourite host.

Vetiveria zizanioides (Linn.) Nash.

Family – POACEAE (Graminae)

Local/Common name(s) – Khas, Ushir, Vettiver, Khus-khus grass.

Distribution – The grass is found throughout the plains in lower hills of India, particularly on the river-banks and in rich marshy soils, ascending to an altitude of 1200 m. In Himachal Pradesh, it has been found to occur in Kangra, Una, Hamirpur, Mandi, Paonta-Nahan; Nalagarh and even in Khaltoo in boggy, marshy nalas. (Chauhan, N.S., 1410).

Description – A densely tufted awnless, wiry, glabrous, perennial grass; is a shy breeder and is considered sterile. The plant grows in large clumps from a much branched spongy rootstock with erect culms, 0.5-1.5 m high. The leaf blades are relatively stiff, long and narrow upto 75 cm long and 8 mm or less in width, glabrous, but, downward rough along the edges. Flowers in conical panicles, 15-30 cm long, narrow, acute, appressed, awnless, sessile and some-what flattened laterally with short, sharp spines. Spikelets are pale or reddish-brown or purplish in colour with 3 stamens and 2 plumose stigmas.

Flowering/Fruiting – July-October.

Parts used – Roots.

Current market rate – Rs. 30/- per kg.

Main ingredients – Roots yield an essential oil, the content of which has been found to vary from place to place from 0.2 to over 1.0 per cent. The ketone fraction varies from 20-90 per cent and the yield of β-vetivone varies from 2-49 per cent (Chopra et al. 1956:254).

Uses – Khas is aromatic, refrigerant, febrifuge, diaphoretic, stimulant, stomachic emmenagogue, expectorant, heart-tonic, nerve sedative and diuretic. Root part is cooling and the essential oil acts as tonic. Useful to allay thirst in fever and inflammatory affections. Indicated in asthma, bilious complaints, hysteria, rheumatism and gout. Infusion is used in several cooling mixtures. (Dastur, 163; Dey, 1980: 156).

Preparations – Ushirasava, Usiradi-kwatha, Ushiradi-churna, Usiradya-taila, Sarvajwarhar-louha etc.

Cultivation notes – The grass is cultivated through slips separated from the clumps, with the roots intact, having 15-20 cm of shoot. Prior to planting the slips/divisions in the field; tops beyond 20 cm from the base are cut off and below the base roots of 8-10 cms in length should be left on the plant. This procedure cuts down the transpiration level, prevents it from drying out and thus gives it a better chance of survival. The slips planted in the nursery can be fertilised with DAP (Diamonium phosphate) to encourage fast tillering. Best planting site for vetiver are the inlets to small dams or tanks, along the banks of streams, brooks and water channels (Greenfield, 1987).

Viola serpens Wall.

Family – VIOLACEAE

Local/Common name(s) – Banfshah, Banaksha, Gule-banaksha, Violet.

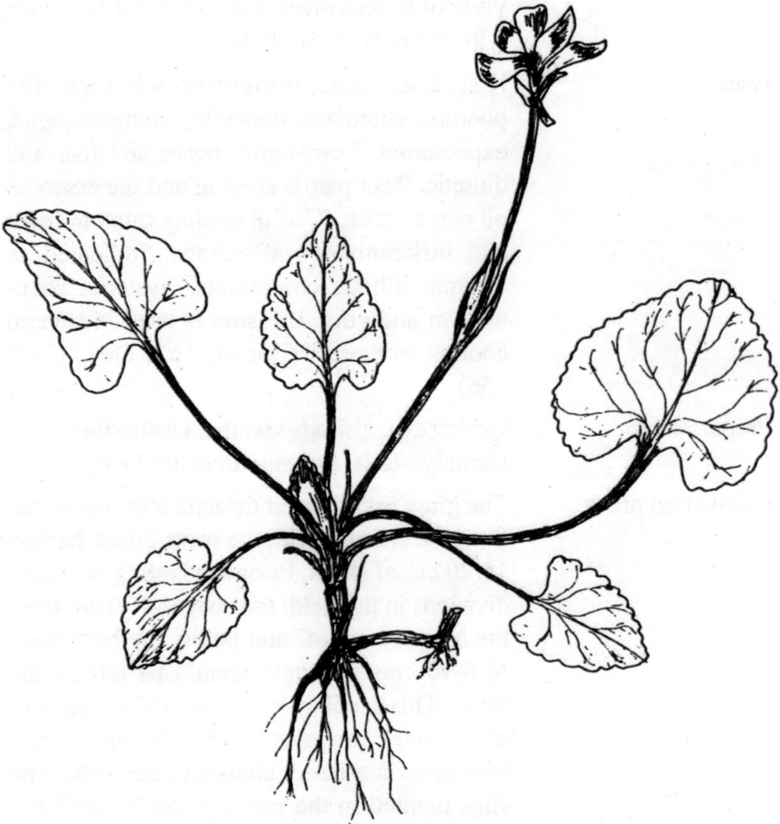

Viola serpens Wall. (Banaksha)

Distribution – A hirsute or glabrous herb, found throughout the temperate Himalayas upto an altitude of 2000 m. In Himachal Pradesh, it is found in Chamba, Kangra, Kinnaur, Kullu, Mandi, Sirmour, Bilaspur, Shimla and Solan districts (Chauhan, N.S., 822, 977, 1359, 1496, 1624, 1751, 2071, 2681).

Description – A small, perennial herb, with short tufted root-stocks. Stem short, but distinct, covered with withered scales, often producing runners. Leaves are ovate-cordate, crenate-serrate. Flowers lilac, blue or sometimes white. Capsule globose, only a few seeded.

Flowering/Fruiting – March-September.

Parts used – Whole plant.

Current market rate – Rs. 140-160/- per kg (Leaves + Flowers). Rs. 325-400/- per kg (Flowers).

Main ingredients – Roots contain glucoside methyle salicylate; yields an alkaloid, violine; a glycoside, viola quercitrin, which is probably identical with rutin and a saponin. Roots contain saponin and an alkaloid. Roots, leaves and blossoms contain methyl-salicylate in the form of a glucoside (Dymock, Vol. I and Dey, 23).

Uses – In action, it is aperient, antipyretic, cooling, demulcent, diaphoretic, diuretic, emetic, emollient, expectorant, febrifuge and purgative. Useful in asthma, bleeding piles, cancer of the throat, constipation, cough, fever, headache and skin diseases. Made into a syrup, it is given in infantile disorders (The Useful Plants of India, 1986:681; Chauhan, N.S., 1984).

Preparations – Banafshdi-kwath, Gulukand-banafsha, Banfasha syrup and Banfasha oil.

Cultivation notes – The plant is propagated by divisions, cuttings or seeds. The plant grows well in a cool and moist climate, but exposure to heavy or frequent rain is fatal for its blooming. Flowers are collected during March-April, air dried, packed, stored and marketed. The soil and climatic conditions of Tehri-Garhwal (U.P.), Shimla, Kullu, Kinnaur, Sirmour and Chamba (H.P.) are conducive for large scale cultivation of this plant.

Other species – *V. biflora*, *V. caneseens* and *V. patrinii*.

***Viscum album* Linn.**

Syn: *V. costatum* Gamble

Family	–	LORANTHACEAE (Viscaceae)
Local/Common name(s)	–	Chulu-ka-banda, Ujjral, Banda, Kahbung, Mistletoe.
Distribution	–	Jaunsar and Tehri-Garhwal upto 2000 m. In Himachal Pradesh, it is parasitising chiefly on Rosaceous trees and shrubs, on elm, walnut, and willows between 1200-2600 m. (Chauhan, N.S., 1007, 1531, 1676, 1977 & 2475).
Description	–	A large, epiphytic, profusely branching shrubby perennial, forming rounded clumps over the host trees. It is green all over; branches dichotomous or whorled. Leaves about 5 by 10 cm, sessile, cuneate-oblong, or oblanceolate, with 3-5 longitudinal, basal nerves. Flowers are dioecious, sessile; in clusters of 3-5, supported by concave bracts. Perianth-segments 3-4, triangular, deciduous. Fruit small, ellipsoid, white, smooth, almost transparent, mucilaginous.
Flowering/Fruiting	–	May-October.
Parts used	–	Leaf, Fruit (Berry), Twigs i.e. the whole plant.
Current market rate	–	Not available.
Chemical contents	–	Main constituents are fatty acids (oleic and palmitic), nicotine alkaloids, amines (Acetyl-choline, choline, histamine and tyramine). Other constituents include flavonoids (quer-cetine flavonol), chaleone derivatives and flavonone derivatives; terpenoids-beta-amyrin, betalinic acid, lupeol and ester combinations; oleanoilic acid, resin acid, ursolic acid, betasitosterol, sterol A, phytosterol glucoside,

viscotoxin (mixture of low molecular weight polypeptides); mucilage, polyoles (e.g. mannitol, dulcitols, inositol, pinitol and quercitol); sugars (fructose, glucose and sucrose), starch, syringin (a phenolic glucoside) and tannin. Also contains a sticky greenish resin, "Viscin", fatty oil, lutein, alpha and beta carotene and vitamin C.

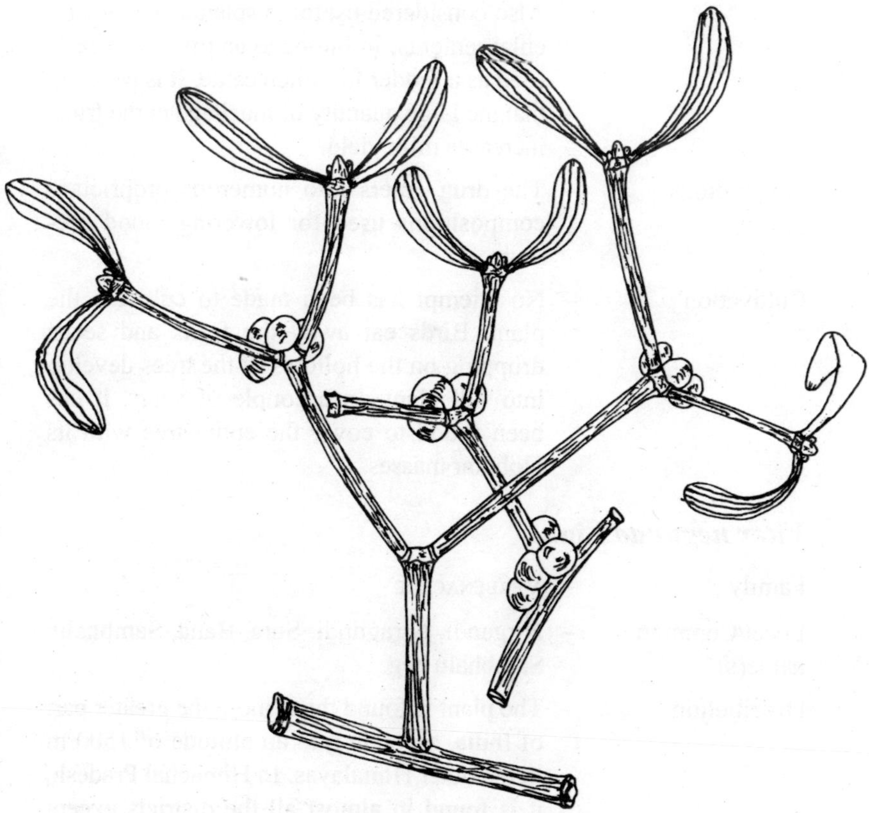

Viscum album Linn.

Properties and uses – The extract and powder of the leaves were official in the French Pharmacopoiea as stabilizers. In Europe, the leaves, stems and berries constitute a drug used for hypertension, arteriosclerosis, tumors and arthrosis. Also used for hysteria, epilepsy and ear diseases. The drug is reputed as a cardiotonic, antiseptic, antispasmodic, diuretic, purgative and emetic with narcotic action. It is employed as a decoction or tincture in intermitted fevers. Also considered useful in splenic and hepatic enlargements; in lumbago and piles. Locally used as a fodder for milch cattle. It is believed that the large quantity of mucilage in the fruits increases milk yield.

Preparations – The drug enters into numerous proprietary compositions used for lowering blood pressure.

Cultivation – No attempt has been made to cultivate the plant. Birds eat away the fruits and seeds dropping on the hollows of the trees develop into full plant. In a couple of years, it has been found to cover the entire tree with its globular masses.

***Vitex negundo* Linn.**

Family – VERBENACEAE

Local/Common name(s) – Nirgandi, Niragundi, Sura, Bana, Sambhalu, Sambhalu beg.

Distribution – The plant is found throughout the greater part of India, ascending to an altitude of 1500 m in the outer Himalayas. In Himachal Pradesh, it is found in almost all the districts except Lahaul-Spiti and has also been planted as hedge and to check soil erosion (Chauhan, N.S., 456, 694, 1121, 1221, 1599).

Vitex negundo Linn. (Niragundi, Bana)

Description	– A shrub, with a thin, grey bark. Branchlets are four-angled, densely white tomentose. Leaves three to five foliolate, petiolate; leaflets lanceolate, coarsely cut and crenate; terminal leaflets quite long; lateral smaller and with much shorter petiolules. Flowers bluish-white to blue in terminal, paniculate racemes. Berry black, the size of a pea.
Flowering/Fruiting	– March-April and June-September.
Parts used	– Whole plant.
Current market rate	– Leaves Rs. 10/- per kg.
Main ingredients	– Fresh leaves yield 0.05% essential oil. Air dried, leaves yield an alkaloid and two new iridoidglycosides-nishindaside and negundoside have been reported. Bark yields flavone glycosides. Seed have been reported to yield vanitin and a mixture of unidentified alkaloids.
Uses	– In action, it is alterative, anodyne, antiarthritic, antiparasitic, appetiser, aromatic, astringent, cardiac demulcent, emmenogogue, expectorant, discutient, febrifuge and nervine tonic. It has been credited with curative effects in asthma, lung diseases, splenic enlargements, urinary troubles, tonsillitis, sciatica, rheumatism etc. Dried leaves are used to protect stored grains and the smoke of leaves repels mosquitoes. (Chauhan, N.S., 1988).
Preparations	– Nirgundi-kalpa and nirgundi-taila.
Cultivation notes	– The shrub can be propagated through cuttings and root suckers. The species has wide adaptability and can be cultivated best in moist situations, open-waste lands and near the deciduous forests.

***Withania somnifera* Dunal.**

Family	–	SOLANACEAE
Local/Common name(s)	–	Ashvagandha, Asgandh, Asugandh, Indian ginseng.
Distribution	–	The plant is found in the drier parts of India, ascending upto 1700 m in the Himalayas. In Himachal Pradesh, it is found sporadic to scattered in Una, Hamirpur, Kullu (Bhunter) Kangra, Mandi, Bilaspur (Bhararighat), Solan and Sirmour districts upto an altitude of 1400 m (Chauhan, N.S., 1165).
Description	–	The plant is an erect, branching perennial under-shrub upto 1.5 m high. All parts are nearly, more or less stellately tomentose and the branches are flexuose, terete and densely tomentose. Leaves are petioled, 5-10 cm long, ovate and sub-acute. Flowers are greenish or lurid yellow, usually about 5 together in sub-sessile umbelliform cymes. Berry is 0.5-0.7 cm in diameter, enclosed in the much enlarged, inflated, somewhat 5 angled, pubescent calyx and is red when ripe. Seeds are many enclosed in pulp.
Flowering/Fruiting	–	June-November.
Parts used	–	Roots
Current market rate	–	Rs. 50/- to Rs. 75/- per kg.
Morphology of the official parts used	–	Roots vary much in size and are light brown in colour and usually occur in small pieces, 10-17.5 cm long and 6-12 mm in diameter, uniform in appearance, smooth and tapering in young roots. Parenchyma contains starch grains, which decrease as the secondary growth starts. The young roots of the wild plants, occurring in the region have the same structure and character of 'Nagori' variety

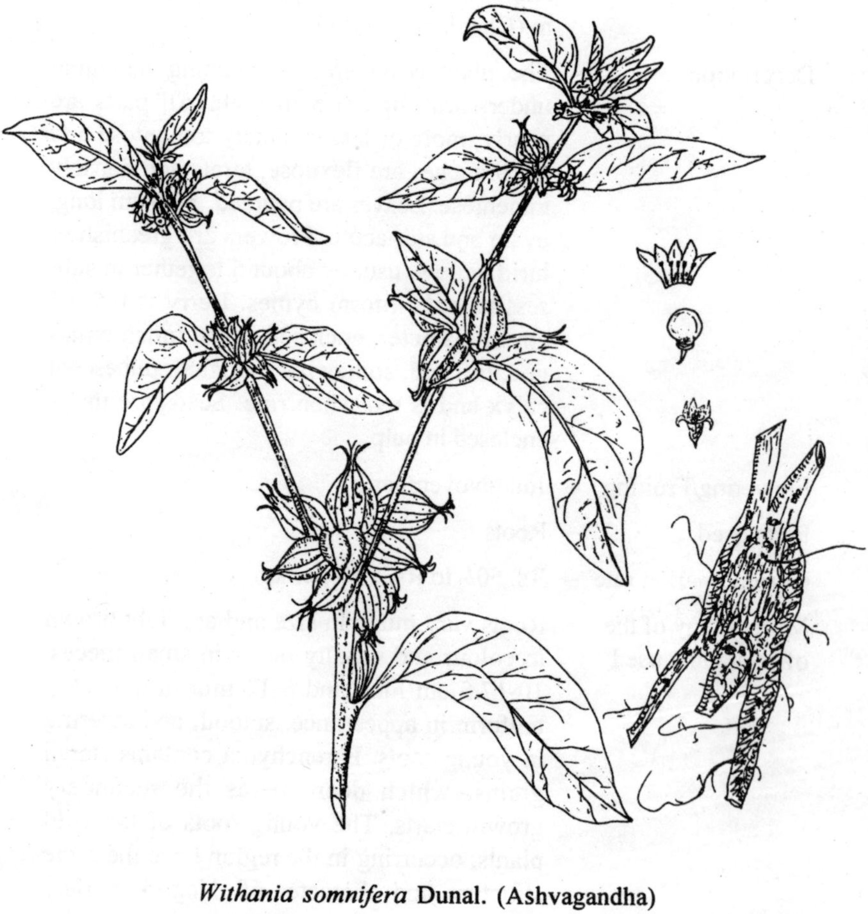

Withania somnifera Dunal. (Ashvagandha)

(growing in Nagore & Madhya Pradesh) i.e. before the secondary growth takes place.

Main ingredients – Ashvagandha contains an essential oil, ipuranol, a crystalline alcohol, whithaniol, hentriacontane, phytosterol and fatty oil. It also contains alkaloids withanine and somniferene (I.P.C.:253).

Uses – The roots are alterative aphrodisiac, tonic, deobstruent, diuretic, narcotic hypnotic, sedative, restorative and abortifacient. These are used in rheumatism, cough, debility from old age, dropsy, emaciation of children, consumption and general weakness. It is valued as a potent tonic, delays ageing, greying of hairs and provides physical as well as mental strength.

Preparations – Ashvagandha-ghrita, Ashvangandharishta, Ashvagandhahadi-pak, Ashwagandha-rasayana.

Cultivation notes – The plant is cultivated in soil that are unsuited for other crops and requires little care. Seeds are sown broadcast in the nursery, just before the onset of the rainy season. Five kilograms of seeds are sufficient to provide seedlings, for an hectare. When the seedlings are sufficiently tall, they are planted in a field at a distance of 30 cm × 5-7 cm row to row and plant to plant to achieve soft roots. Manuring and weeding are not required, Nitrogenous fertilizers promote heavy leafy growth with less yield of roots. Irrigation is hardly required. Excessive rainfall is harmful, as it leads to excessive weed growth. Plant flowers and fruits in December. Harvesting starts from January and continues till March. The entire plant is uprooted for roots, which are separated from the aerial parts by cutting the stem 1-2 cm above the crown. They are then transversely cut into smaller pieces for drying. Occasionally, the roots are dried whole. The yield of the roots is 500 kg per hectare when fresh and 150 kg per hectare, when dried. Yields upto 300 kg of dried roots are also occasionally recorded (The Wealth of India, Vol. X; 1976:582).

Woodfordia fruticosa Kurtz.

Family – LYTHRACEAE

Local/Common name(s) – Dhai, Dhataki, Dhai-ki-phool, Dhataki-pushpa, Woodfordia.

Distribution – Throughout the greater part of India, upto 1600 m in the Himalaya. In Himachal Pradesh, found in all the regions upto 1500 m in open dry slopes, rocky fields and waste-lands. (Chauhan, N.S., 822, 977, 1359, 1496, 1624, 1751, 2071, 2681).

Description – A large, much branched deciduous shrub with bark scaling off in small patches. Leaves are opposite or in whorls of 3, ovate, lanceolate. The flowers and their calyces are bright red, the latter are permanent and retain their colour even after the flowers have faded. Flowers are arranged in short, clustered panicles in the axils of fallen leaves on the old wood and appear before the leaves.

Flowering/Fruiting – March-June.

Parts used – Flowers.

Current market rate – Rs. 15/- per kg (Flowers).

Main ingredients – Flowers as well as practically the whole plant yields tannin upto 20 per cent. Flowers are valued for dyeing.

Uses – In action it is astringent, stimulant, suppurative and styptic. Useful in bowel complaints,

Woodfordia fruticosa Kurtz. (Dhataki, Dhai)

haemorrhages, diarrhoea and dysentery, complaints of the liver, leucorrhoea, menorrhagia, stimulant in pregnancy, skin diseases and vesicular eruptions and foul ulcers. Commercially the flowers are of considerable importance as a dyeing and tanning material (The Useful Plants of India, 1986: 690). Flowers are used in preparation of Asawas and Arishtas in Ayurvedic medicines.

Preparations – Dhatakyadi-churna, Dhatakyadi-taila, Laghu-gangadhar-churna, Kutajarista, Abhyayarista, Pippalayasava, Kanakasava, Asokarista. Much used for colouring and intoxicating various asava and aristas.

Cultivation notes – Propagation is through seeds and cuttings. Flowers are harvested during March.

Xanthium strumarium Linn.

Family – Compositae/Asteraceae

Local/Common name(s) – Adhasisi, Banokra, Chhota dhatura, Musha-ro-khor, Cocklebur, Burweed.

Distribution – Distributed throughout India. Widely distributed throughout the world in tropical and temperate region. Found as a weed in Himachal Pradesh in all parts upto 2500 m elevation (Chauhan, N.S., 938, 1013 and 2393).

Description – An erect, tall, coarse, unarmed weed with triangular, cordate or sub-orbicular, often three-lobed, irregularly toothed, scabrid or hispid leaves. Flowers numerous, white or green in terminal and axillary racemes with both male and female flowers on the same plant; male heads upper most, globose, about 0.75 cm in diameter. Female heads ovoid, 1.25 cm long with hooked bristles. Achenes

		obovoid, thick, enclosed in the hardencd, involucral bracts and clothed with strong hooked spines.
Flowering/Fruiting	–	April-November.
Parts used	–	Roots, leaves and fruits, seed oil.
Current market rate	–	Not available.
Chemical contents	–	The aerial parts of the plant contain a mixture of alkaloids which are said to be toxic i.e. the sesquiterpene lactones-xanthinin, its sterio-isomers-xanthinim and xanthatin. The seeds on solvent extraction yield 30-35% of semi-drying oil, resembling sunflower oil. It has the same taste as other vegetable oils.
Properties and uses	–	The herb is reputed as a medicine in Europe, China, Indo-China, Malaysia and America. The drug is credited with powerful diaphoretic properties. The dose of half to one ounce is recommended in chronic malaria, leucorrhoea and urinary diseases. It is also useful as a styptic and is considered a prophylactic against hydrophobia. In clinical experiments, its pollens have been found to cause asthma and rhinitis; also produce dermatitis in sensitive/susceptible persons especially during autumn when the plant is in pre-fruiting stage. The leaves are given internally in scrofula and herpes in powder form. The root is bitter tonic, useful in cancer and scrofula. Its extract is used locally over ulcers, boils and abscesses. The fruits are used as tonic, diuretic, diaphoretic, sedative, cooling and demulcent and are also given in small-pox. The seeds are used for resolving inflammatory swellings. The herb is reported to be used in snake bite also (Kirtikar and Basu, II: 1355-56; Chopra *et al.*, 1958: 438; Dastur, 252; Wealth of India, XI: 1).

Preparations – Earlier, it used to be used as an effective drug against malaria, before quinine was discovered. Used in various traditional formulations.

Cultivation – Can be cultivated through seeds which the plant produces in plenty.

Zanthoxylum armatum DC.

Syn: *Z. alatum* Roxb.

Family – RUTACEA

Local/Common name(s) – Tumbru, Tejphal, Tirmir, Tejbal, Timbri.

Distribution – Found in Punjab, Kumaon and Khasi Hills. In Himachal Pradesh, it is sporadic to scarce in distribution in slopes, nalahs, field borders and depressions below 2500 m elevation in all parts of the state (Chauhan, N.S., 921, 648, 661, 803, 864, 1455).

Description – An armed, scandent or erect shrub upto 6 m tall with dense foliage. Branches-armed and the prickles are flattened, upto 2 cm long. The bark is pale brown and deeply furrowed. Leaves are imparipinnate or trifoliolate, 5 to 23 cm long, often with lanceolate, entire to glandular, crenate, acute to obtusely acuminate leaflets. Flowers are greenish-yellow in dense, terminal and axillary, sparse panicles. Follicles are generally reddish, sub-globose, glabrous. Seeds solitary in a fruit, globose and shining black.

Flowering/Fruiting – April-June.

Parts used – The bark, fruits and seeds.

Current market rate – Rs. 15-20/- per kg.

Chemical contents – Dried fruits on steam distillation yield 2.3% essential oil which is a pale to olive-yellow

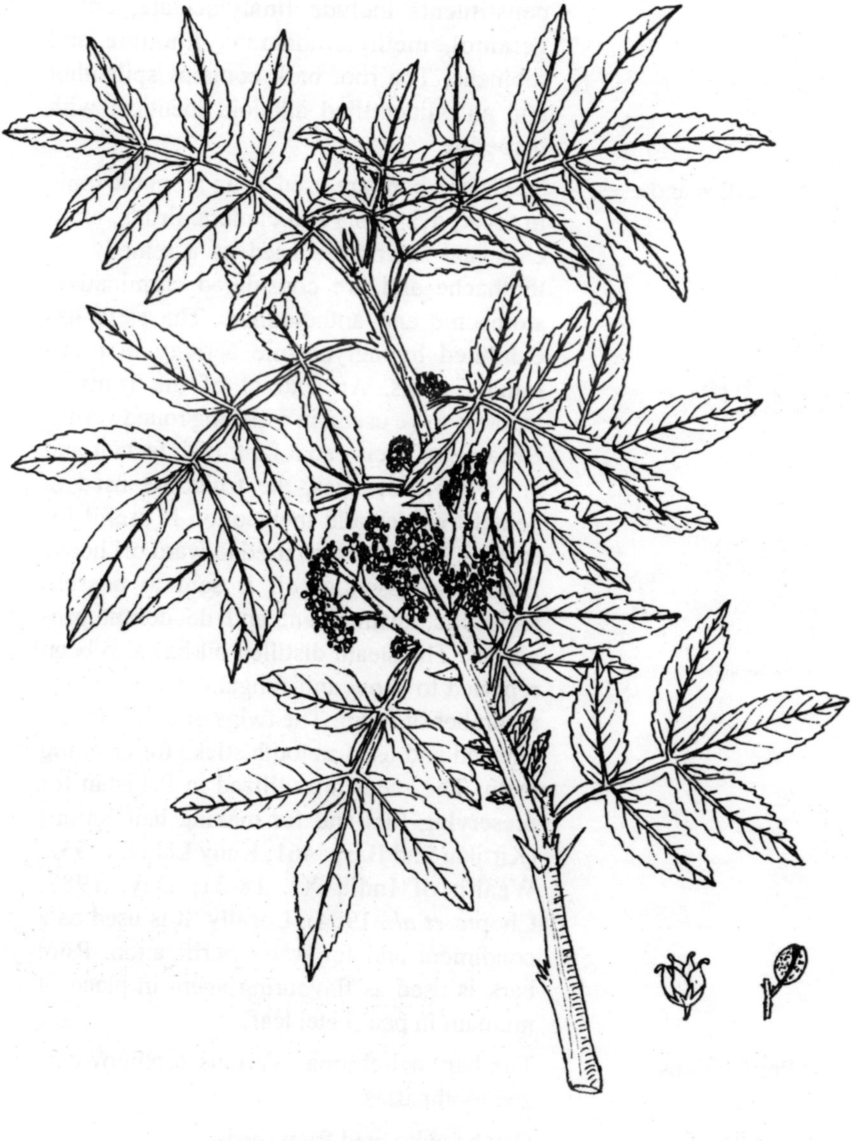

Zanthoxylum armatum DC.

coloured, mobile liquid with a spicy odour resembling cubebs (*Piper cubeba*). The chief content of the oil is linalool, while the other constituents include linalylacetate, citral, geraniol, methylcinnamate, linonine and sabinene. The root bark contains spilanthol and an unidentified alkaloid identical with berberine.

Properties and uses – Seeds and bark are used as an aromatic tonic in fever, dyspepsia and cholera. Fruits, branches and thorns are used as a remedy for toothache and are considered carminative, stomachic and anthelmintic. The stem has exhibited hypoglycaemic activity in preliminary trials. An extract of the fruits is reported to be useful in expelling roundworms. Because of their deodorant, disinfectant and antiseptic properties, the fruits are used in dental troubles and their lotion is useful for scabies. They are also used to ward off house flies. The essential oil is said to possess antiseptic, disinfectant and deodorant properties. The steam distilled oil has also been reported to show anti-fungal activity against a number of fungi. The twigs of branches are chewed and used as tooth sticks for cleaning teeth. The seeds are utilized in Pakistan for preserving furs and for making hair lotions (Kirtikar and Basu: 461; Kany Lal Dey: 332; Wealth of India XI: 18-21; Dey: 1989; Chopra *et al.*, 1954). Locally, it is used as a condiment and for water purification. Root bark is used as flavouring agent in place of mulhatti in pan (betel leaf).

Preparations – Tumbarwadi-churna, Various toothpowders and toothpastes.

Cultivation – Can be cultivated from seeds.

1. *Achillea millefolium*

2. *Aconogonum rumicifolium*

3. *Allium carolinianum*

4. *Aloe barbadensis*

5. *Androsace rotundifolia*

6. *Angelica glauca*

7. *Arctium lappa*

8. *Arnebia euchroma*

9. *Arnebia guttata*

10. *Artocarpus lakoocha*

11. *Asparagus adsendens*

12. *Atropa belladonna*

13. *Bauhinia variegata*

14. *Berberis asiatica*

15. *Berberis lycium*

16. *Bergenia ciliata*

17. *Bergenia stracheyi*

18. *Boerhaavia diffusa*

19. *Bombax ceiba*

20. *Bryonopsis laciniosa*

21. *Bunium persicum* (complete plant showing bulb to flowers)

22. *Butea monosperma*

23. *Calotropis gigantea*

24. *Capparis spinosa* (flowers)

25. *Capparis spinosa* (fruits)

27. *Centella asiatica*

26. *Celastrus paniculatus*

28. *Centratherum anthelminticum*

29. *Codonopsis ovata*

30. *Codonopsis ovata* (showing fleshy, milky roots)

31. *Colchicum luteum*. 32. *Corydalis govaniana*. 33. *Costus speciosus*

34. *Crocus sativus*

35. *Dactylorhiza hatagirea*

36. *Datura fastuosa*

37. *Digitalis lanata*

38. *Digitalis purpurea*

39. *Dioscorea deltoidea*

40. *Dioscorea deltoidea* (specimen showing rhizomes)

41. *Eclipta alba*

42. *Emblica officinalis*

43. *Ephedra gerardiana* (shrub in fruits)

44. *Evolvulus alsinoides*

45. *Fraxinus xanthoxyloides*

46. *Gentianna Kurroo*

47. *Geum elatum*

48. *Ginkgo biloba*

50. *Heracleum candicans*

49. *Gloriosa superba*

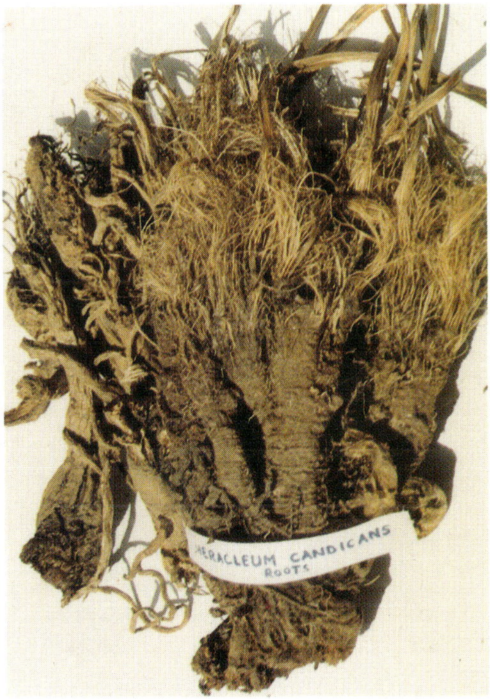

51. *Heracleum candicans* (root stock)

52. *Hippophae rhamnoides* (red fruited type)

53. *Hippophae rhamnoides* (orange fruited type)

54. *Hippophae salicifolia*

55. *Holarrhena antidysenterica*

56. *Humulus lupulus*

57. *Hyssopus officinalis*

58. *Jasminum humile*

59. *Juniperus communis*

60. *Juniperus macropoda*

61. *Jurinea dolomiaea*

62. *Malaxis muscifera*

63. *Malaxis acuminata*

64. *Matricaria chamomilla*

65. *Meconopsis aculeata* (in front and cobra plant in the background)

66. *Morina longifolia*

67. *Mucuna pruriens*

68. *Myrica esculenta*

69. *Nardostachys grandiflora*

70. *Ocimum sanctum*

71. *Oroxylum indicum*

72. *Paeonia emodi*

73. *Physochlaina praealta*

74. *Picrorhiza kurrooa*

75. *Pistacia integerrima* (showing fruits and galls)

76. *Plumbago zeylanica*

77. *Podophyllum hexandrum*

78. *Polygonatum verticillatum*

79. *Prinsepia utilis*

80. *Pyracantha crenulata*

81. *Rauvolfia serpentina*

82. *Rheum australe*

83. *Rheum australe* (showing roots)

84. *Rheum moorcroftianum*

85. *Rhododendron arboreum*

86. *Rhododendron campanulatum*

87. *Roscoea alpina*

88. *Salvia moorcroftiana*

89. *Saussurea costus*

90. *Saussurea gossypiphora*

91. *Saussurea obvallata*

92. *Semecarpus anacardium*

93. *Solanum khasianum*

94. *Solanum surattense*

95. *Spilanthes acmella*

96. *Taraxacum officinale*

97. *Taxus wallichiana*

98. *Thermopsis inflata*

99. *Thymus serpyllum*

100. *Trillidium govanianum*

101. *Typhonium diversifolium*

102. *Valeriana jatamansi*

103. *Viola serpens*

104. *Viscum album*

105. *Waldhemia glabra*

106. *Woodfordia fruticosa*

Collection and Marketing of Plants (Past and Present Scenario)

In Himachal Pradesh, the collection and trade of crude drugs is disorganised and the state has no market of its own. Village people collect the drugs and sell these to the local shopkeepers in return for the items of their domestic needs. These village shopkeepers in fact act as agents of a few business parties from Delhi, Amritsar, Bombay, Dehradun etc. In exchange for the essential commodities, the villagers sell drug plants at throw away prices. In a bid to earn the maximum, merciless exploitation of the natural resources is in evidence resulting in depletion of many valuable herbs as listed in Appendix 3. This system is resulting on one side in exploitation of the rural poor and on the another side it is cutting deep into the state revenue receipts. Though, in Himachal Pradesh, the collection, extraction and marketing of crude drugs and their products was hitherto regulated under the provisions of the Forest/Revenue settlement of erstwhile states or districts, yet these were not uniform in all the forest divisions. In Chamba and Mandi, the collection and export of medicinal and aromatic plants was regulated under the acts i.e. **The Chamba Minor Forest Produce Exploitation and Export Act 2003 Bikrami, and Mandi State Forest Produce Exploitation and Export Act, 1977 Bikrami.** Export royalty was prescribed for certain items under these Acts. In other parts of the state, collection was done on right holding basis and export permit was sanctioned by the concerned Divisional Forest Officer of the State Forest Department at a paltry sum of Rs. 5/- per item irrespective of the quantity. The collected material was and is still stored haphazardly and marketed in plains, where the material is processed for drugs, extracts, powders and various other formulations.

Now, the Government of Himachal Pradesh has notified vide notification No. Fts (A) 3-1/77, dated Shimla-2 the 9th March, 1994, new amendments in regulating the trade and marketing of medicinal and aromatic plants in the state. These rules shall be called—**The Himachal Pradesh Forest Produce Transit (Land Rules) Amendment Rules 1994.** In the amendment of rule 5, 6, 10, 11, 18 and 20 of the Himachal Pradesh Forest (Land Rules) after the words "any forest produce", but before the words "that does not bear, the words and signs" other than fuel wood, khair wood, bamboos, charcoal, medicinal plants and seeds "shall be inserted". Amendment of rule 11 says that the issuing authority shall also determine the other conditions subject to which the pass shall be issued and shall also determine the period for which the pass shall remain valid. However, the validity of the pass shall not under any circumstances exceed a period of six months including any extension allowed. A fee of Rs. 25/- shall be leviable for the issue of such a pass except in case of medicinal plants for which the fee shall be charged as shown in the ensuing Table. As per amendment of rule 18:—No person shall offer any forest produce for export by rail, on any railway station or by post at any post office or by air on any airport within Himachal Pradesh, unless a pass has been issued under these rules and no railway, postal, airport authority shall accept for transport/ transmission, any forest produce by rail, post or air, unless it is accompanied by a valid pass. Fees at varying rates shall be leviable for the issue of such a pass as indicated in the succeeding Table. Certain items which have been categorised as endangered have been restricted/banned for collection and trade. At this juncture, it would not be out of place to present the amended Atc:

The medicinal plants are mostly found in the forest and are included in the forest produce under section-2(4) of the Indian Forest Act which defines the forest produce as under:

"Forest produce" includes the following:

a) whether found in, or brought from a forest, that is to say, timber, charcoal, caoutchouc, catechu, wood oil, resin, natural varnish, bark, lac, mahua flowers, mahua seeds, kuth and myrobalans, and
b) when found in, or brought from a forest, that is to say:
   1) tree and leaves, flowers and fruits, and all other parts of produce not herein before mentioned, of trees,

2) plants not being trees (including grass, creepers, reeds and moss), and all parts of produce of such plants,
3) wild animals and skins, tusks, horns, bones, silk, cocoons, honey and wax, and all other parts or produce of animals, and
4) peat, surface soil, rock and minerals (including lime stones, laterite and mineral oils, and all products of mines or quarries).

The useful part in several medicinal plants is the root portion and unscientific extraction thereof seriously affects the natural regeneration of the plant in forest areas. The forest department permits the export of medicinal and aromatic plants and issues export permits after realising the license fee specified for each item. Many items which are not included in the list are illegally exported causing a great loss to the state ex-chequer on one hand and the biodiversity on the other.

The licencees of such forest produce are generally petty contractors and employ illiterate villagers for extracting useful parts of medicinal plants who are not aware about the process of regeneration and the need to protect it. The pressure on the forest and continuous increase in demand coupled with unscientific exploitation has threatened the survival of many drug plants and several species have become endangered also.

Medicinal and aromatic plants are so far collected following the traditional methods without adhering to the pharmacopoeial standards laid down for the collection of different drugs and their parts. Thus, no control is kept on the quality of the material. Plants are collected as and when the collectors are free from their farming and other activities or any such time when there is a demand from the market. The collection of underground parts like roots, rhizomes, tubers etc. is done by digging out the entire plant by means of *kuti, khilni* or *kudali* as may be available from July-August to March-April. In certain cases, the collectors have been busy collecting these herbs during summers i.e., the growing season, without caring for its survival and regeneration. By doing so, on one side, the collection of immature and undeveloped parts/plant results into a product of inferior quality lacking proper active ingredients and on the other side, the life cycle of the plant is broken as it is not allowed to bear flowers, set fruits

Notification of 17.08.1993, HP Forest Produce Transit (Land Routes, Rules, Ordinary) 1993-Amendment (Rule 11(5) of the Act)

Botanical name	Local name	Pass/export permit fees (in Rs./quintal)
Abies webbiana Lindl.	Talis patra	85.00
Aconitum heterophyllum Wall.	Patish, Atish (Karvi patish)	7,500.00
Aconitum violaceum Jacq. ex Stapf	Patish (Mithi)	1,500.00
Acorus calamus L.	Bach, Bare	130.00
Adiantum lunulatum Burm	Dungtuli, Hansraj	80.00
Ainslaea aptera DC.	Sathjalari	50.00
Angelica glauca Edgew.	Chora	125.00 (Restricted/banned)
Artemisia brevifolia Wall. ex DC.	Senski	50.00
Atropa acuminata Royle ex Lindl.	Belladona (Jharka)	60.00 (Restricted/banned)
Berberis aristata DC.	Rasaunt	500.00 (Restricted/banned)
B. asiatica Roxb.	Rasaunt	500.00 (Restricted/banned)
B. lycium Royle	Kashmal	500.00 (Restricted/banned)
Bunium persicum (Boiss.) Fedts.	Kalajira	2,000.00
Dactylorhiza hatagirea (D. Don) Soo	Salam panja	6,000.00
Dioscorea deltoidea Wall. ex Kunth.	Singhli-Mingli	900.00 (Restricted/banned)
Ephedra gerardiana Wall. ex Stapf.	Somlata	50.00
Girardinia heterophylla Decne.	Bichhu buti	150.00
Hedychium acuminatum Rosc.	Kapoor kachri	70.00
Heracleum candicans Wall.	Padara, Patrala	25.00
Hyoscyamus niger L.	Khurasani ajwain	150.00

Contd.

Botanical name	Local name	Pass/export permit fees (in Rs./quintal)
Jurienea dolomiaea Boiss	Dhoop	500.00
Lichens (Parmelia sp.)	Chharila	225.00
Morchella esculenta (Linn.) Pers.	Guchhi	10,000.00
Nardostachys grandiflora DC.	Jatamansi	690.00
Picrorhiza kurrooa Royle. ex Benth	Karoo	540.00 (Restricted/banned)
Pinus wallichiana A.B. Jackson	Kail cones	1,000.00
Pistacia integerrima Strew. ex Brandis	Kakarsinghi	1,000.00
Podophyllum hexandrum Royle	Bankakri	450.00 (Restricted/banned)
Polygonatum verticilatum (L.) Allioni	Salam Mishri	1,000.00
Potentilla nepalensis Hook.	Dori	40.00
Rheum australe D. Don.	Rewand chini	110.00
Rhododendron companulatum D. Don.	Kashmiri Patha	150.00
Salvia moorcroftiana Wall. ex Benth.	Thuth	180.00
Saussurea costus Clarke	Kuth	300.00
Selinum vaginatum (Edgew.) Clarke	Bhutkesi	400.00
S. tenuifolium Wall. ex Clarke	Muramansi	400.00
Swertia chirayita (Roxb.) Karsten	Chirayata	700.00
Taxus wallichiana Zucc.	Birmi, Talispatri	600.00
Thalictrum foliolosum DC.	Mamiri	335.00
Thymus serpyllum Linn.	Banajwain	100.00
Tinospora sinensis (Lour.) Merr.	Giloye	100.00
Valeriana jatamansi Jones.	Mushak bala	590.00
V. hardwickii Wall.	Nihani	300.00
Viola serpens Wall. ex Roxb.	Banafsha	2,250.00

and shed seeds leading to loss in future regeneration in their natural habitats. As a matter of rule, any plant having underground part, i.e. root, rhizome, tuber, corm and bulb as official part must be harvested when these have borne fruits, shed seeds and aerial parts have started withering. This way before uprooting the plant, the seed shed has taken place, which produces plant, crop in the coming season through natural regeneration. Secondly, the active chemical constituents are also on the higher side at such a dormant stage, when the vegetative growth has ceased and the entire contents are concentrated in the underground parts. Summer collection not only results in breaking the life cycle of these plants, but also in yielding poor quality material.

Drugs like dhoop, karu, dioscorea, atish, salampanja, chora, rhubarb, meda mahameda, valerian etc. fall in this category. Banaksha (*Viola serpens*) and allied spp. are collected by shepherds, graziers and *gwalas* during March-April along with guchhi (*Morchella esculenta*) mostly with entire plant instead of plucking the flowers and leaves only. Leaves of tejpatra (*Cinnamomum tamala*) are collected in winters by ruthlessly cutting the branches, and the fruits of amla, harad and bahera too are collected haphazardly without waiting for the proper maturity stage and often by cutting the branches thereby reducing the fruit-bearing capacity of the trees during the succeeding years. Though people have acquired near expertise in the art of collection, yet owing to the greed of contractors and ignorance of collectors about the quality, the norms are frequently flouted.

Scientific Method of Collection

The following rules prescribed for the collection of different official parts from medicinal and aromatic plants described in Ayurvedic literature and supported by scientific studies should be strictly adhered to:

1. Roots, tubers, rhizomes and barks (root/stem) to be collected during the autumn or early spring (dormant phase).
2. Leaves and flowering tops to be collected at a time when flowers have fully developed and before the maturity of fruits and seeds.
3. Flowers to be collected prior to or just at the time of pollination.
4. Fruits to be collected when fully grown, but unripe.

5. Seeds are collected when fully mature and if possible before the fruits have opened/dehisced.
6. Seed like fruits of coriander, fennel, cumin etc. are to be collectcd at fully developed but green stage to maintain freshness and appeal in the market.

In Himachal Pradesh a nodal agency is required to be established to harness this valuable wealth systematically and judiciously with adequate measures for conservation, protection and production on sustainable basis.

As per records of the state forest department, Govt. of Himachal Pradesh, the quantity of major medicinal plants exported from Himachal Pradesh has declined from 94583 quintals in 1991-92 to 18226 quintals in 1994-95, indicating a fast decreasing trend that comes out to be approximately five times less during 1994-95 as compared to the outturn in 1991-92 (Appendix 2). This gives a fair trend of decrease in the availability of the raw material from the natural resources, indicating thereby the loss to the existing herbal wealth which is an alarm and calls for immediate attention of the scientists, administrators and the public at large to impress upon the Government to devise ways and means for their protection, production and utilisation on rational and sustainable basis.

Cultivation and Conservation of Medicinal Plants used in Indian Systems of Medicines (ISM) and Homoeopathy

Use of plants for medicinal purposes has been known throughout the world from times immemorial. The holistic approach of the ancient systems of medicines has come to be appreciated more and more in the present day when harmful effects caused by administration of drugs of the modern system of medicine are receiving increasing attention of the scientists and the public at large. Herbal remedies with a greater emphasis on curative, promotive, and preventive aspects of health care are the legacies of our ancient sciences which are still to be fully re-discovered for the benefit of the mankind.

About 80 per cent of the raw material for the drugs used in Ayurveda, Unani, Sidha and Homoeopathy are based on plant products. The plants are known in various states, regions and languages by various names. The correct identity of these plants is important for the manufacture of the drugs, galenicals and their use as single drugs. The very credibility of the Indian Systems of Medicines depends on the availability of genuine and authentic raw material for the preparation of standard drugs in accordance with the proportions and procedures laid down in the texts of these systems. In India, around 5,000 plant species have been attributed with medicinal properties. But in all, about 1,100 medicinal plants have been estimated to find regular use in Indian Systems of Medicines. Of these too, 500 medicinal plants are most commonly used in the preparation of Ayurvedic, Unani and Homoeopathic drugs. There are around 7,000 pharmacies of Ayurveda, Sidha, Unani and Homoeopathic for manufacturing the drugs in the country. In addition, many *vaids* and *hakims* prepare their own medicines for dispensing to their patients. The increase in demand for ISM and Homoeopathic drugs has, thus, created problem of supply and availability of sufficient quantities of medicinal plants.

There is no reliable data regarding the quantitative inventory of the medicinal plant's raw material available for use and needed for future requirements. The main source of this raw material has been the forest areas of the country where these are collected as minor forest produce mostly by tribals and local people with whom the rights of collection are vested. The unsystematic and relentless exploitation for many centuries to meet the growing demand of these pharmacies has proved a suicidal trap. Many valuable plants have become vulnerable to extinction in the absence of developmental programmes. The regeneration, protection and preservation of the medicinal plants is a serious challenge for restoring our biological heritage. The strict legislation to regulate the exploitation of our herbal wealth, cultivation, protection and conservation of high-yielding strains of important drug species is, thus, a necessity to provide a sustainable supply of crude drugs on one side and to ease the pressure on natural resources on the other.

Why Cultivate Medicinal Plants?

Medicinal plants have to be cultivated because of the following factors:

- To ease pressure on natural resources.

- To make available fresh, genuine and quality raw material for manufacturing of standardised and efficacious drugs.
- To evolve better strains and high-yielding crops of medicinal plants through improvement programmes and tissue culture techniques.
- To check the use of spurious substitutes and adulterants which have resulted in the deterioration of the standard of drugs of Indian Systems of Medicines.
- To standardise the collection, storage and postharvest technology for medicinal plants.
- To provide regular and alternative source of income to the farmers for amelioration of their economic conditions.
- For introduction and domestication of useful exotic drug plants to minimise import and maximise export.
- To conserve the biological and genetic diversity in medicinal plants for the posterity.

Easing Pressure on Natural Resources

Our forests till date continue to provide major share of raw material to the pharmaceutical industries in India more particularly the sectors manufacturing Ayurvedic, Unani, Sidha and Homoeopathic medicines. Recent clinical reports on *Rauvolfia, Tylophora, Guggulu, Squill, Chirayita, Valeriana, Withania, Spikenard* etc. have brought our forests into world focus. Plant-based drugs have created a sweeping resurgence in the west in their favour and the demand for these has reached a scale never witnessed before. This ever increasing demand for plant drugs has resulted in indiscriminate and avaricious collections from the forests thereby seriously depleting the existing resources. Not only is our country losing revenue because of disappearance of many drug plants, but in the process humanity is also being deprived of the valuable genetic stock of these plants which the mother earth has built up and evolved during the course of centuries. Jain and Sastry (1980) have listed 134 species as endangered plants in India of which 16 belong to medicinal plants of north-west Himalaya. Gupta and Sethi (1981) have identified 8 medicinal plants as endangered, 12 as vulnerable owing to over-exploitation and 8 species as rare, being of very limited distribution. Chauhan (1988) has enumerated 20 species of medicinal plants as endangered/threatened for Himachal Pradesh. Examples of such valuable drug plants for north-west Himalaya are: *Aconitum heterophyllum, Arnebia benthami, A. euchroma, Atropa acuminata, Colchicum luteum, Delphinium vestitum, Dioscorea deltoidea, Ephedra gerardiana, Inula racemosa, Heracleum candicans, Physochlaina praealta, Gentiana kurroo, Nardostachys grandiflora, Onosma bracteatum, Dactylorhiza hatagirea (Orchis latifolia), Paeonia emodi, Podophyllum hexandrum, Rauvolfia serpentina, Gloriosa superba, Costus speciosus, Rheum australe, R. nobile, R. tibetianum, Roscoea alpina, R. procera, Malaxis muscifera, M. wallichii, Saussurea sacra, S. bracteata, Swertia chirayita, Berberis aristata* and *Skimmia laureola*. Similarly *Coptis teeta* is endangered in the eastern Himalayan region. Many others, like Dhoop, Kutki, Tejpatra, Vankasha, Mamira, Tejbal, Bhutkesi, Tagar, Lodhra etc. are entering fast into the threatened list.

It is, thus, a high time that concerted efforts be undertaken to protect, propagate and cultivate these valuable species on large scale. Rich, natural habitats of these herbs have to be identified and declared as 'Biosphere Reserves' with full protection. This reserve can be for a single species or for multiple species depending upon their occurrence in respective areas. In our western Himalayan region, model reserves for *Ephedra gerardiana, Artemisia brevifolia, Arnebia euchroma* (Ratanjot) can be selected in Pooh and Spiti sub-division of Himachal Pradesh. Similar localities can be identified for other species in other parts of the region/country. The *gene pool* protected in these reserves can be utilised for introduction and plantations in new area. Simultaneously, demonstration plots, drug farms and a chain of herbal gardens have to be established for large scale cultivation of the valuable drug plants. Forest departments of the states should encourage the plantation of important drug plants like Neem, Kuchla, Guggulu, Kapur, Tejpatra, Jamun, Shigru, Katphal, Amla, Harad, Bahera, Arjun, Bilve, Karanj, Gambhari, Sirish, Lakuch, Padhal etc. in their plantation programmes. Integration of medicinal species in social forestry, agroforestry and cropping patterns of our farmers will lease a new life to this valuable resource. Priority has to be given to such plants which are either getting extinct or are in short supply.

Availability of Genuine and Quality Raw Material

Because of prevailing confusion and controversies in identification of the pharmacopoeial plant drugs coupled with unsystematic and unscientific collection followed by defective storage and preservation procedures, many a times spurious, adulterated and degraded raw material is utilised for the manufacture of ISM and Homoeopathic drugs. This not only retards the quality of the products, but also adversely affects the status of the very system. There is, thus, neither check on genuineness and quality of the material nor on the prices charged by the practitioners and traders. Most of the times, a great difficulty is also faced by pharmacies in getting adequate supply of drugs.

In order to overcome the maladies being experienced at present by our pharmacies, cultivation of medicinal plants is indispensable. Cultivation involves several checks at every point, i.e. use of the genuine propagating material, optimum use of fertilisers, proper spacings, adequate protection from diseases and pests. The chemical constituents have to be evaluated at different stages of growth to find out the right part of the plant to be used and the right time of collection of different parts of a plant to provide efficacious and standard material for the drug industry.

This venture will, thus, evolve methodology to be adopted regarding their cultural conditions, manurial treatments, method of propagation, harvesting and indicating the best period when they are richest in active constituents and also the storage and preservation techniques before the raw material is finally marketed for consumption. Thus, the standardisation in basic raw material will facilitate manufacturing of good quality products inviting heavy demands from the international markets.

Evolving High-Yielding Varieties

Under cultivation through improvement programmes like grafting, selection, hybridisation, induction of mutagenic variations by radiations and chemical substances, induction of polyploidy through colchicine treatment, many high yielding strains of medicinal plants can be developed. *Cinchona ledgeriana* which is richest in quinine content requires virgin soil for plantation. Same old field can not be utilised for raising this species again. It was also prone to fungal attack which destroyed its roots. The seedlings of *C. ledgeriana* were grafted on the rootstock of *C. succirubra* which is a hardy species and resistant to fungal attack and the problem has been overcome (Mehra, 1986). Selection methods can be used to select high-yielding strains for successive cultivation in *Urginea, Dioscorea, Ipomoea hederacea, Colocynth, Ephedra gerardiana, Artemisia brevifolia, Nux vomica, Camellia* and various others. Hybridisation studies can be done between *Ephedra sinica* (richest in Ephedrine) and *E. gerardiana, E. sinica* and *E. intermedia* (a more robust plant) to produce high ephedrine content on one hand and robustness on the other. Similar studies can be done in *Solanum* spp. for producing solasodine rich strains. Induction of polyploidy can be used to produce tall plants with more biomass and high active contents. Thus, under cultivation not only can the bulk of the produce be increased, but the production of active ingredients can also be enhanced. The application of tissue culture technique to produce large scale quality material for commercial cultivation of valuable and endangered species will prove a most valuable tool in this direction.

Standardisation and Propagation of Potent Substitutes

India has a large number of plant species which are equally potent in contents and efficacy as the exotic species mentioned in Indian, British, Homoeopathic and other official pharmacopoeias. These can be used as substitutes to check the exports e.g. *Lobelia nicotianaefolia* for *L. inflata, Urginea indica* for *U. scilla, Gentian kurroo* for *G. lutea, Valeriana jatamansi* for *V. officinalis, Ipomoea turpethum* for *I. purga, Plantago ovata* for *P. psyllium, Ephedra gerardiana* for *E. sinica, Artemisia brevifolia* for *A. cina* and *Aconitum* sp. for *A. napellus*.

These species can be cultivated to reduce the expenditure on their import and a sizable foreign exchange can be saved. There are still a large number of unworked plants which can further supplement the resource after thorough investigations.

Standardisation of Collection, Storage and Postharvest Technology

Cultivation is followed by harvesting, collection, storage and marketing of the produce. Collection and preservation of official parts of a drug plant are the important aspects on which depends the entire process of quality products. It is, thus, to be done on scientific lines through skilled persons at appropriate season. The standards have been laid down for collection of different parts of plants in different seasons in different pharmacopoeias.

After the material is collected either from forest or farmers fields/ herb farms, it has to be dried under optimum conditions, garbled, graded and packed into suitable air-tight containers and stored in appropriate places. It is said that plant drugs have to be used fresh and invariably within a year of collection, otherwise they lose their efficacy. By the use of solar dryers, solar kilns, improved packing materials and air conditioned storages, the material can be dried fast and stored effectively for a considerable period of time before it is transported to the consumer industry. This work can only be handled by an independent agency adequately funded by the state or the Union Government.

Introduction and Domestication of Exotics

Many useful exotic plants have been introduced and domesticated successfully in the country. Common examples are *Belladona, Lavender, Rosemary, Salvia officinalis, Pyrethrum,* Periwinkle/ Sadabahar (*Catharanthus roseus*), *Datura inoxia, Hyoscyamus niger, Digitalis, Dubosia myoporides, Ammi majus, A. visnaga, Solanum laciniatum, Glaucium flavum, Glycyrrhiza glabra* etc. India possesses diverse agro-climatic conditions ranging from tropical to dry alpine nature. Plants from any region of the world can successfully be cultivated here in respective conditions. The requirement of such herbs which is so far met from exports can be met internally and the foreign exchange thus saved can be used for other developmental purposes.

Conservation of Biological/Genetic Diversity

Conservation has been defined as the management of resources of the biosphere for the benefit of all life including human, so that sustainable benefit may be derived by the present with potentials to meet the needs and aspirations of the future generations. Biological diversity refers to the richness of species i.e. number of species in a community of living systems while genetic diversity is a heritable diversity within and between the species of a genus. Basic rationale behind the conservation of biological diversity is to ensure sustainability in the utilization of species and ecosystems by maintaining essential ecological processes in the life-support system. Indian region is the centre of diversity of 152 economic species including medicinal species like Cardamoms, Nutmeg, Dioscorea, Pepper, Ginger, Rhododendrons, Jasmines, Betel leaf etc.

The concept of biosphere reserve is a method wherein specific ecosystems are conserved and managed for posterity. The germ plasm collection in the herbal gardens/drug farms of valuable medicinal species can be a major tool in conserving genetic/biological diversity. It will provide the basic material for future cultivation of desired plants.

Government's Efforts

It is estimated that the world demand for medicinal and aromatic plant products is growing at the rate of 7 per cent per annum and at this trend India's export by 2000 AD would be of the value of Rs. 3,500 million. The demand of these plant products in the home market may be 3-5 times of the export figure at raw material level. Thus, considering the demand for export and material use, the annual growth rate in the production of medicinal and aromatic plants had been fixed at 10 per cent during the 8th Plan.

The world market for herbal medicines is fast increasing with the increasing faith of the people, particularly in western countries towards holistic systems of medicine like Ayurveda, Unani, Sidha, Chinese, Tibetan and Yoga. It is reported that although US spends Dollar 800 billion per year on health care, yet, every third American is suffering from chronic disease. This has attracted people in the developed western countries towards alternative systems and it is estimated that one out of three Americans now uses alternative medical treatments such as Ayurveda, acupuncture and faith healing. The US spends over dollar 25 billion on these therapies. In the UK, Canada, Germany and Netherlands too, alternative medicine is booming.* It is also estimated that world trade in medicinal plants is over US dollar 60 billion and in India it is estimated to be of the order of Rs. 1,000 crores per year. It is predicted that world trade in medicinal plants would rise to US dollar 5 trillion (approximately Rs. 175 lac crores) by the year 2050 AD.** It is apprehended that with the commercialisation of medicinal plants, there could be a great pressure on herbal wealth of the state like Himachal Pradesh, which is not only a repository of several valuable medicinal plant species, but also a cradle of several systems of medicines including Ayurveda, Tibetan and other traditional systems basically based on plant medicines. Thus, the next 50 years are not only a challenge, but also a great opportunity for India. However, the present share of India in the world trade in the medicine is hardly 2% and other Asian countries are far ahead. China is controlling about 80% of the world trade. A major policy intervention is, therefore, urgently needed to encourage and facilitate investments into comprehensive cultivation of medicinal plants and production of high quality herbal drugs.

Keeping the significance being attached to medicinal and aromatic plants, the Ministry of Agriculture, Government of India has embarked upon a plan to develop the medicinal and aromatic plants in the country. The state agricultural universities and institutes under Council of Scientific and Industrial Research have evolved a number of high-yielding varieties with better quality traits. Under this Central Sector Scheme, the Ministry has set targets for development and cultivation of medicinal and aromatic plants under the following categories.

During the 9th Plan period, the Ministry of Agriculture and Co-operation is implementing all components of the above scheme on year to year basis.

The Ministry of Health and Family Welfare, Government of India too, is giving grant to various institutions for the cultivation of

---

* Sunday Review December 29, 1996.

** World Bank Report, 1996.

Development programmes/components for implementation

Objective	Physical target (Rs. in lac)	Financial outlay (Rs. in lac)
A. Aromatic plants		
Production and distribution of quality planting materials	230	80.50
Financial assistance for setting up modern distillation units	126	63.00
B. Medicinal plants		
Establishment of herbal gardens	60	129.00
Establishment of nursery centres	16	44.50
Establishment of demonstration-cum-seed production	2,100	31.50
Setting up of Regional Analytical Laboratories for quality testing	5	75.90
Staff and infrastructure	–	75.60
Total	2,537	500.00

medicinal and aromatic plants especially for such species which are in great demand by the Indian Systems of Medicines. A list of 45 such species has been circulated for the purpose. Ministry of Environment and Forests, Govt. of India has undertaken a co-ordinated programmes to propagate/multiply endangered species of medicinal and aromatic plants through the use of bio-technology. The departments of Science and Technology/Bio-technology are also funding such activities. Publication and Information Directorate (PID) of Council of Scientific and Industrial Research (CSIR), New Delhi is maintaining data base on medicinal and aromatic plants of the country. National Bureau of Plant Genetic Resources (NBPGR), New Delhi, Central Institute of Medicinal and Aromatic Plants (CIMAP), Lucknow, Tropical Botanical Gardens and Research Institute (TBGRI), Parode, Thiruvanantapuram, Kerala, have initiated steps for long term preservation of germplasm of medicinal and aromatic plants. Foundation for Revitalisation of Local Health Traditions (FRLHT), Bangalore is also engaged in *ex situ* conservation and preparation of data base for agro-technology of medicinal and aromatic plants especially for three state i.e. Tamil Nadu, Kerala and Karnataka. Jawahar Lal Nehru Herbal Garden at Pune is maintaining 320 species of medicinal and aromatic plants in its 8 hectare farm. Pioneer institutions like, Central Drug Research Institute, Lucknow, Regional Research Laboratories under CSIR, various Medical/ Ayurvedic/Homoeopathic/Unani/Tibetan/Sidha, research institutions and universities are also engaged in researches on different aspects of medicinal and aromatic plants.

These programmes have been undertaken with the objectives of collection, cultivation, multiplication, and preservation of the valuable medicinal and aromatic plants so as to meet the demands of the Pharmaceutical industries on a sustained basis and also to provide scientific, technological, pharmacological, toxicological, clinical and therapeutic inputs in a bid to bring the efficacy of medicinal and aromatic plants of the country into world focus.

State Policies

Himachal Pradesh has been a natural habitat of a large variety of medicinal plants, some of which are on the verge of extinction due to unscientific exploitation from the wild to meet the ever increasing demand of medicinal herbs. As per the state industrial policy, medicinal plants have been kept under first priority list and incentives have been provided to encourage entrepreneurship development in the state. The HP Khadi Board provides loans for collection, marketing and processing of medicinal plants ranging from Rs. 75,000/- to Rs. 1,000,000/-. HP State Science, Technology & Environment Council has sanctioned small schemes to the NGOs functioning in Himachal Pradesh for documentation of traditional knowledge on medicinal plants.

The department of India Systems of Medicine has proposed an action plan (Negi, 1997) with the following objectives which are under active consideration of the government.

1. To identify and estimate the natural plant wealth of the state including trees, roots, stems, bark, leaves, flowers, seeds, fruits, resins and other useful parts of medicinal and cosmetic use.

2. To harness the plant wealth to promote industries to produce phyto-chemicals, essential oils, medicines and cosmetics etc.
3. To identify the commercially viable species in each agro-climatic zone and to standardise the agro-practices for cultivation and preservation of these medicinal plants.
4. To transfer the agro-technology to the farmers and to encourage them to take up the cultivation of medicinal plants particularly in forests, wastelands and orchards as an additional income-generating activity.
5. To reduce the pressure on forests to conserve the biodiversity and protect the natural environment of the state.
6. To strengthen the capacity for undertaking research and development in evolving new herbal products and techniques for their manufacture by utilising the existing capacity within and outside in an optimal manner. It is intended to strengthen this capacity through the World Bank project on Capacity Building for Public Health.
7. To standardise intermediate and final products using the medicinal plants grown or to be grown in the state and to introduce better technology in the existing pharmacies.
8. To provide professional and technical assistance to the public, private and joint sector enterprises to manufacture high quality products at competitive prices for sale in the domestic and international market.
9. To ensure adequate supply of quality medicines in Indian Systems of Medicines and Homoeopathy for use in government hospitals and dispensaries and to sell the surplus in open market.

Tie-up with other developmental departments of Rural Development, Agriculture, Horticulture, Forests, Science and Technology can also play a significant role in promoting the cultivation of medicinal plants. The Department of Rural Development, Govt. of India gives some assistance for innovative schemes and some of the Deputy Commissioners proposed and got sanctioned some projects for medicinal plants. This source could be tapped with the help of Rural Development Department by working out joint strategy for cultivation of medicinal plants. Agriculture and Horticulture departments are already getting some of the staff trained in Floriculture and

Herbiculture in the Dr. Y.S. Parmar University of Horticulture and Forestry. Their services would be utilised for extension work. Horticulture Department has some schemes for certain crops like Kalazeera and Kesar. Forest department has taken some steps for conserving medicinal plants in the forests and are also planting some plants of medicinal value. Assistance from Govt. of India will be sought for plantation of medicinal plants on a large scale in forest areas. The Department of Science & Technology, Govt. of India is also encouraging development and cultivation of medicinal plants. It is proposed to seek central assistance from there also. Ayurveda Department is taking help of CSIR, Universities in the State and CIMAP Lucknow for developing agro-technology and dissemination of agro-technology of medicinal plants.

In this policy document, dissemination of know-how, skill and information to farmers of the state to cultivate and propagate medicinal plants, preparation and use of essential oils, alkaloids and other related activities, it is proposed to undertake training activities on a large scale. The Department of Ayurveda will organise its own training programmes through farmer camps, conferences, workshops, demonstration plots and distribution of publicity material containing relevant information. The department will also take the help of functionaries of other departments like Forests, Agriculture, Horticulture, Rural Development and Panchayats by imparting them training through its own training programmes and also in coordination with other institutions like Dr. Y.S. Parmar University of Horticulture and Forestry.

In order to encourage the farmers to take up the cultivation of medicinal plants, a scheme for providing subsidies on various inputs will be launched. It is essential to have such type of schemes to match with similar scheme of other departments for other crops and also to cover the risks in exploring a new area for cultivation.

The department proposes to tie up buy-back arrangements for the produce of the farmers. It is proposed to achieve this goal by government efforts and involvement of private sector and industry with department playing a role of a facilitator.

The department will buy-back those herbs at competitive prices which are required for production of medicines in the pharmacies of the department. The department will explore the possibility of entering into joint venture for upgradation of these pharmacies and thus try to increase the consumption of herbs. However, the department does not intend to introduce any support price mechanism.

The department will get in touch with leading pharmaceutical companies and other bulk customers and keep them in touch about the activities undertaken by the farmers. The transactions for sale and purchase would be done by the parties themselves. The department will play the role of facilitator by providing two-way information, assist in quality control and offer other services without getting involved in the export transactions.

The department will also encourage cooperative societies, NGOs and other voluntary agencies in collecting herbs to be produced by farmers and supply them to the manufacturers and other customers. The department will also launch training programmes for herb collectors to educate them about the need for scientific extraction and methods thereof to preserve and conserve the medicinal flora.

The Department of Ayurveda has proposal to establish drug testing-cum-research centre. Ayurveda and other Indian Systems of Medicines have many uses of medicinal plants in the manufacture of medicines for treatment of various diseases and cosmetic use. At present, they are used in classical and proprietary medicines manufactured by the pharmacies of Ayurveda department and private pharmacies numbering about 54 in the state. Substantial quantity of medicinal herbs are exported out of the state which are consumed by large pharmacies. However, to provide a stable market for medicinal plants grown by the farmers, it is necessary to diversify the range of intermediate and finished products of common medicinal and cosmetic use. The classical texts have prescribed the methods of manufacture which are basically for small quantity of medicines. However, the large scale manufacturing requires introduction of modern technology and standardisation of products. The weakest area of Indian Systems of Medicines is quality control and in order to ensure that the products manufactured are competitive and acceptable in the national and international markets, standardisation and uniform quality has to be ensured. The new products will have to be developed based on local raw materials and methods of manufacture and technology to be used are to be standardised and upgraded. It is, therefore, proposed to establish a well equipped 'Drug Testing Cum Research Centre' at Jogindernagar by suitably upgrading the Drug Testing Laboratory of

Ayurveda Department. A project for assistance of World Bank was submitted to the Govt. of India and the World Bank pre-appraisal mission has already indicated financial assistance for the development of this facility. Incidentally, this would be the first project to be financed by the World Bank in the field of ISM in the country and would open new horizons of assistance and collaboration in the field of herbal medicines based on the principles of Indian Systems of Medicines with the following objectives:

1. Screening and analysis of medicinal plants and their products.
2. Exchange of research activities with central, state institutions and reputed companies.
3. Manufacturing of standardised drugs.
4. Manufacturing of herbal cosmetics.
5. Extraction of essential oils and alkaloids.

Role of Dr. Y.S. Parmar University of Horticulture and Forestry

As a step towards conservation of biological diversity of medicinal and aromatic plants of Himalayan region, the Y.S. Parmar University of Horticulture and Forestry, Nauni-Solan has done systematic survey of medicinal plant wealth of the state to assess the quantity and quality of available and commercially viable species and their useful parts followed by *ex situ* conservation by developing/establishing Arboreta/Botanic Garden/Herbal Gardens at different agro-climatic zones of the state. Presently, such activities are in progress at the main campus Solan, Dhaulakuan, Jachh, Bajaura, Manali, Rahla, Sharbo and Tabo. In the botanical garden at the campus, 300 species belonging to 75 families have been planted, whereas Dhaulakuan arboretum has 64 species. Nearly 220 species of medicinal and aromatic plants have been raised in herbal gardens and a large number of species as per mandate of the funding agency are under cultivation under the Central Sector Scheme "Cultivation of Medicinal and Aromatic Plants" funded by the Ministry of Agriculture and Co-operation and Ministry of Health and Family Welfare, GOI, New Delhi. Under the ICAR's coordinated research project on breeding and improvement of medicinal and aromatic plants, agro technology for a dozen species has been developed. Many projects for undertaking

Herbal gardens in different agro-climatic zones in Himachal Pradesh (Indicating important species only)

Nauni (1270 m)

Acorus calamus
Abelmoschus moschatus
Centella asiatica
Costus speciosus
Digitalis lanata
Dioscorea deltoidea
Gloriosa superba
Matricaria chamomilla
Psoralea corylifolia
Rosmarinus officinalis
Salvia officinalis
Withania somnifera

Jachh (900 m)

Abrus precatorius
Achyranthes aspera
Aloe barbadensis
Bryonopsis laciniosa
Boerrhavia punarnava
Silybum marianum

Tabo (3600 m)

Arnebia euchroma
Capparis spinosa
Dactylorhiza hatagirea
Ephedra gerardiana
Heracleum candicans
Hippophae rhamnoides
Hyoscyamus niger
Hyssopus officinalis
Rheum australe

Dhaulakuan (750 m)

Eclipta prostrata
Holarrhena antidysenterica
Mentha spp.
Ocimum sanctum
Rouvolfia serpentina
Solanum laciniatum
Terminalia chebula
Withania somnifera

Sharbo (2600 m)

Artemisia brevifolia
Angelica glauca
Atropa belladona
Bunium persicum
Crocus sativus
Carum carvi
Hippophae salicifolia

Rahla (3000 m)

Aconitum heterophyllum
Angelica glauca
Chrysanthemum cinerariaefolium
Dactylorhiza hatagirea
Jurinea dolomiaea
Picrorhiza kurrooa
Podophyllum hexandrum
Skimmia laureola
Valeriana jatamansi

Herbal gardens in different agroclimatic zones in Himachal Pradesh under Dr. Yashwant Singh Parmar University of Horticulture and Forestry, Nauni-Solan 173 230 (HP)

propagation and conservation of germplasm of medicinal and aromatic plants through tissue culture and cryopreservation, and maintaining the herbal garden of the endangered, threatened and rare medicinal and aromatic plants and their mass scale propagation are lying for financial approval with the funding agencies. Efforts are also in progress to preserve germplasm for long term in collaboration with NBPGR, New Delhi and micro-propagation through tissue culture with Department of Biotechnology, New Delhi.

As per policy of the State Govt., each of the 72 developmental blocks in the state will raise herbal gardens of the species particular to that block. The Horticulture Development Officers posted in these blocks are to be trained in the University for this purpose. A batch of

36 Horticulture Development Officers and 11 Ayurvedic Doctors posted in rural dispensaries were imparted training in the University for a fortnight during July 1996. Seed material is being supplied to State Agriculture Universities and other institutions in the country including National Pharmacopoeial Laboratory, Gaziabad, National Bureau of Plant Genetic Resources (NBPGR), New Delhi and Central Institute of Medicinal and Aromatic Plants (CIMAP), Lucknow. Free seed packets and technical know-how is provided to the farmers by the University. It is a pioneer institution in the country having well established training facilities supported by Herbarium, Herbal Garden, Laboratory and a strong faculty.

Future Action Plan

Forests have been exploited, butchered and assaulted for a variety of economic gains for many centuries. Their production capacity is at the lowest ever. Many valuable plant species have disappeared and quite a large number are on the fringe margin. It is, therefore, a compelling obligation on the society that this valuable treasure responsible for our health and longevity is protected, replenished and properly managed. Cultivation on large scale is the most viable factor that can substantially supplement this erosion of genetic diversity.

The Ministry of Health and Family Welfare has a great role in co-ordinating the multi-pronged efforts put in by the Central Councils for Indian Systems of Medicines and Homoeopathy, State Governments, Universities, research institutions and other autonomous organisations connected with this enterprise. Sufficient funds have to be created in the central pool to finance the research activities undertaken by different organisations.

The following points need immediate attention:

- Creation of protectorates/biosphere reserves to conserve the genetic stock of endangered species.
- Large scale cultivation of rare and endangered species used in Indian Systems of Medicines and Homoeopathic Pharmacopoeias to be initiated through the use of appropriate technology and its transfer to farmer's fields. Transplant and tissue culture techniques should be adopted for such species which are either

shy seed producers or their propagation through seeds is not easy.

- Establishment of demonstration plots for herbal drugs at institutional level and a chain of herbal gardens in all states rich in medicinal plants followed by establishment of Herbal Farms and a well stocked central herbal garden. Integration of medicinal and aromatic plants in social/agro-forestry and cropping patterns of our farmers will provide more economic gains per unit area per unit time.
- As a result of thorough survey, an inventory of medicinal plants including the plants used in local folklore, has to be prepared for judicial exploitation and sustainable management of our resources both for the present and for posterity.
- Establishment of nodal agencies both at the centre and the states for coordination of cultivation, collection, extraction and utilisation of the herbal resources by those involved in the vocation.
- Generation of trained and skilled manpower to handle all aspects of medicinal plants.
- Public awareness programmes, specifically conservation consciousness need to be intensified among school and college students through the introduction of topics on medicinal plants in their curricula. To start with, this can be initiated by including most common plants in the form of botanical names, local/common vernacular names, brief description and broad medicinal uses. The National Council of Educational Research and Training, New Delhi has already devised syllabus (the author being one of the expert members of the course curriculum development committee) under vocationalisation of education in the trade "Medicinal and Aromatic Plant Industry", which can also be introduced in Himachal Pradesh to achieve tangible results.

Conclusion

The production, processing and use of medicinal and aromatic plants has great potential for employment generation particularly at village level. The recent emphasis on rural development and particularly on tribal development shows that medicinal and aromatic plants if properly managed have a prominent role to play in this direction.

Nature's bounty of medicinal plants which can cure even the rare diseases is getting destroyed with fast depleting forests. The future of hundreds of plant species used for medicinal purposes in India which once grew abundantly in our forests and elsewhere are at risk as they often are picked indiscriminately from the wild, leaving little scope for their regeneration. Another reason for the disappearance of many species is ignorance on our part with regard to knowledge of the identity and use of such species. In our ignorance, we are invariably treating these otherwise useful species as useless weeds and destroying them and leaving no scope for their regeneration.

The growing popularity of Ayurveda, Unani, Homoeopathy and Sidha systems of medicines is putting an increasing pressure on the plant population. Conservationists blame pharmaceutical industries for depletion of wild flora.

Ninety-five per cent medicinal plants in use today are obtained directly from the wild and are not yet cultivated. Manufacturers and traders admit this, and they acknowledge that the time has come to go in for cultivation of medicinal plants in a big way to ensure a constant supply of raw material. Some have already made a start in this direction, but most have found it easier to collect their requirements from the wild rather than making farm investments. The traditional pattern of collection and utilisation of medicinal plants is breaking under the changing social and economic structure.

Earlier, the middleman had a hey-day, but with the growing consciousness, the farmers and the tribals have realised/visualised the great potential in growing medicinal and aromatic plants. Setting up of small pharmaceutical units and cottage industries will not only provide them employment at their doorsteps, but will also create increased employment opportunities for rural youths, women and entrepreneurs engaged in collection, processing, marketing etc.

India is no exception to the global phenomenon of environmental problems and depletion of natural resources. Nature has provided us a rich diversity in medicinal and aromatic plants which need to be exploited judiciously without disturbing its ecological balance. Undoubtedly, resources need to be harnessed for economic development, but, simultaneously we have to be concerned about their regeneration, preservation and propagation, so that these are available to us for use on a sustainable basis. Thus, the restoration and preservation of our biological heritage is a challenge to our planners, administrators, scientists and farmers alike.

In these contexts, it is of paramount importance that the herbal resources of the state are scientifically identified, inventorised, documented and a database prepared. It is only possible to harness these resources on sustainable basis without jeopardising their ecological equilibrium, if we actually recognise and know their utilisation potential. For this, scientific skill followed by conservation consciousness has to be created in the minds of the people involved in the trade, local populace, officials of the departments of Forests, Ayurveda, Industry, Agriculture, Horticulture, Tribal and Rural Development. Course contents have to be introduced for school, college and university students to educate, motivate and activate them to care and respect this valuable treasure the nature has bestowed upon our state. The collection and harvesting has to be strictly regulated and market infrastructure developed within the state. The pharmacies and other units consuming medicinal and aromatic plants should guarantee the consumption of the produce locally available, ensuring remunerative prices to the collectors and cultivators. This will boost commercial cultivation and ease pressure on natural habitats, thereby contributing to the ecological stability and bio-diversity conservation. It is emphatically suggested that no material be allowed to go out of the state in crude form. If we add value to it by establishing small processing units at central and salient places in herb rich areas, it will not only create a chain of employment to our women and unemployed youths at their door step, but will also fetch increased returns for their products, thereby bringing an economic revolution. To achieve success in this direction, the government should, as a matter of policy incentive, channelise the produce to be consumed by the manufacturing units (including governmental pharmacies) in order to effect optimum utilisation of the production capacity of these units, provide immediate returns to the collectors/cultivators, improve the quality of formulations of Ayurvedic drugs using fresh, genuine and quality material, become self sufficient in meeting the drug requirement of over 1000 Ayurvedic dispensaries/hospitals existing in the state, and above all enter into the national and international market by floating standard drugs into the world market. If judicious exploitation coupled with the government efforts is ensured, the day will not be far off when our state can boast of being a pioneer in the enterprise. Having earned the fame as an apple bowl of the country, it can earn similar status in the field of medicinal and aromatic plants too.

Appendices

Appendix 1: Commercial plants for trade and drug industry

There is a long list of such species growing in different parts of the region which are collected regularly on small scale or often on large scale as and when there is a demand from the market. Some important drug plants are enumerated hereunder:

Sr. No.	Name of species	Local name	Official part(s)
1.	*Abies pindrow* Royle	Tosh, Talispatri	Leaves/needles
2.	*Abrus precatorius* Linn.	Ratti, Gunja	Seed, Root
3.	*Abutilon indicum* Linn.	Atibala	Seeds, Leaves
4.	*Acacia catechu* Willd.	Khair	Katha (extract of heartwood)
5.	*A. nilotica* (L.) Willd. ex Del.	Kikar, Babul	Bark, Gum
6.	*Achillea millefolium* Linn.	Biranjasif	Herb
7.	*Achyranthes aspera* Linn.	Apamarga	Entire plant
8.	*A. bidentata* Blume	Puthkanda	Entire plant
9.	*Aconitum chasmanthum* Stapf. ex Holmes	Banbalang	Root
10.	*A. deinorrhizum* Stapf.	Vatsnabh, Mohrabish	Root
11.	*A. heterophyllum* Wall.	Atis (Ativisa), Patis	Root
12.	*A. violaceum* Jacq.	Dudhi Mohra Mitha Patish	Root
13.	*Acorus calamus* Linn.	Bach, Bare	Rhizome
14.	*Adiantum lunulatum* Burm.	Hansraj, Kalijhant	Fronds (Plant)
15.	*A. pedatum* Linn.	Maiden hair	Plant

Contd.

Sr. No.	Name of species	Local name	Official part(s)
16.	*Aegle marmelos* Corr.	Bel, Bilve	Fruit rind, leaves and bark
17.	*Aesculus indica* Colebr.	Horse-chestnut, Khanor	Seed
18.	*Agave americana* Linn.	Kantala	Root, Leaf juice
19.	*Agrimonia eupatorium* Linn.	Bajardant	Root
20.	*Ailanthus altissima* (Mill.) Swingle	Maharukh	Bark
21.	*Ajuga bracteosa* Wall. ex Benth.	Nilkanthi	Plant (Leaves)
22.	*Albizia lebbek* Benth.	Siris	Bark
23.	*A. odoratissima* Benth.	Siras Ki Chhal	Bark
24.	*A. procera* (Roxb.) Benth.	Safed siris	Bark
25.	*Aleurites montana* (Lour.) E.H. Wils.	Tungnut	Oil
26.	*Allium govanianum* Wall. ex Baker	Duno, Farna	Leaves and Bulbs
27.	*A. carolinianum* DC.	Laot, Jangli Lahsun	Flowers, Leaves & Pseudobulbs
28.	*A. jaquemontii* Kunth	Koche, Pharna	Flowers, Leaves & Pseudobulbs
29.	*A. sativum* Linn.	Lahsun	Bulbs
30.	*Aloe barbadensis* Mill.	Ghrit-kumari	Leaf pulp
31.	*Althaea officinalis* Linn.	Tukhm khatmi	Root fibres
32.	*Amaranthus caudatus* Linn.	Chaulai, Rajagiri	Plant (leaves and seed)
33.	*A. spinosus* Linn.	Kanta-chaulai	Root, leaves
34.	*Amomum subulatum* Roxb.	Bari ilaichi	Fruit

Contd.

Sr. No.	Name of species	Local name	Official part(s)
35.	*Amphicome emodi* Royle ex Lindl.	Kaur	Plant
36.	*Anagalis arvensis* Linn.	Jonkmari	Herb
37.	*Angelica glauca* Edgew.	Chora	Root
38.	*Anogeissus latifolia* (DC.) Wall. ex Bedd.	Dhava, Chhal	Bark
39.	*Arctium lappa* Linn.	Jangli kuth, Burdock	Root
40.	*Ardisia solanacea* Roxb.	Gurhbheli	Root
41.	*Arenaria serpyllifolia* Linn.	Chiki, Gandial	Herb
42.	*Argemone mexicana* Linn.	Swaran khshiri	Yellow Juice, Root, Seed
43.	*Arnebia benthami* (Wall. ex G.Don.) Johnston	Ratanjot	Root
44.	*A. euchroma* (Royle) Johnston	Ratanjot	Root
45.	*A. guttata* Bunge	Ratanjot	Root
46.	*Artemisia brevifolia* Wall. ex DC.	Santonin plant, Seinski	Flowering shoots
47.	*A. dracunculus* Linn.	Teragon	Floral shoots
48.	*A. scoparia* Waldst. & Kit	Donojhan, Jhau	Seed, Flowering fruiting tops
49.	*A. vulgaris* Linn.	Chhambra	Floral shoots
50.	*Artocarpus lakoocha* Roxb.	Lakucha	Fruit and Bark
51.	*Asclepias curassavica* Linn.	Kaktundi	Root, Leaf juice
52.	*Asparagus adscendens* Roxb.	Shatavari, Safed musli	Root (fasiculated)
53.	*Astragalus strobiliferus* Royle	Garmezu	Gum

Contd.

Sr. No.	Name of species	Local name	Official part(s)
54.	*Atropa acuminata* Royle ex Lindl.	Sag-angur	Root and leaves
55.	*Azadirachta indica* A. Juss	Neem	Root bark, leaves and seed oil
56.	*Bacopa monnieri* (Linn.) Pennell.	Brahmi, Jalneem	Plant
57.	*Baliospermum montatum* Muell.-Arg.	Dantimool	Root bark
58.	*Barleria cristata* Linn.	Jhinti, Nilabansa	Plant
59.	*Bauhinia purpurea* Linn.	Khairwal, Kaliar	Bark, Flowers
60.	*B. vahlii* W. & A.	Maljhan, Taur	Seed, Leaves and Bark
61.	*B. variegata* Linn.	Kachnar	Bark (stem)
62.	*Berberis aristata* DC.	Daruharidra	Rasont (Extract)
63.	*B. asiatica* Roxb. ex DC.	Chothar	Root bark/extract
64.	*B. chitria* Lindley	Totoli	Root bark/extract
65.	*B. lycium* Royle	Kashmal	Root bark/extract
66.	*Bergenia ciliata* (How.) Sternb.	Pashanbhed	Rhizome
67.	*B. stracheyi* (Hook f. Thoms) Engl.	Pashanbhed	Rhizome
68.	*Betula utilis* D. Don.	Bhojpatra	Papery bark & fungal growth
69.	*Bidens pilosa* Linn.	Chor pushpi	Infusion, Seed
70.	*Boerhaavia diffusa* Linn.	Punarnava	Root
71.	*Bombax ceiba* L.	Simal	Root of young seedling
72.	*Brassica campestris* Prain	Sarson	Seed

Contd.

Sr. No.	Name of species	Local name	Official part(s)
73.	*B. juncea* Linn.	Rai	Seed
74.	*Bridelia retusa* Spreng.	Khaja	Root and Bark
75.	*Bryonopsis laciniosa* (L.) Naud.	Shivlingi	Seed
76.	*Buchanania lanzan* Spreng.	Chironji	Seed oil, Gum
77.	*Butea monosperma* (Lamk.) Taub.	Plash	Seed, flowers, root bark
78.	*Bunium persicum*	Kalazira	Seed like fruit
79.	*Buxus wallichiana* Baill.	Papri	Leaves, Bark
80.	*Caesalpinia crista* L.	Kat-Karanj	Leaves, Bark, Seed, Oil
81.	*Calendula officinalis* Linn.	Zergul	Plant with flowers
82.	*Callicarpa macrophylla* Vahl.	Priangu, Daya	Leaves, Root oil
83.	*Calotropis gigantea* (Linn.) Dryand	Aak, Akanda	Root
84.	*C. procera* (Ait.) R. Br.	Aak, Madar	Root
85.	*Caltha palustris* Linn.	Mamiri	Root
86.	*Cannabis sativa* L.	Vijaya, Bhang	Resin, Fruiting top, Seed
87.	*Capparis spinosa* Linn.	Kabra, Caper	Root bark, Flower buds
88.	*Careya arborea* Roxb.	Kumbhi	Bark, Flowers, Fruit
89.	*Carica papaya* Linn.	Papita	Fruit, Seed, Milky Juice
90.	*Carum carvi* Linn.	Kalajira	Fruit
91.	*Cassia absus* Linn.	Chaksu	Leaves, Seed

Contd.

Sr. No.	Name of species	Local name	Official part(s)
92.	*C. fistula* Linn.	Amaltas	Pods, Roots and Bark
93.	*C. occidentalis* Linn.	Kasmard	Seed
94.	*C. tora* Linn.	Chakramard	Seed
95.	*Cedrus deodara* (Roxb.) Loud.	Devdaru	Wood and Cedar Oil
96.	*Celastrus paniculatus* Willd.	Malkangni	Seed, oil
97.	*Centella asiatica* (L.) Urban	Mandookparni, Brahmi	Herb
98.	*Centratherum anthelminticum* Kuntz.	Kalijiri, Ghorajiri	Seed
99.	*Chenopodium album* Linn.	Bathu	Plant
100.	*C. ambrosioides* Linn.	Sugandh-vastuk	Essential oil
101.	*Cichorium intybus* Linn.	Kasni, Chikori	Seed/root
102.	*Cinnamomum camphora* Nces. & Ebrm.	Kapur	Camphor oil
103.	*C. tamala* Nces.	Tejpatra	Leaves
104.	*Cissampelos pareira* Linn.	Patha	Root, Leaves
105.	*Citrullus colocynthis* Schrad	Indrayan	Fruit
106.	*Citrus* spp.	Nimbu, Jamiri, Galgal etc.	Fruit
107.	*Clerodendrum phlomidis* Linn.	Agnimanth (Arni)	Root and Leaves
108.	*C. serratum* Roxb.	Bhrangi	Root and Leaves
109.	*Codonopsis ovata* Benth.	Sardanda	Root
110.	*Colchicum luteum* Baker	Hirantutiya	Bulb

Contd.

Sr. No.	Name of species	Local name	Official part(s)
111.	*Colutea nepalensis* Sims.	Braa	Leaves, Seed
112.	*Convolvulus arvensis* Linn.	Hiranpadi	Root, Leaves
113.	*Cordia dichotoma* Forst.	Lasurha	Fruit, Bark, Leaves
114.	*Coriandrum sativum* Linn.	Dhaniya	Fruit
115.	*Corydalis govaniana* Wall.	Bhutjata	Root
116.	*Corylus colurna* Linn.	Bhotia-badam	Nut
117.	*Costus speciosus* (Koen. ex Retz.) Sm.	Keu, Kemuk	Rhizome
118.	*Crataegus oxycantha* Linn.	Hawthorn	Fruit
119.	*Crataeva nurvala* Buch.-Ham	Varun	Bark, Fresh leaves
120.	*Crocus sativus* Linn.	Kesar	Stigma and Styles (dried)
121.	*Cryptolepis buchanani* Roem. & Schult.	Kali Sariva	Root
122.	*Curculigo orchioides* Gaertn.	Kalimusli	Root
123.	*Curcuma longa* Linn.	Haldi	Rhizomes
124.	*Cuscuta europaea* Linn.	Amarvel	Herb
125.	*Cydonia oblogna* Mill.	Bihidana	Fruit
126.	*Cymbopogon jwarancusa* Shult.	Lamjak	Herb, Essential oil
127.	*C. martinii* (Roxb.) Wats.	Makora ghas	Essential oil
128.	*Cynodon dactylon* Linn.	Durwa	Herb
129.	*Cyperus rotundus* Linn.	Nagarmotha	Bulbous root

Contd.

Sr. No.	Name of species	Local name	Official part(s)
130.	*C. scariosus* R. Br.	Nagarmotha	Rhizomes/stolons
131.	*Dactylorhiza hatagirea* (D. Don.) Soo	Salam Panja	Tuber
132.	*Datisca cannabina* Linn.	Akalbir	Root, Herb
133.	*Datura fastuosa* L.	Sadah Dhatura	Whole plant
134.	*D. metel* Sims.	Kala dhatura	Whole plant
135.	*D. stramonium* Linn.	Safed dhatura	Pnaching/seed/leaves
136.	*Delphinium denudatum* Wall.	Nirbisi	Root
137.	*Desmodium gangeticum* DC.	Shalparni	Plant
138.	*Didymorcarpus pedicellata* R. Br.	Patharphori	Leaves
139.	*Digitalis lanata* Ehrh.	Foxglove, Hritpatri	Leaves
140.	*Dioscorea bulbifera* Linn.	Varahikand	Tubers
141.	*D. deltoidea* Wall. ex Kunth.	Singli-Mingli	Rhizomes
142.	*Dolichos uniflorus* Lam.	Kulthi	Seed
143.	*Drocera peltala* Sm.	Mukhajali	Plant
144.	*Eclipta alba* (Linn.) Hassk.	Bhringraj	Plant
145.	*Embelia tsjeriam-cottom* A. DC.	Bai vidang	Fruit
146.	*Emblica officinalis* Gaertn.	Aonla	Fruit
147.	*Ephedra gerardiana* Wall.	Somlata	Twigs
148.	*E. intermedia* Schrank & May	Somlata	Twigs

Contd.

Sr. No.	Name of species	Local name	Official part(s)
149.	*E. major* Host.	Soma	Twigs/branches
150.	*Equisetum arvensis* Linn.	Horse tail, Harhjori	Herb
151.	*Eucalyptus citriodora* Hook.	Safeda	Essential oil
152.	*Eulophia compestris* Wall.	Amrita	Rhizome
153.	*Euphorbia hirta* Linn.	Dudhi	Plant
154.	*E. pilosa* Linn.	Saptala, Satla	Root
155.	*E. royleana* Linn.	Thoohar	Latex
156.	*E. thymifolia* Linn.	Chhoti-dudhi	Plant
157.	*Evolvulus alsinoides* Linn.	Shankhpushpi	Plant
158.	*Fagopyrum* spp.	Ogal, Chabru, Phafra	Young leaves, Seed
159.	*Ferula jaeschkeana* Vatke	Jangli Hing	Gum, resin
160.	*Ficus benghalensis* Linn.	Vat, Bargad	Aerial roots, latex
161.	*F. carica* Linn.	Anjir	Fruit
162.	*F. racemosa* Linn.	Udumbar, Gular	Bark, Fruit
163.	*F. religiosa* Linn.	Pipal, Ashwath	All parts
164.	*Foeniculum vulgare* Mill.	Mithi sonf	Fruit
165.	*Fraxinus xanthoxyloides* (G.Don) DC.	Pipjul, Thuma	Bark, Twigs
166.	*Fritillaria oxypetalum* Royle	Lahsunia salep	Bulb
167.	*F. roylei* Hook.	Kakoli	Bulb

Contd.

Sr. No.	Name of species	Local name	Official part(s)
168.	*Fumaria parviflora* Lamk.	Pitpatra	Herb
169.	*Gentiana kurroo* Royle	Karu	Root
170.	*G. tenella* Fries	Teeta	Herb
171.	*G. tianschanica* Rupr.	Tikta	Root
172.	*Geranium nepalense* Sweet	Laljari	Root
173.	*G. wallichianum* D. Don.	Laljari	Root
174.	*Geum elatum* Wall. ex G. Don.	Masreen	Root
175.	*Gerardinia diversifolia* (Link.) Friis	Bara bichhua, Karal	Root
176.	*Gloriosa superba* Linn.	Kalihari, Langli	Tuber
177.	*Gmelina arborea* Linn.	Gambhari	Bark
178.	*Habenaria edgeworthii* Hook. f.	Varidhi	Tubers
179.	*H. intermedia* D. Don.	Ridhi	Tubers
180.	*Hedychium acuminatum* Rosc.	Kapurkachri, Banhaldi	Rhizomes
181.	*Helicteres isora* Linn.	Marorphali	Twisted fruits and Bark
182.	*Heliotropium indicum* Linn.	Hathi-sundi	Plant
183.	*Hemidesmus indicum* R. Br.	Sariva, Anantmool	Root
184.	*Heracleum candicans* Wall. ex DC.	Padara, Patrala	Root
185.	*Hippophae rhamnoides* Linn.	Chharma, Sarla	Fruit, Bark, Oil
186.	*H. salicifolia* D. Don.	Chuk, Chuma, Amli	Fruit, Bark, Oil

Contd.

Sr. No.	Name of species	Local name	Official part(s)
187.	*Holarrhena antidysenterica* Wall.	Kutaj, Kura, Indrajau	Bark, Seed
188.	*Humulus lupulus* L.	Hops	Female cones (dried)
189.	*Hydrocotyle rotundifolia* Roxb.	Mandukparni	Herb
190.	*Hyoscyamus niger* Linn.	Khurasani ajwain	Leaves, Seed
191.	*Hypericum perforatum* Linn.	Basant	Plant
192.	*Hyssopus officinalis* Linn.	Jufah, Tengu	Flowering tops
193.	*Inula racemosa* Hook. f.	Pushkarmool	Root
194.	*Iris germanica* Linn.	Safed Bach	Rhizomes
195.	*Jasminum humile* Linn.	Pitmalti	Leaves, flowers
196.	*J. officinale* Linn.	Chameli	Leaves, flowers
197.	*Jatropha curcas* Linn.	Byaghrerand	Seed, Twigs, Juice
198.	*Juglans regia* Linn.	Akhrot	Bark, Nut
199.	*Juniperus communis* Linn.	Hauber	Fruit (Berry)
200.	*J. macropoda* Boiss.	Dhupi, Leur	Fruit, Twig
201.	*J. recurva* Ham.	Hauber, Bethar	Twigs
202.	*Jurinea dolomiaea* Boiss.	Dhoop	Root
203.	*Justicia adhatoda* L.	Basuti, Vasaka	Leaves
204.	*Kalanchoe pinnata* Pers.	Zakhmehayat	Leaves
205.	*Lactuca sativa* Linn.	Salad	Leaves/plant

Contd.

Sr. No.	Name of species	Local name	Official part(s)
206.	*Lannea coromandelica* (Houtt.) Merr.	Jhingan	Gum, Bark, Leaves
207.	*Lawsonia inermis* Linn.	Mehndi	Leaves, Bark, Oil
208.	*Lepidium sativum* L.	Hallo, Chandrashoor	Seed
209.	*Leucas cephalotes* (Roth.) Spreng.	Dronpushpi	Herb
210.	*Lilium polyphyllum* D. Don.	Kakoli	Bulbs
211.	*Linum usitatissimum* Linn.	Alsi	Seed
212.	*Lippia nodiflora* Mich.	Jalpipli	Plant
213.	*Litsea chinensis* Lam.	Meda-lakri	Bark, Oil
214.	*L. umbrosa* (Nees) Hook. f.	Chirindi	Seed, Oil
215.	*Malaxis acuminata* D. Don.	Rishbhak	Green horn like rhizome
216.	*Malaxis muscifera* (Lindley) Kuntz.	Jeewak	Bulb
217.	*Mallotus phillippinensis* Muell.-Arg.	Kamila	Fruit powder
218.	*Malva rotundifolia* Linn.	Khubbasi	Herb, leaves/seeds
219.	*M. verticillata* L.	Mridupatra	Herb, leaves/seeds
220.	*Malvastrum coromandelianum* (L.) Garcke	Piliabuti	Plant
221.	*Mangifera indica* Linn.	Am ki guthli	Kernel
222.	*Martynia annua* Linn.	Kaknasa	Fruit
223.	*Matricaria chamomilla* Lamk.	Babuna	Flowers, Essential oil
224.	*Melissa officinalis* Linn.	Billi Lotan, Baam	Herb, Oil

Contd.

Sr. No.	Name of species	Local name	Official part(s)
225.	*Mentha longifolia* Huds.	Jangli Pudina	Aerial parts
226.	*M. piperita* Linn.	Pipermint	Pepermint oil
227.	*M. spicata* Linn.	Pudina	Menthol
228.	*Milletia auriculata* Baker ex Brand.	Gauj	Bark, Seed
229.	*Mimosa pudica* L.	Lajwanti	Seed, Root
230.	*Mirabilis jalapa* Linn.	Gulabans	Root and Leaves
231.	*Monochoria vaginalis* (Burm.) f. Presl.	–	Leaves and roots
232.	*Morchella esculenta* (Linn.) Pers.	Guchhi	Fruiting body
233.	*Morina coulteriana* Royle	Bish kandara	Root
234.	*M. longifolia* Wall. ex DC.	Dhoop	Root
235.	*Moringa oleifera* Gaertn.	Sahinjan	Seed, Oil
236.	*Mucuna pruriens* Hook.	Kaunch	Seed
237.	*Murraya koenigii* (L.) Spreng.	Mithinim, Curry leaf	Leaves
238.	*Myrica esculenta* Linn.	Kaiphal	Fruits, bark
239.	*Myrsine africana* Linn.	Bidang	Fruit
240.	*Nardostachys grandiflora* DC.	Jatamansi	Root (hairy part)
241.	*Nerium indicum* Mill.	Karvir, Kaner	Root bark, leaves
242.	*N. odorum* Soland.	Kanermool (Rakta)	Root
243.	*Nyctanthes arbor-tristis* Linn.	Harsingar	Leaves

Contd.

Sr. No.	Name of species	Local name	Official part(s)
244.	*Ocimum basilicum* Linn.	Tukhm Tulsi	Herb
245.	*O. sanctum* Linn.	Tulsi	Herb
246.	*Olea europea* Linn.	Olive, Jaitum	Oil
247.	*O. ferruginea* Royle	Kahu, Wild olive	Oil, Leaves, Bark
248.	*Onosma bracteatum* Wall.	Kahu, Wild olive	Roots
249.	*O. hispidum* Wall. ex G. Don.	Gojihva, Ratanjot	Leaves & Roots
250.	*Operculina turpethum* Silva-Manso.	Nishoth	Root
251.	*Origanum vulgare* Linn.	Sathra	Herb, Oil
252.	*Oroxylum indicum* Vent.	Shyonak, Arlu	Bark, Seed
253.	*Oxalis corniculata* Linn.	Changeri	Herb
254.	*Paeonia emodi* Wall.	Udsalib	Rhizome, Seed, Flowers
255.	*Paris polyphylla* Smith	Kshir-kakoli	Rhizome
256.	*Parmelia perlata* Ach.	Chharila	Thallus
257.	*Peucedanum graveolens* Linn.	Soa (shatpushpa)	Fruit
258.	*Phytolacca acinosa* Roxb.	Jharka, Jalga	Leaves, Root
259.	*Picrorhiza kurrooa* Linn.	Kutki	Rhizomes
260.	*Pinus gerardiana* Wall.	Chilgoza	Kernels
261.	*Pistacia integerrima* Stewart ex Brandis	Kakadsinghi	Insect gall
262.	*Plantago major* Linn.	Bartang, Jangli Isabgol	Seed/husk

Contd.

Sr. No.	Name of species	Local name	Official part(s)
263.	*Plumbago zeylanica* Linn.	Chitrak	Root
264.	*Podophyllum hexandrum* Royle	Bankakri	Root
265.	*Polianthes tuberosa* Linn.	Tuberose, Rajnigandha	Flowers
266.	*Polygonatum verticillatum* Allioni	Mahameda	Rhizome
267.	*P. cirrhifolium* (Wall.) Royle	Meda	Rhizome
268.	*Polygonum aviculare* Linn.	Anjubar	Root
269.	*Pongamia pinnata* Vent.	Karanj	Seed,leaves
270.	*Potentilla fulgens* Wall.	Bajradanti	Root
271.	*P. nepalensis* Hook.	Laljari	Root
272.	*Premna latifolia* Roxb.	Arni, Bakar	Leaves, Bark (Juice)
273.	*Prinsepia utilis* Royle	Bhekhal	Oil
274.	*Prunella vulgaris* Linn.	Ustakhadus	Herb
275.	*Prunus amygdalus* Batsch.	Maghz Badam	Kernels
276.	*P. cerasoides* D. Don.	Patmakh, Pajja	Bark
277.	*Psoralea corylifolia* Linn.	Vakuchi beej, Babchi	Seed
278.	*Pueraria tuberosa* Linn.	Vidarikand	Tubers
279.	*Punica granatum* Linn.	Darim	Seed, Fruit and Root bark
280.	*Putranjiva roxburghii* Wall.	Jiyapota	Bark, Leaves, Stone (Fruits)
281.	*Randia dumetorum* Lam.	Rada, Madanphal	Fruit

Contd.

Sr. No.	Name of species	Local name	Official part(s)
282.	*Raphanus sativus* Linn.	Mooli	Leaves and Root
283.	*Rauvolfia serpentina* Benth. ex Kurge	Sarpagandha	Root
284.	*Rheum australe* Wall ex Meissn.	Rhubarb, Revandchini	Root
285.	*R. moorcroftianum* Royle	Rhubrab	Root
286.	*Rhododendron anthopogon* D.Don	Talispatra	Leaves
287.	*R. campanulatum* D.Don	Kashmiripatha	Leaves
288.	*Ricinus communis* Linn.	Erand	Seed and Oil
289.	*Rosa damascena* Miller.	Gule Gulab	Petals
290.	*Roscoea alpina* Royle	Kakoli	Tuberous roots
291.	*R. purpurea* Smith	Safed Musli, Kakoli	Root
292.	*Rosmarinus officinalis* Linn.	Rosemari	Oil
293.	*Rubia manjith* Roxb. ex Fleming	Manjishtha	Root
294.	*Sapindus mukorossi* Gaertn.	Ritha, Saopnut	Fruit
295.	*Sausurea costus* Clarke	Kuth	Root
296.	*Selinum tenuifolium* Wall. ex DC	Bhutkesi	Fibrous underground stem
297.	*S. vaginatum* C. B. Clarke	Muramansi	Fibrous underground stem
298.	*Semecarpus anacardium* Linn.	Bhilava	Fruit
299.	*Sesamum indicum* Linn.	Til tail	Seed/oil
300.	*Sesbania grandiflora* Linn.	Agast ke pushpa	Flowers

Contd.

Sr. No.	Name of species	Local name	Official part(s)
301.	*Shorea robusta* Burm. f.	Safed rall (Resin)	Gum-oleoresin
302.	*Sida acuta* Burm. f.	Bariara, Kharenti	Plant
303.	*S. cordifolia* Linn.	Kharenti	Plant
304.	*S. rhombifolia* L.	Sahdevi, Swet-Barela	Plant
305.	*S. veronicaefolia* Laud.	Nagbala	Plant
306.	*Skimmia laureola* Sieb. & Zucc.	Ner, Kasturipatra	Leaves
307.	*Solanum indicum* Linn.	Badi Kateri	Whole plant/Berries
308.	*S. nigrum* Linn.	Makoi	Whole plant/Berries
309.	*S. surattense* Burm. fl.	Chhoti Kateri	Whole plant/Berries
310.	*S. viarum* Bunal.	Bari Kantkari	Aerial parts/Berries
311.	*Sphaeranthus indicus* Linn.	Gulmundi/Gorakhmundi	Herb
312.	*Spilanthes acmella* Murr.	Akarkara	Herb, Root, Seeds
313.	*Spondias pinnata* Kurz.	Ambara	Fruit
314.	*Stephania glabra* (Roxb.) Miers.	Bishkhapra, Gindaru	Tuber
315.	*Stereospermum suaveolens* DC.	Padhal, Patla	Bark
316.	*Swertia chirayita* Buch.-Ham.	Chirayata	Whole plant
317.	*S. angustifolia* Buch.-Ham.	Chirayata	Whole plant
318.	*S. purpurascens* Wall.	Nepali chirayata	Whole plant
319.	*Symplocos paniculata* Wall.	Lodhra	Bark

Contd.

Sr. No.	Name of species	Local name	Official part(s)
320.	*Syzygium cumini* Skeels	Jamun guthli	Seed
321.	*Tanacetum longifolium* Wall. ex DC.	Tansy	Herb
322.	*Taraxacum officinale* Weber	Kanphul	Root
323.	*Taxus wallichiana* (Zucc.) Pilger	Talispatri	Needles
324.	*Teramnus labialis* Spreng.	Mashparni	Herb, Seed
325.	*Terminalia alata* Heyne ex Roth.	Badam	Fruit (Kernel)
326.	*T. arjuna* Bedd.	Arjun	Bark
327.	*T. bellirica* Roxb.	Bahera (Baheda)	Fruit rind
328.	*T. chebula* Retz.	Harad, Haritaki	Fruit rind
329.	*T. tomentosa* W. & A.	Asan	Bark
330.	*Thalictrum foliolosum* DC.	Pilijari, Mamiri	Root
331.	*Thevetia peruviana* (Pers) Shum.	Pila Kaner	Root
332.	*Thymus serpyllum* Linn.	Banajwain	Herb
333.	*Tinospora malabarica* (Lam.) Miers.	Giloye, Guduchi	Twigs/branches
334.	*Trachyspermum ammi* Linn.	Ajwain	Fruit
335.	*Tribulus terrestis* Linn.	Gokhru	Fruit
336.	*Trichosanthes cucumerina* Linn.	Patol	Fruit
337.	*T. dioica* Roxb.	Patol patra	Fruit
338.	*Trigonella foenum-graecum* Linn.	Methi	Seed
339.	*Tylophora indica* (Burm. f.) Merr.	Damabuti	Leaves
340.	*Uraria lagopoides* DC.	Prishniparni	Plant

Contd.

Sr. No.	Name of species	Local name	Official part(s)
341.	*U. picta* Desv.	Prishniparni	Plant
342.	*Urginia indica* Kunth.	Jangli piaz	Bulb
343.	*Valeriana hardwickii* Wall.	Nihani, Tagar	Rootstock
344.	*V. jatamansi* Jones	Mushkbala	Rootstock
345.	*V. pyrolaefolia* Decaisne	Nihani	Rootstock
346.	*Vanda roxburghii* R. Br.	Rasna	Plant
347.	*Verbena officinalis* Linn.	Karaita	Herb
348.	*Vernonia cinerea* Less.	Sahdevi	Plant
349.	*Vetiveria zizanioides* (L.) Nash.	Khas	Root, essential oil
350.	*Viburnum nervosum* D. Don.	Tilani	Bark
351.	*Viola canescens* Walliach	Banaksha	Flowers, Leaves
352.	*Viola serpens* Wall. ex Roxb.	Gule banafsha	Flowers, Leaves
353.	*Viscum album* Linn.	Banda, Mistletoe	Plant
354.	*Vitex negundo* Linn.	Nirgundi, Sambhalu	Leaves
355.	*Vitis vinifera* Linn.	Draksha, Angur	Fruits
356.	*Withania somnifera* Dunal.	Ashwagandha	Roots
357.	*Woodfordia fruticosa* Kurz.	Dhai, Dhataki	Flowers
358.	*Wrightia tomentosa* Roem. & Shultz.	Mitha-Indrajau	Bark, Seed
359.	*Xanthium strumarium* Linn.	Bara-Gokhru	Plant, Fruit
360.	*Zanthoxylum armatum* DC.	Tirmir, Tejbal	Fruit
361.	*Zizyphus mauritiana* Lamk.	Ber	Fruit

Appendix 2. Annual outturn of medicinal and aromatic plants in Himachal Pradesh for trade and commerce (1991-92 to 1994-95)

The quantity of major medicinal plants exported from Himachal Pradesh has declined from 38,124.72 quintals during 1991-92 to 17,826 quintals during 1994-95. (Source: Deptt. of Forest, H.P.)

Sl. No.	Latin name	Common name	Year/quantity in quintals 1991-92	1992-93	1993-94	1994-95
1	2	3	4	5	6	7
1.	*Aconitum heterophyllum*	Karvi Patish	2.71	33	17	25
2.	*Aconitum violaceum/*	Mithi Patish	48.24	60	30	–
3.	*Acorus calamus*	Bach, Bare	–	180	18	9
4.	*Angelica glauca*	Chora	–	199	46	66
5.	*Betula utilis*	Bhojpatra	–	305	–	15
6.	*Berberis lycium*	Berberis roots	12,824.00	8,154	3,946	–
7.	*Bunium persicum/ Carum carvi*	Kalajira	5.45	5	–	–
8.	*Chaerophyllum villosum*					
9.	*Cinnamomum tamala*	Tejpatra	–	663	745	675
10.	*Dactylorhiza hatagirea*	Salampanja	–	50	18	–
11.	*Dioscorea deltoidea*	Dioscorea, Singli-mingli	380.00	315	533	5
12.	*Ephedra gerardiana*	Somlata	–	738	1,892	–
13.	*Jurinea dolomiaea*	Dhoop	4939.83	2,519	2,166	3,260

Contd.

1	2	3	4	5	6	7
14.	*Morchella esculenta*	Guchhi	2,800.89	1,209	475	490
15.	*Picrorrhiza kurrooa/ Gentiana kurroo*	Kutki/Karu	2899.50	610	321	343
16.	*Pinus gerardiana*	Neoza	600.00	116	47	403
17.	*Pistacia integerrima*	Kakarsinghi	278.25	138	15	17
18.	*Podophyllum hexandrum*	Bankakri	–	513	65	122
19.	*Potentilla nepalensis*	Dorighas	–	790	385	25
20.	*Rheum australe*	Chukri	–	478	120	94
21.	*Saussurea costus*	Kuth	667.92	575	212	321
22.	*Selinum tenuifolium*					
23.	*Selinum vaginatum/*	Bhutkesi, Mathosal	–	157	82	74
24.	*Taxus wallichiana*	Birmi, Talishpatra	335.50	9,361	5,538	782
25.	*Valeriana hardwickii*	Nihani	–	–	–	1,159
26.	*Valeriana jatamansi*	Mushkbala	2,014.43	1,840	7,084	483
27.	*Viola serpens*	Banafsha	2,800.89	97	47	71
	Others		10,328.00	18,441	7,519	9,387
	Total	38,124.72	47,546	31,321	17,826	

Over a dozen items are extracted under local names, the identity of which has to be ascertained and hence have not been included. Items pilfered and used for bonafide local use like Rattanjot, Mohra, Pharna, Kasturi patra, Nesar, Brahmakamal, Ninayin, Banajwain, Sardanda, Titre, Cedar wood, Rasont, etc. have also not been included. Major items like Harde, Bahera, Amla too are not included. Kaurtumba, Kamila, Chirayata, Karvibuti (Gallonkri), Shivlingibeej, Vijayabeej, Dhatura, Kantkari etc. too do not figure in the list.

Appendix 3. Endangered species of medicinal and aromatic plants

As a result of continuous and relentless extraction during the course of many decades, dozens of valuable species are facing danger to their survival in their natural abode in the absence of management practices for their regulated and scientific exploitation, protection, conservation and multiplication. Unregulated exploitation and disorganised trade practices are responsible for the sharp decline of the herbal wealth of the state. Local villagers in their off time collect the herbs and sell them to local shopkeepers in exchange for their daily need items. In the process, the collection rules are frequently flouted, the life cycle of the plant is broken and the quality of the material so collected gets deteriorated. A comprehensive list of threatened/endangered species of medicinal and aromatic plants in Himachal Pradesh is given below:

Sr. No.	Name of species	Common/ Vernacular name	Distribution in Himachal Pradesh
1.	*Aconitum chasmanthum* Stapf. ex Holmes	Mohra-bish	Quite rare in alpine meadows between 3500-4000 m in Chamba and Kullu district.
2.	*A. deinorrhizum* Stapf.	Mohra, Vatsnabh	Sparse in Shimla & Kinnaur around 3500 m.
3.	*A. heterophyllum* Wall. ex Royle	Atis, Patish	Temperate to alpine areas between 2400-4500 m.
4.	*A. voilaceum* Jacq. ex Stapf.	Mitha patish	Alpine slopes between 3600-4800 m in Kullu, Shimla, Kinnaur and Lahaul-Spiti.
5.	*Allium stracheyi* Baker	Jimmu, Farna	Occasionally met with in dry open slopes, temperate meadows between 2600-3000 m.
6.	*Angelica glauca* Edgew.	Chora, Chorak	Shimla, Kullu, Chamba, Kangra between 2400-3800 m in forest shades.
7.	*Aquilegia fragrans* Benth.	Ludud-dorje	Cold desert regions of Spiti between 3500-4500 m, often growing in rock crevices.

Contd.

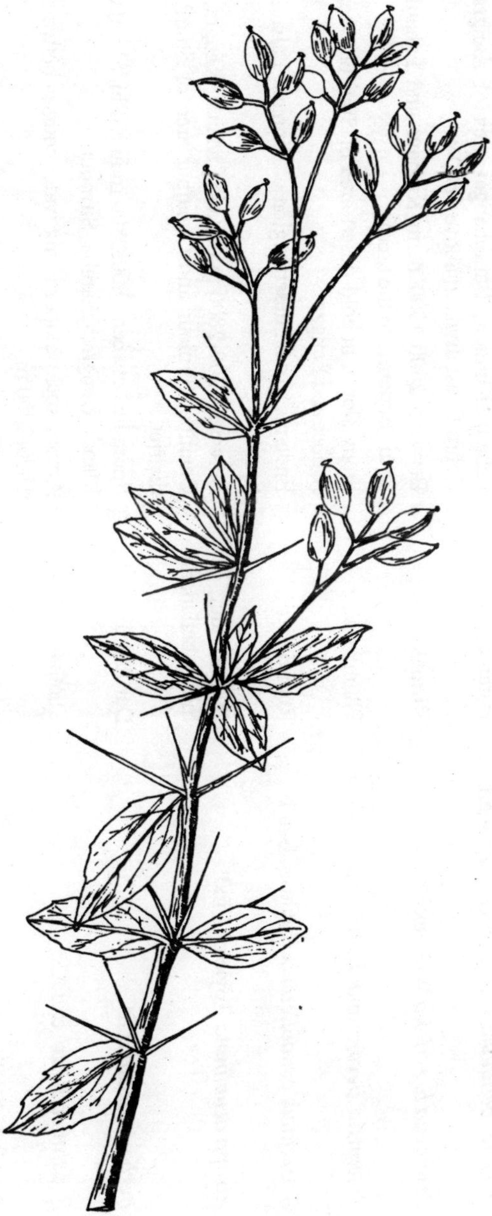

Berberis chitria

Sr. No.	Name of species	Common/ Vernacular name	Distribution in Himachal Pradesh
8.	*Aquilegia pulbiflora* Wall. ex Royle	Bishloghas	Scattered in temperate forest slopes between 2000-3200 m.
9.	*Arnebia benthami* (Wall. ex G.Don) John.	Ratanjot	Lahaul, Kinnaur, Barabhangal area of Kangra, Rohru forest division above 3500 m.
10.	*Artemisia brevifolia* Wall. ex DC.	Kirmala	Between 2600-4500 m in Kinnaur and Lahaul-Spiti (Endemic to the area).
11.	*Artemisia sieversiana* Wild.	Phumbasa	Stony slopes in cold desert area of Spiti between Kibber and Langza.
12.	*Astragalus condolleanus* Royle ex Benth.	Rudravanti	Endemic to cold desert region of Kinnaur, Lahaul-Spiti.
13.	*Atropa acuminata* Royle ex Lind.	Jharka, Jalgi, Indian belladona	Between 2200-2800 m in Kullu, Mandi, Kangra, Chamba, Sirmour and Chhajpur area of Shimla district.
14.	*Berberis aristata* DC.	Daruhaldi	Sporadic between 1800-2800 m in Kullu, Shimla, Mandi, Kangra, Chamba, Sirmour.
15.	*B. asiatica* Roxb. ex DC.	Chothar	Among rocky slopes along water courses between 1800-2800 m.
16.	*B. chitria* Lindl.	Kashmal	Sporadic between 1800-2800 m in Kullu, Shimla, Mandi, Chamba, Sirmour.
17.	*B. pseudoumbellata* Parker	Kushmol	Rare between 2400-2800 m in forest shade.

Contd.

Sr. No.	Name of species	Common/ Vernacular name	Distribution in Himachal Pradesh
18.	*Betula utilis* D. Don.	Bhojpatra	In snowline areas in patches or scattered/dilapidated form.
19.	*Bunium persicum* (Boiss.) Fedesch.	Kalajira	Lahaul, Kinnaur and Pangi-Bharmour.
20.	*Carum carvi* L.	Caraway	Dry temperate areas in Kinnaur, Lahaul-Spiti and Pangi-Bharmour.
21.	*Cinnamomum tamala* (Buch.-Ham.) T. Nees and Eberm.	Tejpatra	Drahal (Jogindernagar) and Darang forest in Palampur area.
22.	*Colchicum luteum* Baker	Suranjan	Udaipur range in Lahaul division (Endemic).
23.	*Costus speciosus* (Koen. ex Roten.) Sm.	Kemuk	Shivalik zone i.e., Nahan division, Una, Kangra, Bilaspur, Mandi, Solan below 1200 m.
24.	*Crocus sativus* Linn.	Kesar	Cultivated in Sangla farm, Kinnaur.
25.	*Curculigo orchioides* Gaertn.	Kalimusli	Sal forests is Paonta Range, Una and Kangra forest divisions.
26.	*Dactylorhiza hatagirea* (G. Don.) Soo	Salampanja	Kinnaur, Kullu, Lahaul-Spiti, Kangra, Shimla, Chamba above 3000 m.
27.	*Dioscorea deltoidea* Wall. ex Kunth	Singli-mingli Ganj, Dioscorea	Temperate zone between 2000-2800 m in Kullu, Shimla, Mandi and Sirmour.
28.	*Ephedra gerardiana* Wall. ex Stapf.	Somlata	Cold deserts in Kinnaur, Lahaul-Spiti.
29.	*Eulophia campestris* Wall. ex Lindl.	Salep	Pine forest and in hill slopes between 1600-2500 m.

Contd.

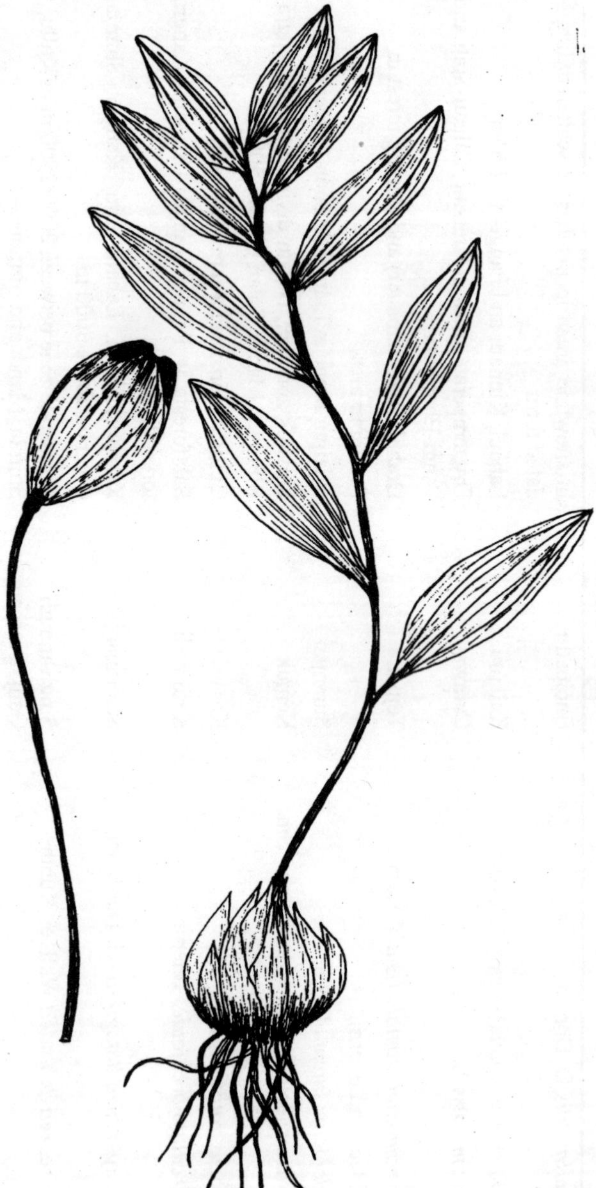

Fritillaria oxypetala

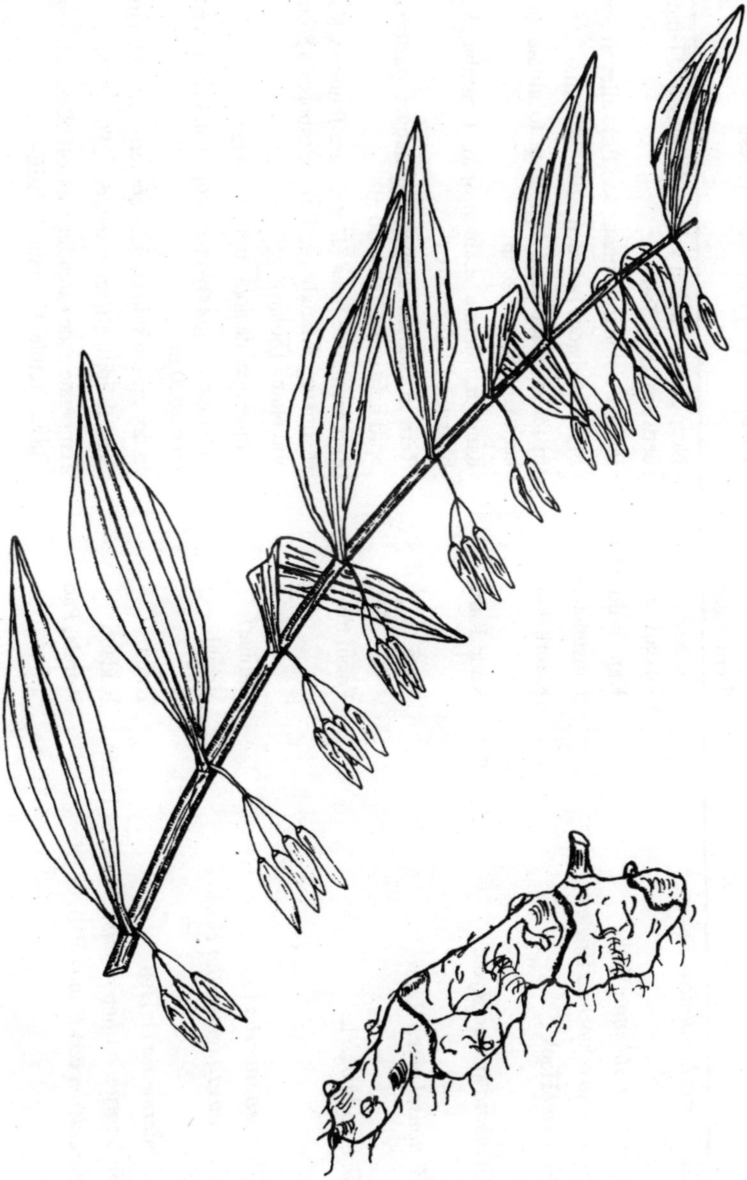

Polygonatum multiflorum

Sr. No.	Name of species	Common/ Vernacular name	Distribution in Himachal Pradesh
30.	*Eremurus himalaicus* Baker	Himalayan Foxtail-lily	Near glaciers in Pin valley, Keylong and Udaipur between 2800-3800 m.
31.	*Ferula jaeshkeana* Vatke	Kait, Wild Hing	Pooh division in Kinnaur and Pin valley in Spiti.
32.	*Fritillaria oxypetala* Royle	Lahsunia salep	In forest shades between 2200-2800 m.
33.	*F. roylei* Hook	Kshirkakoli	In forest shades and temperate to alpine slopes between 2800-3800 m.
34.	*Gentiana kurroo* Ryole	Kaur, Karu	Kinnaur, Shimla, Lahaul-Spiti, Chamba, Solan between 1800-3800 m.
35.	*G. tianschanica* Rupr.	Tikta	Rare in Sagnam slopes and Langza pastures in Spiti between 3200-4500 m.
36.	*Ginkgo biloba* L.	Fossil tree	Representative trees (1 or 2) are found in Kalpa (Kinnaur), Manali (Kullu), Mashobra (Shimla) and Nauni (Solan).
37.	*Gloriosa superba* L.	Kalihari	Sub-temperate areas upto 1300 m.
38.	*Habenaria edgeworthii* Hook. f.	Vridhi	In temperate grassy open slopes and field borders upto 2800 m.
39.	*H. intermedia* D. Don.	Ridhi	In dry grassy slopes, field borders upto 2600 m.
40.	*Hedysarum cashmerianum* Benth. ex Baker.	Rudanti	Lahaul-Spiti, Kinnaur between 2800-4200 m.
41.	*Heracleum candicans* Wall. ex DC.	Patrala, Padara	Temperate zone extending upto 4200 m in Kangra, Kullu, Shimla, Kinnaur, Lahaul-Spiti.

Contd.

Sr. No.	Name of species	Common/ Vernacular name	Distribution in Himachal Pradesh
42.	*Inula racemosa* Hook. f.	Pushkarmool	Pattan valley and Triloknath area of Lahaul-Spiti.
43.	*Jurienea dolomiaea* Bioss.	Dhoop	Alpine slopes in Chamba, Kullu, Kangra, Kinnaur, and Shimla.
44.	*Lilium polyphyllum* D. Don.	Kakoli	Temperate forest between 2400-3200 m.
45.	*Malaxix acuminata* Lindl.	Rishbhak	Sub-temperate areas, at Gallu forest of Joginder-nagar and Jutogh area of Shimla.
46.	*M. muscifera* (Lindl.) Kuntz.	Jeevak	Temperate and alpine areas in Shimla, Kullu, and Kangra districts.
47.	*Nardostachys grandiflora* DC.	Jatamansi, Balchharh	Jakha-kanda, Tangan khai and Manjiban areas bordering Kinnaur, U.P. and Great Himalayan National Park, Kullu.
48.	*Onosma hispidum* Wall.	Gajwanbhed	Open slopes and shrubberis between 2800-4000 m.
49.	*Physochlaina praealta* Hook. f.	Laltang	Pooh division in Kinnaur and Lahaul division in Lahaul-Spiti.
50.	*Picrorhiza kurrooa* Royle ex Benth.	Kutki	Sub-alpine areas alpine rocky slopes above 3500 m.
51.	*Pinus gerardiana* Wall.	Chilgoza	Endemic to Kinnaur and Holi Range in Bharmour.
52.	*Podophyllum hexandrum* Royle ex Camb.	Bankakri	Kullu, Kinnaur, Shimla, Lahaul-Spiti, Chamba, between 2800-3800 m.

Contd.

Sr. No.	Name of species	Common/ Vernacular name	Distribution in Himachal Pradesh
53.	*Polygonatum verticiliatum* (L.) All.	Mahameda	Temperate forest between 2400-2800 m in Shimla, Sirmour, Kullu, Kangra & Chamba.
54.	*P. cirrihifolium* (Wall.) Royle	Meda	Temperate forest between 2400-2800 m in Shimla, Sirmour, Kullu, Kangra & Chamba.
55.	*P. multiflorum* (L.) All.	Salam-mishri	Forest shades and temperate pastures between 2800-3800 m.
56.	*Rauvolfia serpentina* (L.) Benth. ex Kuntz.	Sarpagandha	Only in Paonta Range of Nahan division.
57.	*Rheum australe* D. Don.	Revandchini,	Above 3000 m on rocky boulders in Shimla, Kullu, Kangra, Chamba, Kinnaur, Lahaul & Spiti division.
58.	*R. moorcroftianum* Royel	Leechu	In high alpine slopes between 3500-4500 m.
59.	*Rhododendron anthopogon* D. Don.	Talsar	Alpine slopes and snowline areas between 3200-4500 m.
60.	*R. lepidotum* Wall. ex G. Don.	Talispatra	Temperate to alpine forests between 2800-4200 m.
61.	*Roscoea procera* D. Don.	Kakoli, Safed-musli	Temperate forests, open grassy slopes between 2400-2800 m.
62.	*Saussurea bracteata* Decne.	Chhota doda	Alpine slopes near glaciers, 3800-4800 m.
63.	*S. costus* (Falc.) Lipsch	Kuth	Pangi-Bharmour, Lahaul, Sangla Dogri and Tali-Thatch.
64.	*S. gossypiphora* D. Don.	Ghuggi Badshah	Bara & Chhota bhangal, Mani-Mahesh, Rupin Pass, 3800-5000 m.

Contd.

Sr. No.	Name of species	Common/ Vernacular name	Distribution in Himachal Pradesh
65.	*S. obvallata* (DC.) Edgew.	Brahm Kamal	Near glaciers and alpine rock boulders between 3800-4800 m.
66.	*S. sacra* Edgew.	Jogi badshah	Alpine slopes above timber line in Chamba, Kangra, Lahaul areas.
67.	*Semecarpus anaçardium* L. f.	Bhilava	Sporadic in Paonta range in Nahan forest divisions.
68.	*Skimmia laureola* Hook f.	Kasturi patra	Temperate forest between 2800-3200 m.
69.	*Staphylea emodi* Wall.	Nagdaun	Temperate forest between 2500-2800 m.
70.	*Swertia chirayita* (Roxb. ex Flem.) Karst	Chirayata	Mandi, Kullu, Kinnaur, Kangra, Kotgarh and Rohru division (rare).
71.	*Symplocos racemosus* Roxb.	Lodhra	In broad leaved temperate forest shades between 2400-2800 m.
72.	*Taxus wallichiana* (Zucc.) Pilger	Talispatri	Mixed in fir and spruce forest, between 2600-3200 m.
73.	*Thermopsis inflata* Camb.	Lamo	Fossilized sand areas in Spiti, 3500-4500 m.
74.	*Urginea indica* (Roxb.) Kunth.	Jangli Piaz	Shivalik range upto 1300 m.
75.	*Valeriana hardwickii* Wall.	Tagar	Forest shades and temperate slopes between 2200-3200 m.
76.	*V. pyrolaefolia* Decne	Nihani	Alpine meadows between 2800-3800 m.
77.	*Vanda tessellata* (Roxb.) Hook. ex G. Don	Banda	Lower valley of the state as an epiphyte on mango trees.
78.	*Zanthoxylum armatum* DC.	Tejbal, Tirmir	Warmer areas of the state up 1400 m.

Appendix 4. Plant species having fluctuating demand

There is a long list of such species growing in different parts of the region which are collected on small scale or often on large scale as and when there is a demand from the market. Some important drug plants are enumerated hereunder:

Sr. No.	Botanical name	Trade name	Parts used
1.	*Abies pindrow* Royle	Tosh, Talispatri	Leaves/ needles
2.	*Abrus precatorius* Linn.	Ratti, Gunja	Seeds, Root
3.	*Acacia nilotica* (L.) Willd ex. Del.	Kikar, Babul	Bark
4.	*Achillea millefolium* Linn.	Biranjasif	Herb
5.	*Achyranthes aspera* Linn.	Apamarga, Puthkanda	Herb
6.	*Aconitum deinorrhizum* Stapf	Vastsanabh,	Roots
7.	*A. chasmanthum* stapfex Holmes	Mohrabish, Banbalong, bish	Roots
8.	*Adiantum lunulatum* Burm.	Hansraj, Kalijhant	Fronds
9.	*Aesculus indica* Colebr	Horse-chestnut, Khanoor Kuttakhar	Seeds
10.	*Albizia lebbek* Benth.	Sirish	Bark
11.	*A. procera* (Roxb.) Benth	Safed siris	Bark
12.	*Allium carolinianum*	Laot (Spiti)	Flowers, Leaves & pseudobulbs.
13.	*A. jaquemontii*	Koche, Pharna	Flowers, Leaves & pseudobulbs.
14.	*Aloe barbadensis* Mill.	Ghrit kumari	Leaf pulp
15.	*Arctium lappa* Linn	Jangli kuth, Burdock	Roots
16.	*Arnebia benthami*	Ratanjot	Roots
17.	*A. euchroma* (Royle) Johnston	Ratanjot	Roots
18.	*Artemisia brevifolia*	Santonin plant, Nurcha	Flowering shoots

Contd.

Sr. No.	Botanical name	Trade name	Parts used
19.	*A. maritima* var. *thomsoni*	Seinski	Flowering shoots
20.	*Artemisia dracunculus* Linn.	Teragon	Floral shoots
21.	*A. scoparia* Waldst. & Kit.	Donajhan, Barna	Fruiting shoot
22.	*A. vulgaris* Linn.	Chharmar	Floral shoots Essential oil
23.	*Artocarpus lakoocha* Roxb.	Lakuch	Fruit & Bark
24.	*Asparagus abscendens* Roxb.	Sahasrapal, Shatavari, Safed musli	Roots
25.	*Azadirachta indica* A. Juss	Neem	Leaves, Bark, Seed and oil
26.	*Bauhinia variegata* Linn.	Kachnar	Bark
27.	*B. vahlii* W & A	Maljhan, Taur	Seeds, leaves bark
28.	*Betula utilis* D. Don	Bhojpatra	Papery bark
29.	*Boerhavia diffusa* Linn.	Punarnava	Roots
30.	*Bryonopsis laciniosa* (Linn) Naud	Shivlingi	Seeds
31.	*Butea monosperma* (Lam Kuntze)	Palash	Seeds
32.	*Callicarpa macrophylla* Vehl	Priyangu	Seeds
33.	Calotropis procera (Ait) R.Br	Aak	Roots
34.	*Cassia fistula* Linn.	Amaltas	Pods and Roots Bark
35.	*C. occidentalis* Linn.	Kasmard	Seeds
36.	*C. tora* Linn	Chakramard	Seeds
37.	*Cedrus deodara* (Roxb) Loud.	Devdaru	Wood, dust and cedar oil
38.	*Celastrus paniculatus* Willd.	Malkangni	Seeds

Contd.

Sr. No.	Botanical name	Trade name	Parts used
39.	*Centella asiatica*	Mandookparni, Brahmi	Herb
40.	*Centratherum-anthelmin-ticum* (Wild.) Kuntze	Kalijiri, Ghorajiri	Seeds
41.	*Citrullus-colocynthis* Schrad	Indrayan	Fruit
42.	*Corydalis govaniana* Wall	Bhutjata	Roots
43.	*Cryptolepis buchanani* Roem & Schult	Kali Sariva	Roots
44.	*Curculigo orchioides* Gaertn.	Kali Musli	Roots
45.	*Cymbopogon martinii* Var. 'Motia'	Makora ghas	Aerial parts
46.	*Cyperus rotundus* Linn.	Nagarmotha	Roots
47.	*Datura metel* Linn.	Kala dhatura	Whole plant
48.	*D. stramonium* Linn.	Safed dhatura	Panchang or seeds
49.	*Desmodium gangeticum* DC.	Shalparni	Plant
50.	*Dioscorea bulbifera* Linn.	Varahikand	Tubers
51.	*Eclipta prostrata* Syn. *E. alba* Hassk.	Bhringraj	Plant
52.	*Fumaria parviflora*	Pitpepra	Herb
53.	*Gentiana kurroo* Royle	Karu	Roots
54.	*Geranium nepalense* Sweet	Laljari	Root
55.	*Gloriosa superba* Linn.	Kalihari, Angishikha	Tubers/Seeds
56.	*Hedychium acuminatum* Rosc.	Kapurkachri, Shati, Banhaldi	Rhizomes
57.	*Helicteres isora* Linn.	Marorphali	Twisted fruits & bark
58.	*Hemidesmus indicus* R. Br.	Sariva, Anantamool	Roots
59.	*Holarrhena antidysenterica* Wall.	Kutaj, Kura, Indrajau	Bark, seeds

Contd.

Sr. No.	Botanical name	Trade name	Parts used
60.	*Hyoscyamus niger* Linn.	Khurasaniajwain	Fruiting branches, leaves and seeds
61.	*Hyssopus officinalis* Linn.	Jufah, Tengu	Flowering tops
62.	*Inula racemosa* Hook. f.	Pushkarmool	Roots
63.	*Iris germanica* Linn.	Safed Bach	Rhizomes
64.	*Jasminum humile* Linn.	Pitmalti	Leaves, flowers
65.	*J. officinale* Linn.	Chameli	Leaves, flowers
66.	*Juglans regia* Linn	Akhrot	Bark, nut
67.	*Juniperus communis* Linn.	Hauber	Fruits (Berry)
68.	*Justicia adhatoda* Syn. *Adhatoda vasica* Nces	Adusa, Basuti, Vasaka	Leaves
69.	*J. macropoda* Boiss.	Dhupi, Leur	Fruits, twigs
70.	*J. recurva* Ham.	Hauber, Bethar	Twigs
71.	*Leucas cephalotes* Spreng.	Dronpushpi	Herb
72.	*Litsea umbrosa* (Nces) Hook. f.	Chirindi	Seeds/ Fatty oil
73.	*Lilium polyphyllum* Don.	Kshir Kakoli	Bulbs
74.	*Mallotus philippinensis* Muell-Arg.	Kamila	Fruit powder (Kamila powder)
75.	*Martynia annua* Linn.	Kaknasa	Fruits
76.	*Mentha longifolia* Huds.	Jangli pudina	Aerial parts
77.	*Milletia auriculata* Baker ex-Brand	Gauj	Bark, seeds
78.	*Mucuna pruriens* Hook.	Kaunch	Seeds
79.	*Murraya koenigii* (L.) Spreng.	Mithinim, Curry leaf	Leaves
80.	*Myrica esculenta* Ham.	Katphal	Bark
81.	*Nardostachys grandiflora* DC.	Jatamansi	Roots

Contd.

Sr. No.	Botanical name	Trade name	Parts used
82.	*Olea ferruginea* Royle	Kahu, Wild olive	Fruits/oil
83.	*Onosma echioides*	Gojihva, Ratanjot	Leaves and roots
84.	*O. bracteatum* Wall	Gojihva, Ratanjot	Roots/plant
85.	*Oroxylum indicum* Vent.	Shyonak, Arlu	Bark, Seeds
86.	*Paris polyphylla* Sm.	Kshir-kakoli	Tuber
87.	*Pistacia khinjuk* stocks	Kakarsingi	Galls
88.	*Plumbago zeylanica* Linn.	Chitrak	Roots
89.	*Podophyllum hexandrum* Royle	Bankakri	Roots
90.	*Polianthes tuberosus*	Tuberose, Rajnigandha	Flowers/oil
91.	*Polygonatum verticillatum* Allioni	Mahameda	Rhizome
92.	*P. cirrifolium* Royle	Meda	Rhizome
93.	*Polygonum aviculare* Linn.	Anjubar	Roots
94.	*Potentilla fulgens* Hook.	Bajradanti	Roots
95.	*Pueraria tuberosa* DC	Vidarikand	Tubers
96.	*Punica granatum* Linn	Darim	Seeds, fruit and root bark
97.	*Randia dumetorum* Lam.	Rada, Madanphal	Fruits
98.	*Rauvolfia serpentina* Benth ex Kurge.	Sarpagandha	Roots
99.	*Rheum australe* Wall. ex Meissn	Chukri, Archa, Rhubarb, Revandchini	Roots
100.	*R. moorcroftianum* Royle	Rhubarb	Roots
101.	*Rhododendron campanulatum* D. Don	Kashmiripatha	Leaves
102.	*R. anthopogon* D. Don	Talispatra	Leaves
103.	*Ricinus communis* Linn.	Erand	Seeds & Oil
104.	*Roscoea purpurea* Royle	Safed Musli, Kakoli	Roots
105.	*Rubia manjith* Roxb.	Manjishtha	Roots
106.	*Sapindus mukorossi* Gaertn.	Ritha, Soapnut	Fruit

Contd.

Sr. No.	Botanical name	Trade name	Parts used
107.	*Solanum viarum* Dunal	Bari Kantkari	Aerial parts/ berries
108.	*S. surattense* Burm f.	Laghu Kantkari	Aerial parts
109.	*S. nigrum* Linn.	Makoi	Aerial parts
110.	*Spondias pinnata* Kurz	Ambara	Fruit
111.	*Swertia chirayita*	Chirayata	Whole plant
112.	*S. angustifolia* Buch-Ham	Chirayata (local)	Whole plant
113.	*S. purpurascens* Wall	Nepali chirata	Whole plant
114.	*Symplocos paniculata* Wall	Lodhra	Bark
115.	*Terminalia arjuna* W & A	Arjun	Bark
116.	*T. tomentosa* W & A	Asan	Bark
117.	*Thalictrum foliolosum* DC	Pilijari mamira	Roots
118.	*Thevetia peruviana* (Pers) Schum	Pila kaner	Root bark
119.	*Tinospora cordifolia* (Willd) Miers	Giloye	Twigs
120.	*T. malabarica* (Linn) Miers.	Guduchi, Gulje	Twigs
121.	*Trichosanthes cucumerina* Linn	Patol	Fruit
122.	*Uraria lagopoides* DC	Prishniparni	Plant
123.	*U. picta* Desv.	Prishniparni	Plant
124.	*Valeriana hardwickii* Wall.	Nihani, Tagar	Root stock
125.	*V. pyrolaefolia* Decaisne	Nihani	Roots
126.	*Viburnum nervosum* Don.	Tilani	Bark
127.	*Withania somnifera* Dunal	Ashwagandha	Roots
128.	*Woodfordia fruticosa* Kurz.	Dhai, Dhataki	Flowers
129.	*Zizyphus mauritiana* Lamk.	Ber	Fruit

Appendix 5. Unexploited rare species of medicinal and aromatic plants

There are a large number of herbs which have potential for their commercial exploitation. Many herbs are used in household remedies, which have not been worked upon, but local *Vaids* in rural and far flung areas are using them from generation after generation with great efficacy in a variety of ailments. There exist many verses, lores and mores; songs and phrases which signify the importance of these plants. Uniyal and Chauhan (1971, 1972, 1973) documented the traditional uses of medicinal and aromatic plants of Kangra and Kullu forest circles. Chauhan (1984) has described 85 plants of ethnobotany from Pabbar valley of Shimla district and same author (1988, 1989) has presented ethno-medico-botany of 30 selected herbs from different parts of the state and about 2 dozens from Spiti respectively. During recent studies by Chauhan and his students, valuable information about the traditional local uses of the plants from Spiti, Nauni-Solan, Churdhar, Rohtang, Chhota and Bara Bhangal areas has since been documented. Some rare and interesting plants are enumerated hereunder:

Sr. No.	Botanical name	Local name	Use
1.	*Achyranthes aspera* Linn.	Puthkanda	The local people observe fast on 'Rishi Panchmi' and they use it as 'Datun' on that day. The juice obtained by crushing the leaves is massaged on the waist of the patient to get relief from waist pain. This massage is applied by making the patient sit in sun for 2-4 hours. It is also used for snake bites; the vegetative buds are crushed and applied as paste on the wound and also given internally. Panchang (entire plant) is used in asthma, as it is smoked in hooka and also taken with honey.

Contd.

Sr. No.	Botanical name	Local name	Use
2.	*Achyranthes bidentata* Bl.	Umblokando, Umblakanta	The root of the plant is given in snake bite. About three mashas of root powder is mixed and powdered with equal quantity of smoked tobacco residue from hooka (hubble-bubble) and is applied in the form of a paste to the snake bitten/ swollen organ after giving a proper cross cut. The patient is not allowed to sleep for 12 hours atleast. Sufficient ghee is given to the patient orally.
3.	*Aconitum deinorrhizum* Stapf.	Mohra, Mouro bish	Roots are highly poisonous and instances of suicides and deaths are accredited to it. Roots are kept hidden and out of reach of ladies and children and taken out very cautiously by the old/matured persons for medicinal purposes in small pieces. The roots are smoked in hooka (hubble-bubble) in acute gastric pain, toothache and severe headache; used as massage in rheumatic joints, paralytic parts etc. It is never applied on abrased skin for fear of poisoning.
4.	*Aconitum violaceum* Jacquem. ex. Stapf.	Dudhi-mohra	The roots are rubbed and dissolved in nursing mother's milk to be administered to the infants

Contd.

Sr. No.	Botanical name	Local name	Use
			suffering from severe diarrhoea, stomatitis and allied gastric complaints.
5.	*Aconogonum rumicifolium* (Royle)	Goronthu, Choarh,	The leaves are cooked and squeezed. The water is thrown away and the cooked leaves are prepared as vegetable by the shepherds in Dodra-Kwar and Jakha-Pandar area. The taste is said to be acidic. The root is regarded as an antidote in Aconite poisoning and is chewed to satiate the thirst. It is reported to be an antidote to snake bites and scorpion stings. The paste of the leaves is applied locally in abscesses and boils.
6.	*Acorus calamus* Linn.	Bach, Bare	The roots are dried, crushed and put in a cloth which is tied to the chest of a patient suffering from cough. It helps in expulsion of phlegm and gives relief. Decoction of the roots is made and gargles are done to get relief from toothache.
7.	*Adiantum venustum* *G.* Don.	Dhumtolo, Bai-re-barum	The paste of the aerial parts of the fern is made into six tablets of the size of the almond nut and one tablet, two times a day is given for three days in cases of leucorrhoea. Its juice can be extracted and

Contd.

Sr. No.	Botanical name	Local name	Use
			two spoons each in the morning and evening are given for three days. Only butter and chapati should be taken as food during this period. Chillies, potatoes, urad, rajmash, oily and spicy items should be avoided at all costs for the drug to be more effective.
8.	*Aesculus indica* Colebr.	Kanor, Khanor	The fruits are dried and beaten into flour, washed several times in water to remove the bitter taste, dried and kept for use as tonic for ladies; considered useful in colic and stomach problems. The dried fruits are beaten into coarse particles alongwith shell cover and mixed with fresh or fried soyabean, are given to young sheep and goats as tonic during winters. Leaves are used as dry fodder. The oil expressed from the fruits is used in healing of wounds and the bark is applied in the form of a paste in dislocated joints.
9.	*Ainslaea aptera* DC.	Karvi-buti	The powder of the dried root in the dose of two mashas (Approx. 2 gms) with luke warm water is an effective cure for acute gastric pain. The root is extremely bitter and may be useful as a wormicide.

Contd.

Sr. No.	Botanical name	Local name	Use
			To avoid the bitter taste, a piece of jaggery is taken if required. Shepherds have been using it effectively in hill diarrhoea also. Fresh roots can be chewed in emergency and its effect in acute stomach pain and diarrhoea is quite encouraging.
10.	*Ajuga bracteosa* Wall. and *A. parviflora* Benth.	Ratpacho, Nilkonthe	The plant is considered as bitter tonic. The leaves are used to erase deposition on tongues of children suffering from stomach complaints and fever. The infusion or the juice is used in fevers and gastric pains. Leaf juice is dropped into the ear to treat earache and otorrhoea. The pounded leaves in the form of pea size pill are given in pneumonia and typhoid fever with satisfactory results. Three tablets a day are given in fever, bronchitis and in the cases where blood clots appear in phlegm while coughing.
11.	*Allium carolinianum* DC.	Laot, Jangli Lahsun	Leaves are cooked as vegetables and flower heads are used as condiment. It is a beautiful plant for ornamental purpose. The dried, pink flower heads are sold in the market.

Contd.

Sr. No.	Botanical name	Local name	Use
12.	*A. przewalskianum* Regel.	Koche, Pharna, Saanp-Saperia	Leaves form a flavoured condiment. Bulbs and flower heads are also used as vegetable and spice. Dried, cut parts of leaves are marketed from the area. In case of snake bite, the bulbs are eaten with milk.
13.	*A. rubellum* Beib	Neulagu	Pseudobulbs are used as vegetable and the aerial parts are used as a prized spice. Useful as anti-inflammatory and anti-bacterial.
14.	*Androsace rotundifolia* Hardw.	Golpatte-phulnu, Nirodhak-buti	The leaves of the herb alongwith a few needles of deodar are powdered and formed into tablets of the size of apricot kernel with the addition of gur (Jaggery). As per one informant, six such tablets are prepared and given in the dose of one tablet three times a day alongwith about 25 gm of ghee and same quantity of gur (juggery) for two days from the first day of the onset of menses for affecting birth control. As per the practice of a late lady, one tablet a day is given for four days from the first day of menses for similar effect.
15.	*Aquilegia fragrans* Benth.	Lunded dorje, Lamo	The herb is diuretic, diaphoretic, antiscorbutic and

Contd.

Sr. No.	Botanical name	Local name	Use
			tranquilizing. In homoeopathy, it is employed for hysteria, somnolency, dysmenorrhoea, and chronic skin troubles. Roots are used for removal of kidney stones. Seeds are used as an oxytosic and for jaundice; contain a fatty oil and resin.
16.	*Arctium lappa* Linn.	Jangli kuth, Burdock	Roots contain an essential oil, inulin (45%) and is used as diuretic, diaphoretic, in gout and skin affections. Seeds contains arctin and seed tincture is used for psoriasis and toothache.
17.	*Arnebia euchroma* (Royle) Johnston	Khamet, Ratanjot	Plant contains Ac-Shikonin and related compounds. Red dye from the roots is used for dyeing silk and wool. The roots are used against tootache, ear-ache and has anti-cancer activity. It has been found to have antibacterial properties as well.
18.	*A. guttata* Bunge	Laljarhi	Root are used as Ratanjot, yield red dye and the paste is applied to cuts and wounds and also in fire burns.
19.	*Artemisia dracunculus* Linn.	Barma, Tarragon, Mugworty	Aromatic leaves are used as stomachic, stimulant, and febrifuge and also as spice. The herb yields an essential oil, used for

Contd.

Sr. No.	Botanical name	Local name	Use
			flavouring vinegar, sauces, salads, canned soups and liquors; also employed as cheap perfume.
20.	*Artemisia scoparia* Waldst. & Kit.	Jhau	The leaves are dried, powdered and mixed with ghee and is massaged on joints to get relief from joint pains.
21.	*Asparagus abscendens* Roxb.	Bankhelti, Sainsarbuti	The roots are used to increase the milk yield in cattle and also to get germ free milk.
22.	*Atylosia mollis* Benth.	Ban-kulthi	The paste of the root is applied against snake bites to remove the poison.
23.	*Bauhinia vahilii* Wright & Arn.	Tour, Jall	Root bark is powdered, made into paste in water and applied to fire burns as a fine layer. The paste is not allowed to dry otherwise it will squeeze the skin. To avoid this problem, the watery juice from the bark is sprinckled over the paste. It is claimed that with its application, no scar of fire burn is left behind.
24.	*B. variegata* Linn.	Kachnar, Karyala	The bark is dried and powdered and the decoction is used for garling to relieve throat diseases.
25.	*Begonia picta* Linn.	Khatta patta	Used in Thaneda (pain related to navel)-the leaves are crushed and consumed with cold water.

Contd.

Sr. No.	Botanical name	Local name	Use
26.	*Bergenia ciliata* Wall.	Pathan Belg	In backache; the roots are debarked and crushed and mixed with milk and consumed. On burns; it is applied after mixing with curd.
27.	*Bidens pilosa* Linn.	Kando, Kumbro, Chorpushpi	The leaves and young shoots are used as a vegetable curry in skin diseases including leprosy. The leaf juice is used in ulcers and wounds. The achenes are used as anthelminitc for domestic cattle.
28.	*Bistorta macrophyllum* (D. Don.) Sonak	Chhoti ninayin, Kanda-reninai	The purplish coloured, recurved, rhizomatic rootstock is sweetish in taste and the paste of it is given to children and infants as a cure for stomach problems. Older patients chew the roots as such for similar problems. The dried roots are arranged in a string of thread and offered by Kinners to their favourite persons as a token of regards and good wishes. The roots are eaten in dry state also. The flowering spikes are arranged in scented *Pleurospermum* garlands to make it more showy and attractive.
29.	*Boerhaavia punarnava* Linn.	Itsit	A paste of the entire plant is applied to get relief

Contd.

Sr. No.	Botanical name	Local name	Use
			from pain resulting from poisonous insect bites.
30.	*Bupleurm falcatum* Linn.	Kaligewar	Plant yields saponins (2.37% on dry weight basis), rutin (0.3-0.4%) and siakosaponins (2.06-3.02%) from the rhizomes. Herb is used as diaphoretic and for stomach and liver complaints.
31.	*Caltha palustris* Linn.	Pipling-tsa, Munire	Poisonous herb. Plant contains coumarins, a pigment calthaxanthin. Flower buds are kept in vinegar and used as capers.
32.	*Canna orientalis* Rosc.	Shivdarshan	Leaf juice is extracted and 1-2 drops are put in the ear to relieve ear ache.
33.	*Carduus nutans* Linn.	Rucharg, Ruchoa, Musk Thistle	Plant contains Kaempferol, -3-alpha-L- Rhamnofuranoside and acacetin-7-beta-D- glucoside. Flowers are used as febrifuge and are edible alongwith thick pith of the stem. Roots are often eaten raw and the root extracts can be used to kill round worms.
34.	*Cassiope fastigiata* D. Don.	Hieunshelo, Hieunsheli	The leafy twigs are ground into a paste and applied to fire burns. It exerts immediate cooling effect and is effective in healing the wound also. It is kept in houses as an emergency medicine for this purpose.

Contd.

Sr. No.	Botanical name	Local name	Use
35.	*Celtis australis* Linn.	Khouruk, Kharik, Khirak	The bark of the tree is made into a fine paste and is applied to the hurt (as a result of physical injury) muscular part; where the part of the body has become blue-back with the skin intact and is quite painful. The application of the paste is said to declot the blood and subsides the pain and helps in healing the wound. Fruits are sweet and eaten in amenorrhoea and colic. Bear enjoys the fruits during July-September.
36.	*Centratherum anthelminticum* O. Kuntze	Kalijiri	The fruit is taken with water to relieve cough and stomach-ache.
37.	*Chenopodium album* Linn.	Bathu	The young leaves are used as vegetable and given to patients suffering from leucoderma. The juice of the leaves is also applied on the depigmented skin spots for about a month. It is said to be effective in early stages of the disease. The decoction of seeds is given in large doses to induce abortion in women. In small doses, it is supposed to be effective in cough and coryza. The flour prepared out of its seeds is utilized as a catalyst for preparing

Contd.

Centratherum anthelminticum O. Kuntze

Sr. No.	Botanical name	Local name	Use
			"soor" a local alcoholic beverage. In far flung areas; the seeds are cooked along with cheena (*Panicum milliaceum*) and shathi rice as staple food occasionally.
38.	*Chrysanthemum leucanthemum* Linn.	Chhambar	The leaf juice is extracted and applied on wounds and cuts for early healing.
39.	*Cissampelos pareira* Linn.	Patindu, Batindu	In case of heat in children, the leaves are crushed and applied on the palate. For dysentry in children-roots are dried and powdered. If it is due to heat, it is given with milk or if due to other reason, it is given with honey. Root is used in snakebite, tooth-ache and leucorrhoea.
40.	*Citrus aurantifolia* Swingle	Nimbu	The fruits are dried, burnt and made into ash. Ash is mixed with greater cardamom and taken with honey to relieve cough and cold.
41.	*C. jambhiri* Lush	Khatta Jhamtu	The leaves are crushed and juice extracted, which is used for washing hair and later 'amla' oil is applied to remove dandruff.
42.	*Cirsium arvense* (L.) Scop.	Bhungchee, Bhungcho	The root is eaten fresh for stomach complaints, especially useful in pains due to roundworm infestation. The taste is agreeable with a faint peculiar

Contd.

Sr. No.	Botanical name	Local name	Use
			aroma. The juice extracted from roots is given to relieve the burning sensation in the liver in snake bites. It exercises a cooling effect.
43.	*Clematis nutans* Royle	Kauni-bel	Leaves are highly pungent, chewed in tooth-ache and infusion of leaves is used as errhine as it causes sneezing.
44.	*Codonopsis ovata* Benth. and *C. rotundifolia* Benth.	Sardandi, Khiri, Sardanda	The milky roots are cut into slices and added to goat milk to make it viscous and gel type. The roots can be eaten raw, are sweetish in taste, considered as a good physical and sexual tonic. The unbranched radish like root acts as aphrodisiac and stimulant and is known as sardanda. Sardandi has thick robust, long, branched roots and is attributed with feminine properties and its use is claimed to be of cooling nature.
45.	*Colocacia antiquorum* (L.) Schott.	Arbi, Kuchawari, Ghandiali	The cooked and peeled corms are cut into piece in the evening and kept on roof or open places uncovered overnight to allow dew drops to fall on it. In the morning, the cut pieces are given with honey to the patient suffering from jaundice to effect relief.

Contd.

Sr. No.	Botanical name	Local name	Use
46.	*Colutea nepalensis* Sims.	Braa	Leaves are purgative and also used as an adulterant of senna leaves (Cassia senna). Seeds are emetic.
47.	*Cotinus coggygria* Seop.	Tung	It is used as 'Datun' (tooth stick) commonly and the sap is applied on wounds for healing.
48.	*Cuscuta reflexa*	Sutrawlo, Akashbel	The plant is used in piles. 20-35 gm of crushed herb, 5-7 seeds of *illaichi* (lesser cardamon) and about 25 gm of kuje-misri are all put in a glass of water in the evening and the decanted solution is given for drinking regularly for eight days in the morning to cure piles. The paste of the plant with purified butter is applied effectively in sinus. In the area, legend goes that who-so ever finds the thread like white root of this plant and keeps it in milk in his house and changes in fresh milk pot daily, is reputed to bring prosperity. As a parasite, there is no scientific basis of its bearing the roots, but even then this lore is much talked about).
49.	*Cynodon dactylon* Pers.	Joob, Doob	The grass shoots are used for sprinkling holy water during poojah. The shoots with token money is offered to the parents of

Contd.

Sr. No.	Botanical name	Local name	Use
			new born male issue as a symbol of good will and blessings called as *jubadali*. The parents concerned happily return double the amount to the offerer. The juice of the plant is used to check nose bleeding, called *nathar* or *nakseer*. A few drops are dropped into the nostrils and it acts as styptic. The whitish roots, cleaned and washed are used as memory sharpner.
50.	*Cynoglossum glochidiantum* Wallich. ex. Benth.	Kumbro, Khodre, Kuri	The juice from the properly washed leaves is applied like eye drops in conjunctivitis and reddening of the eyes. The juice is purplish in colour. The flowers are collected and added in a small bottle containing distilled water and tightly corked. After a few days, the solution so obtained is filtered and used as eye drops with efficacy.
51.	*Delphinium vestitum* Wall.	Salyan, Nirbisha	The Tibetan use the decoction of the plant in fevers and as an astringent in hill diarrhoea. The flower powder mixed with bland oil is applied to head (hairs) to kill lice. The powder of the roots is reported useful in healing ulcers and wounds in cattle.

Contd.

Sr. No.	Botanical name	Local name	Use
52.	*Dicliptera bupleuroides* Nees	Pipri, Ludra-buti, Sam-ki-buti	A fine thick paste of the herb (leaves and top shoot) is applied on the wounds in snake bite. After a few minutes of its applications, yellowish secretion is reported to ooze out in the form of poison. The juice of leaves is given to children suffering from fever and stomach troubles in the dose of half to one tea spoon. In bone fractures, the pounded leaves are given in the form of a tablet. It is claimed to be effective in cases of cattle, sheep and goats as well.
53.	*Dioscorea kamoonensis* Syn. *D. kumaonensis* Kunth	Magaiyun	The tubers are eaten as vegetable by the women suffering from leucorrhoea.
54.	*Duchesnea indica* Andr.	Kaphal	The leaves are crushed and applied as paste on skin diseases and also on wounds and cuts.
55.	*Echinops niveus* Wall.	Oontkatara	Root bark is powdered and mixed with honey to make 2-3 g tablets and taken internally to cure cough and asthma.
56.	*Eleusine coracana* Gaertn.	Koda, Kodra, Mandua	The pudding prepared of its flour with gur is given in colds, coughs and coryza. The flour is also utilized in preparing local beverage called as "soor"

Contd.

Sr. No.	Botanical name	Local name	Use
			and good quality product is used as medicine in acute stomachache, in whooping cough, common colds etc.
57.	*Equisetum arvensis* L.	Harjori	The infusion of the herb in the form of a raw tablets or juice is given orally in bone cracks and fractures for 7-8 days. The paste made by boiling the herb in cow's urine is also applied externally on the affected part after fixing the bone properly. It helps in early unification of broken bony pieces, provided, the part is properly bandaged. The juice expressed from the stem and branches is given for 2-3 days with milk and sugar in leucorrhoea.
58.	*Eremurus himalaicus* Baker	Tache, Himalayan	Hordenine is isolated from the plant. Young leaves are used as vegetable. It is a very attractive plant when in bloom for its long tail like, stout, creamy white, spicate inflorescence.
59.	*Euphorbia pilosa* Linn.	Shodhan	The plant has got both emetic as well as purgative properties. Shoots are emetic; whereas the root is purgative. The milky latex has also the similar properties in respective parts. If a part of branch/stem is plucked, the latex oozing

Contd.

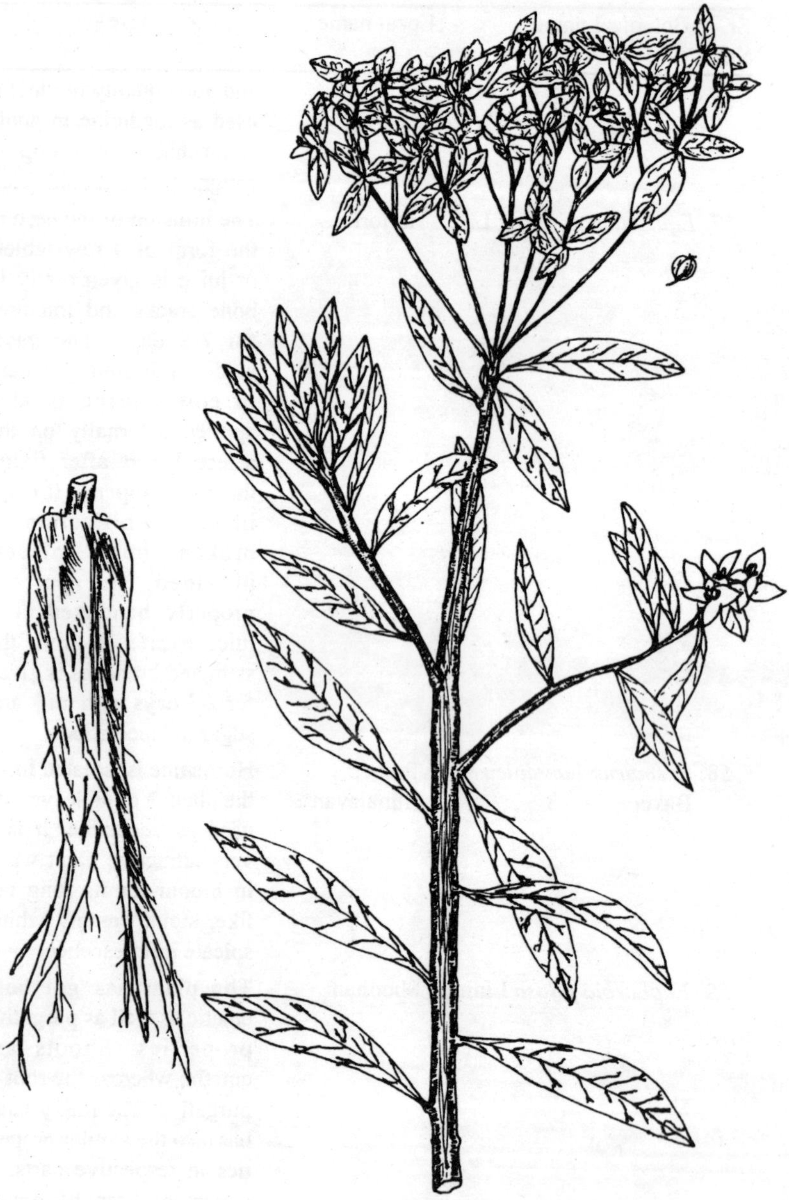

Euphorbia pilosa Linn.

Sr. No.	Botanical name	Local name	Use
			out from the separated part is considered as purgative. It is mixed in *sattu* or *gur* and stored properly for use. The latex from the broken top of the plant which oozes upwards is collected in the same way and used as emetic at the time of need. The root part, when ground into a paste or powder by moving the grinding stone only in one direction repeatedly, causes purgation and if as a normal process, the stone is moved backward and forward while grinding, the use of this material causes both purgation and vomiting (detailed investigation are suggested).
60.	*Euphorbia royleana* Linn.	Danda Saro	Its milky latex is applied to fire burns on legs, but precaution is taken that water does not touch the affected area. After drying of the sap, it is washed after 5-6 hours. In case of snake bites, the decoction prepared from small roots (6 masha) and kali mirch 6 masha) is given to the patient as a medicated drink.
61.	*Fraxinus xanthoxyloides* (G. Don.) DC.	Pipjul, Thuma	The bark is an esteemed cattle medicine. The decoction of the stems/twigs is

Contd.

Sr. No.	Botanical name	Local name	Use
			applied to wounds and bone fractures. The green paste obtained as a result of extraction is locally applied to the wounds caused by horse bites, wherein part of skin & flesh is removed as a result of individual fights.
62.	*Galium rotundifolium* Linn.	Manjith	Glycoside asperuloid is present in the plant. It is used against colic, sore throat and chest complaints. Pounded leaves are applied to check bleeding from fresh cuts.
63.	*G. verum* Linn.	Safed manjith	Roots yields asperuloside and salicylic acid. Yields yellow dye used for colouring butter and cheese and for dyeing woolens. Plant juice is used for curing epilepsy and hysteria.
64.	*Gentiana tianschanica* Rupr.	Karu, Kour, Nilakant, Tikta	Roots are used as Indian Gentian for *Gentiana kurroo* as bitter tonic, stomachic, liver stimulant, blood purifier, antipyretic and useful in urinary affections. Also used as a Masala for fattening horses in Spiti and Ladakh.
65.	*Gentianella moorcroftiana* (Wall. ex. G. Do) Airy Shaw	Tikta	The decoction of the whole plant is used as febrifuge, antipyretic, blood purifier; in cough as expectorant and also to cure bile complaints.

Contd.

Sr. No.	Botanical name	Local name	Use
66.	*G. paludosa* Hook.	Tikta	The whole plant is used in fevers and as blood purifier, either singly or in combination with other herbs.
67.	*Geranium wallichianum* D. Don. ex. Sweet	Polo, Purlomendo	Herb contains bitter substance geranin and as astringent is used against toothache and eye troubles. Roots yield tanning material/red dye.
68.	*Geum elatum* Wall.	Masreen	The fine powder of the dried roots with ghee is applied on deep cuts and wounds. It binds the wound together and effects quick healing even by a single application. The best procedure adopted is to wash the wound with mild-hot water properly and dry it by the use of cotton or cloth piece. Apply luke warm *desi ghee* all over the wound and the fine root powder is applied on it in a thick layer. Proper bandaging of the wound is done and the healing results within three to seven days time. It is very effective home remedy even for deep cuts.
69.	*Girardinia diversifolia* Syn. *G. heterophylla* Dcne	Bichhubuti, Karal, Bhabar	The forced contact of the stinging shoots is made with swollen joints to activate blood circulation.

Contd.

Sr. No.	Botanical name	Local name	Use
			The blisters produced as a result thereof, give out watery juice and ultimately result in subsiding of the swellings. The paste of the root is applied to boils to enhance suppuration. The fibre from the plant forms a good quality material for ropes, cords and 'Pula' (fibre shoes).
70.	*Habenaria pectinata* D. Don.	Safed Musli	The leaves are crushed and applied as paste in snake bites. Its tuber is mixed with 'khoya' and greater cardamom and eaten, to get relief from joint pains.
71.	*Humulus lupulus* Linn.	Hops	Herb contains an essential oil, a bitter substance, choline, asparagine and lupulin. Used for providing aroma and flavour to beer. Fiber from stems is used for twining and for textile.
72.	*Hydrocotyle javanica* Thumb.	Jangli ajawain	The seeds are boiled with black salt (kala namak). The amount of water should be such, that it is completely soaked up and is then taken with cold water, to relieve stomach-ache.
73.	*Jasminum dispermum* Linn.	Chameli	For acidity and gastric troubles, the green leaves are made into tablets of 1 g mixed with gur, and consumed with cold water.

Contd.

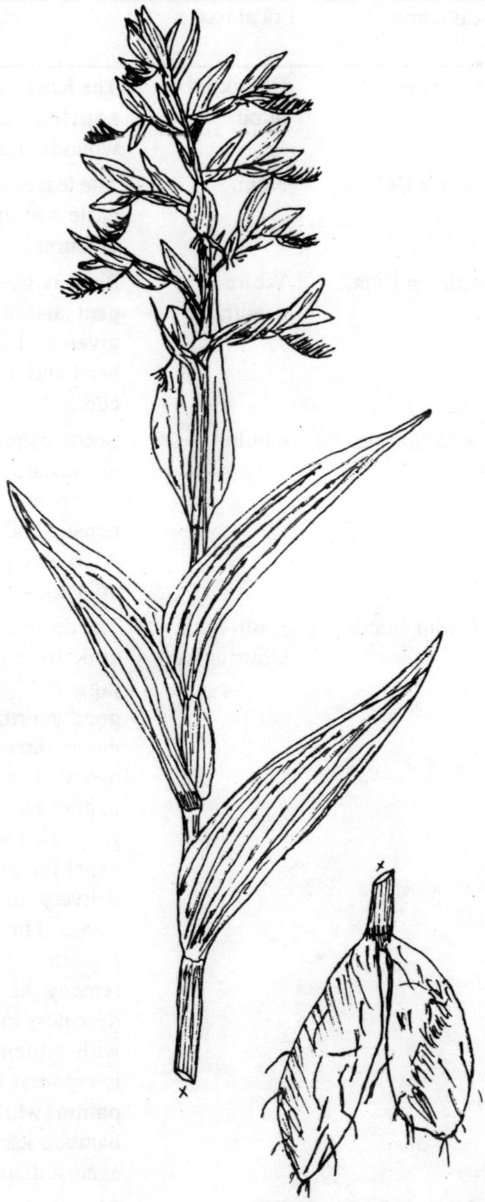

Habenaria pectinata D. Don.

Sr. No.	Botanical name	Local name	Use
74.	*Kalanchoe pinnata* Pers.	Dhank ka nunu	The leaves are crushed and applied as paste on wounds and burns.
75.	*K. spathulata* DC.	Nunu	The leaves are made into a paste and applied as paste on burns.
76.	*Lamium album* Linn.	White deadthisle	Plant is used as an astringent and its decoction is given in haemorrhages of head and nose. Leaves are edible.
77.	*Lepidium latifolium* Linn.	Chulti	Seeds contain a glycoside of mustard oil. Infusion of the plant is given in hepatic and renal troubles. Also used against skin affections.
78.	*Leucas lanata* Benth.	Kathi buti, Dhurlughas	The decoction of the leaves and apical shoots with jaggery (gur) acts as a good abortifacient upto or under three months pregnancy. The roasted leaves in ghee have been successfully tried as a remedy to expel the placenta as after delivery in bufalloes and cows. The plant is also reported to be a good remedy in diarrhoea and dysentery in cattle. If given with buttermilk (lassi), it is reported to treat constipation while given with bamboo leaves, it is a cure against diarrhoea.
79.	*Litsea chinensis* Lamk.	Rihan	The bark is applied on

Contd.

Sr. No.	Botanical name	Local name	Use
			fractured bones and tied with cloth for joining the bones.
80.	*Lonicera angustifolia* Wall.	Pidoo, Bandakh	The fruits are quite sweet and make a delicious desert. The paste of the leaves is said to be usefully applied to cure boils and sores.
81.	*Mallotus philippinensis* Muell.-Arg.	Kamila	Bark and leaf powder are mixed with butter and applied on sores and boils.
82.	*Malva rotundifolia* Linn.	Chambasa	The plant contains sterols, terpenes, alkaloids, tannins and sugars. Seeds are used against cough and inflammation of bladder. Its mucilaginous extract is given in leucorrhoea.
83.	*Malva verticilata* Linn.	Shorshotle, Shochlo, Shochle	The juice expressed from the roots is said to be useful in leucorrhoea. The crushed root in water is given to cows and buffaloes to facilitate detachment and expulsion of placenta after delivery. The infusion of the root is used in washing of woollen garments. It gives foam due to saponin content and helps in making the garments thick and stuffy, and aids in cleaning as well.
84.	*Mangifera indica* Linn.	Aam	The mango seeds are powdered and mixed with 'Mishri' and dried 'Bras''

Contd.

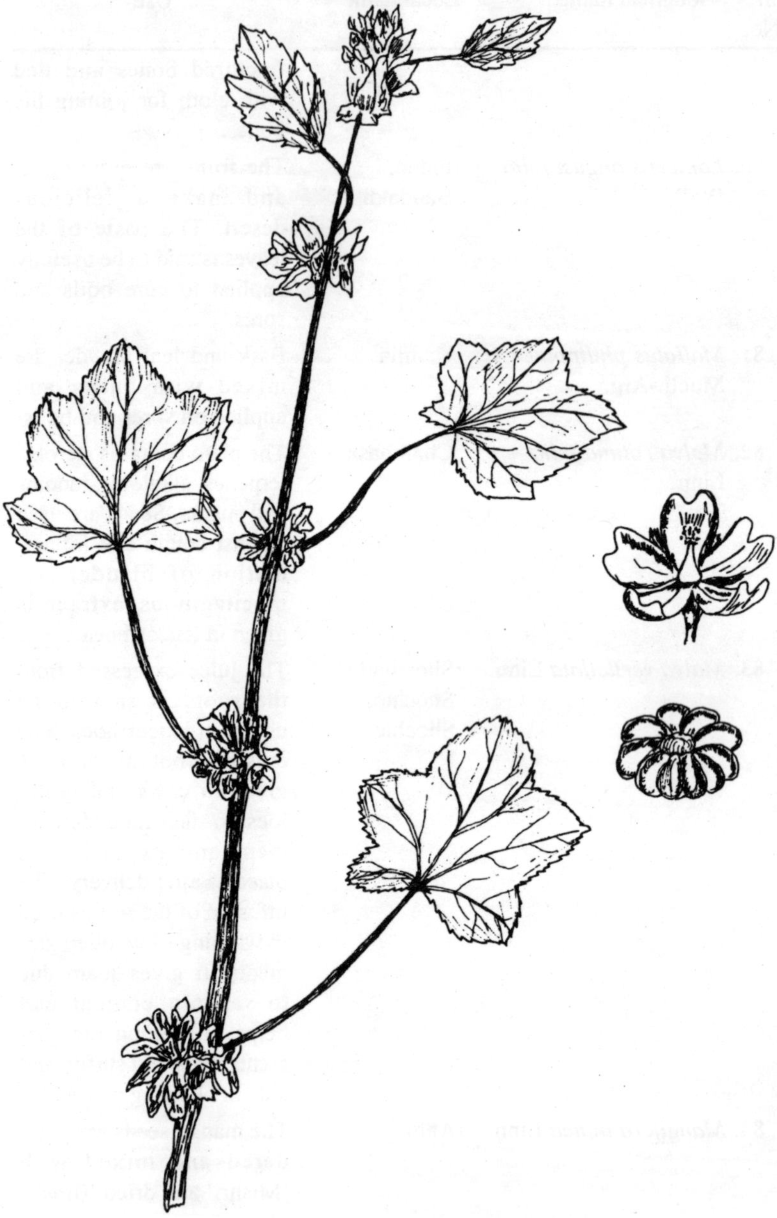

Malva verticilata Linn.

Sr. No.	Botanical name	Local name	Use
			(*Rhododendron abroreum*) flowers and given in dysentery.
85.	*Melilotus officinalis* (Linn.) Pall	Bhuksup	Dried leaves and flowering tops yield coumarin, are used in the preparation of cigarettes which are smoked for relief in asthma.
86.	*Mentha arvensis* Linn.	Pudina	Mix pudina leaves, greater cardamom and 'gur' to make tablets and take it with water to get relief from stomachache. Crush the leaves, extract juice and give it to the child, to stop vomiting.
87.	*Mentha longifolia* (L.) Hudson	Chachri, Chachar	The thin fine paste of the leaves is applied to rheumatic and painful swollen joints. If a thick paste is applied, blisters may be produced on the skin. The watery juice oozing out from these blisters helps in subsiding the swellings and gives permanent relief from the painful problem. Thin paste is advisable to be on the safer side as excessive blisters may lead to wounds.
88.	*Meriandra strobilifera* Benth.	Murta, Murte	Leaves are spread where the cattle sits to get rid of fleas. If 'leech' affects the cattle, juice of leaves is extracted and put in their nose to ward off leech.

Contd.

Sr. No.	Botanical name	Local name	Use
89.	*Micromeria biflora* Benth.	Indrapan, Indian wild thyme	The leaves are dried and powdered and the paste is applied on wounds and also eaten in stomachache, to get relief. Essential oil similar to camphorcitral is obtained from the plant. It is used for disinfecting worm infected wounds of cattle. The leaves are mixed with 'Brahmi' and 'Pathan bel' and eaten by diabetic patients.
90.	*Nardostachys grandiflora* DC.	Machhi, Mansi	Local shepherds and old people keep the roots in their possesion. A small piece of the root is powdered and mixed with tobacoo and smoked in cases of palpitation of heart and mental tension. The roots kept in oil is applied by woman as an aromatic oil and to promote the luxuriant growth of hairs.
91.	*Nasturtium officinale* R.Br.	Chala, Chhuchha	The young shoots are used as leafy vegetables and its use is said to act as appetizer, laxative and diuretic. It is believed to increase blood circulation.
92.	*Nepeta ciliaris* Benth.	Pissu-buti	The brushed leaves are spread on the floor of the room or on the cot, under bed cover to act as repellent for fleas, bed dugs and other harmful insect pests.

Contd.

Sr. No.	Botanical name	Local name	Use
			The plant has a strong odour, which helps as repellent and keeps the pests away.
93.	*Nepeta podostachys* Benth.	Ribuksu, Ribhuksu	Herb yield terpenes, tannins and sugars and is used in cough and fevers.
94.	*Opuntia elatior* Mill.	Suro	The sap is applied on forehead and temples during headache.
95.	*Origanum vulgare* Linn.	Basloghas, Jakham-buti	The plant is regarded an important house hold remedy for various purposes. The pasts of leaves and terminal shoots is applied to boils, ulcers, wounds, cuts and weeping eczima. Two to three fruits of black pepper are mixed while making the paste. The paste of the leaves is reported to be highly useful in healing the wounds caused by fire-burns and in abscess of the breast. A tablet made of its leaves alongwith the leaves of *Plectranthes coesta* is given with flour or folded bread to animals in cases of bone fractures. The root piece of the plant is bound in a cloth piece and tied to the necks of infants as a protective measures against conjuctivitis.
96.	*Oxalis corniculata* Linn.	Umlori, Alori	A fine paste of the aerial parts of the herb alongwith

Contd.

Sr. No.	Botanical name	Local name	Use
			a few seeds of black pepper is applied to boils, abscesses, wounds and weeping eczema. This application results in oozing out of juice from the affected parts and helps in drying and healing of the wounds. This paste is reported quite useful in abscesses of the mammary glands. Leaf juice is regarded useful in curing cataract of the eyes and also used externally to remove warts.
97.	*Oxyria digyna* (Linn.) Hill	Suma, Chucha	The herb is considered as antiscorbutic and refrigerant.
98.	*Physochlaina praealta* (Walp.) Miers.	Laltang	Yield alkaloids like Atropine and hysocyamine which enter into various modern formulations as antidote, sedative, narcotic, anodyne and in dilation of the pupil.
99.	*Picris hieracioides* Linn.	Chava	Leaves are bitter and used as febrifuge.
100.	*Picrorrhiza kurrooa* Royle.	Kutki	In the cases of nose bleeding, leaves are crushed and 1-2 drops of the juice are put in the nose to stop bleeding.
101.	*Pinus wallichiana* A.B. Jackson	Cheel, Kail	The resin from the young sapling is used as a lotion for healing of the cuts and wounds. The bark taken

Contd.

Sr. No.	Botanical name	Local name	Use
			out in the form of long circular cylinders is used to bandage the dialocated/ fractured organs both in cattle, sheep and goats and even on human beings after setting the dislocated or broken bone in proper place. This helps as a plaster of paris in keeping the organ in resting position and facilitates calcification.
102.	*Pistacia integerrima* Stewart	Kakkar singhi	The gall is burnt and ash is taken with honey to relieve whooping cough. Leaves are dried and powdered and mixed with kali mirch (black pepper) and taken in headaches.
103.	*Plantago depressa* Linn. Syn. *P. tibetica* Hook. f. & Thoms.	Musal-ninai	A paste of the plant, including leaves and roots is applied to the muscular abscess (where the skin has cracked and watery secretion and shooting pain occurs). The paste is applied two to three times a day. The paste of leaves is also applied to skin eruptions, boils and rashes. The root of the plant is tied to the button hole of the infants as a protective measure against infection of stomach diseases.
104.	*Plantago major* Linn.	Yakram, Jharam naram	Seeds are used in gastric complaints, burning sensation in stomach and dysen-

Contd.

Sr. No.	Botanical name	Local name	Use
			tery. Leaves are applied to bruises.
105.	*Pleurospermum brunonsis* Benth.	Nesar	The powder of the flowering shoot is mixed with cow's butter and massaged over the entire body to allay fever. The dried herb is used as a preserving agent against the attack of moth and silver fish to protect woollen garments kept in boxes. It can be of great value in perfumery industry. It is used as a condiment in Churdhar area of Sirmour district.
106.	*Plumbago zeylanica* Linn.	Chicha	Roots are crushed, made into a paste and applied externally on the neck to subside tonsils. The root paste is applied on the palate and later massaged with oil to get relief from headache.
107.	*Polygala abyssinica* R. Br.	Pagal buti	The entire plant is crushed and mixed with butter or ghee and is given to the animal/dog as a cure against rabies (rabid dog bites).
108.	*P. crotalarioides* Linn.	Ghandeli	If children suffer from Whitlow (Galeki) the entire plant is crushed and mixed with butter and fed to the child.
109.	*Polygonum hydropiper* Linn.	Mirchighas	The juice of the leaves is reported to be useful as a

Contd.

Sr. No.	Botanical name	Local name	Use
			cure for pussing ear. It is often used as a fish poison.
110.	*P. recumbens* Royle	Vishalyakarni	The paste of the leaves and top shoots with a few seeds of black pepper is applied to cuts, wounds, boils, scabies and skin rashes. Its paste is also useful in weeping eczema and effective in healing the ulcerations between the fingers of the toe—a problem which the barefooted hill people usually suffer during rainy season.
111.	*Potentilla eriocarpa* Wall. ex. Lehm	Penma	Used in the form of tea in diarrhoea, kidney stones; arthritis and intestinal troubles.
112.	*P. nepalensis* Hook.	Laljarhi	The root powder or the paste is applied in cuts, burns and wounds as a healing agent. The root is also substituted for ratanjot as a red colouring matter for hair oils.
113.	*Primula nivalis* D. Don. var. *macrophylla*	Nili-ka-phool	The beautiful flowers are given preference over all flowers of the alpine regions by the local people. As a legendary tale, it is said that he, who has this flower in possession can not be affected by any magic, Spirit (evil) and tantric tricks. He can even detect the tricks played by

Contd.

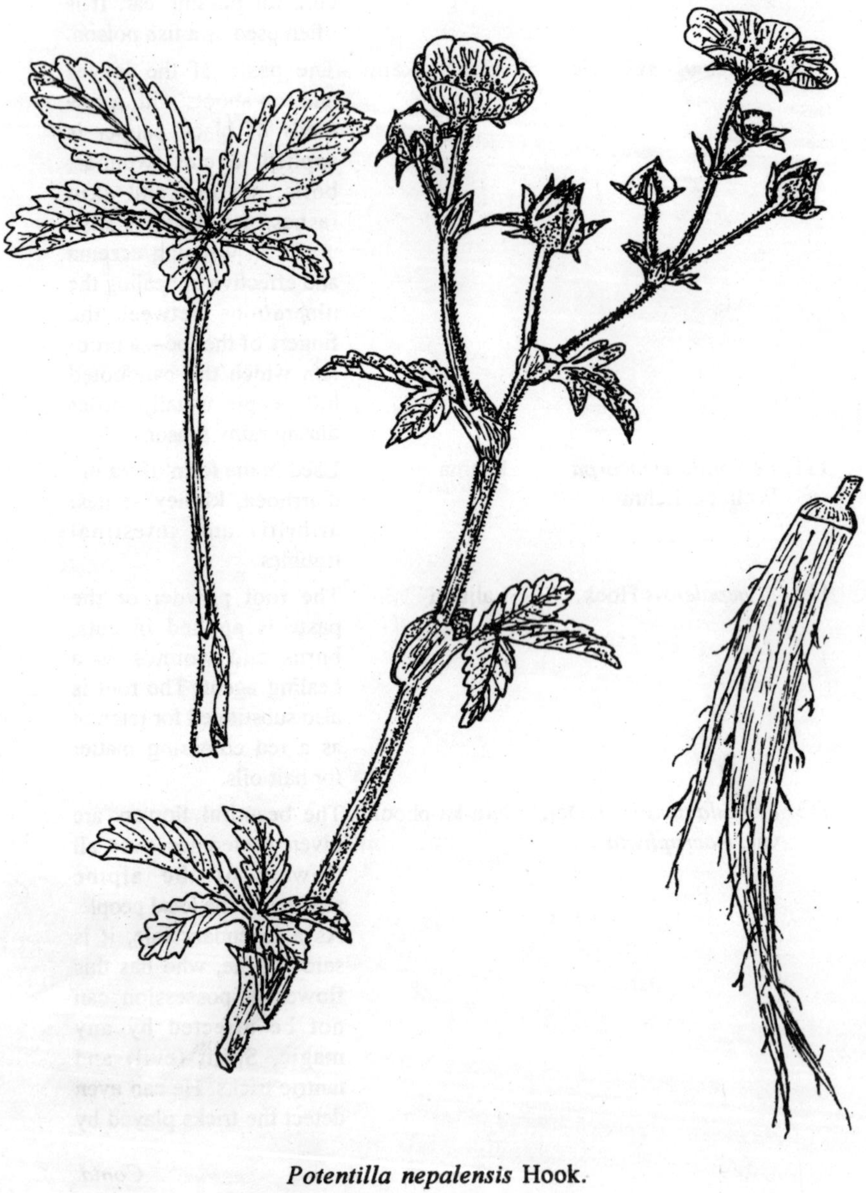

Potentilla nepalensis Hook.

Sr. No.	Botanical name	Local name	Use
			the magicians which otherwise none can visualize under ordinary conditions.
114.	*Prinsepia utilis* Royle.	Bhekhali, Bhekhal	The oil extracted from the seeds is taken internally as tonic and is considered useful in general debility and theumatism. The oil is also massaged on rheumatic joints for relief. The extract of bhekhal root is given in bloody dysentery. The hollow branches are made into flutes and toy pistols by the children. The long hollow tubes of its, branches/stems are also used as pipes for smoking tobacco by fitting it in the outlet of hubble-bubble (Hukka).
115.	*Prunus armeniaca* L.	Chulu, Sharhi, Wild apricot	The fleshy part of the fruit is eaten. The ripe fruit as a whole are dried and kept in storage for use during winters. The fresh fruit is beaten into pulp and preserved in the form of slabs or circular cakes called as "chulottu" in Pabbar valley, Shimla hills and later used as sour ingredient in chutneys. It is also considered very useful for washing the silver and gold ornaments. The edible part of the fruits is also used in higher altitude

Contd.

Sr. No.	Botanical name	Local name	Use
			areas for preparing local alcoholic beverage '*soor*' or '*soro*'. The oil from the kernels is highly priced for domestic use i.e. cooking, eating in lieu of ghee and for external application like hair oil. It is regarded equivalent in properties to almond oil. The hard shell of the kernel is used as a coal. The charred oily seeds are rubbed into a paste (black), which is a highly efficaceous ointment in healing of the cuts, wounds and sores. Chuliphant an watery extract of fruit pulp is taken to reduce obesity and cleaning of the stomach of gases and other problems in Kinnaur.
116.	*Prunus cerasoides* D. Don.	Paja	It is used in marriage ceremony, as it is considered holy. The twigs are kept in the water and the bride and groom take bath in that water.
117.	*Prunus persica* (L.) Batsch.	Arua, Aru	Dried and powdered flowers are given as a cure in menstrual disturbances to teenaged girls. Fruits are edible.
118.	*Punica granatum* Linn.	Daru	The fruit rinds are dried and powdered and taken with cold water to relieve cough. When children (new borne) start cutting

Contd.

Sr. No.	Botanical name	Local name	Use
			out teeth, the peels are powdered mixed with Kashmal (*Berberis lycium*) roots, made into a paste and applied on the palate to ease the process of emergence of teeth.
119.	*Rhamnus triqueter* Wall.	Karu	Half a glass of leaf juice is given early morning in empty stomach for 3 consqutive days against malarial fever and to kill intestinal worms.
120.	*Rhododendron anthopogon* D. Don	Simrish, Talsar	The powder of the dried flowers is mixed in any bland oil and massaged over the entire body in complications like fevers, coughs and colds. The aroma of the plant is reported to cause giddiness, nausea and vomiting in its natural habitat.
121.	*R. arboreum* Smith	Buras, Burans	The fresh petals are used in chutneys. The flowers and leaves are tied in long ropes made of Munja grass and tied around the houses including temples as decorations during Baisakhi (Bishoo) festival. The powder of the dried flowers is used as an efficaceous drug to check bloody and chronic dysentery.
122.	*Rubia cordifolia* Linn.	Chipdua, Manjistha	The leaf juice followed by applications of leaf paste is

Contd.

Sr. No.	Botanical name	Local name	Use
			applied to check bleeding from cuts and hastens healing.
123.	*Rumex hastatus* D. Don.	Khatti-mithi	The bark of roots is used to cure burnt portion and is applied as 'lep' (paste). Leaves are used in chutneys, being acidic in taste.
124.	*R. nepalensis* Spreng.	Mareli	The roots are crushed, made into paste and applied on skin, affected with 'Jaaj' (white blotches). The root paste is also applied to remove warts.
125.	*Sapium insigne* Trimen.	Chhirna	Due to some infection, if the hair start falling suddenly, the milky sap is applied, to stop hair fall as in the case of alopacia.
126.	*Saussurea gossypiphora* D. Don.	Ghuggi-badshah, Bhergandha (Jakha) Loskarch	The plant is regarded as sacred. The wool of the herb is applied to the cuts, where it sticks compactly, seals the wound, stops bleeding and facilitates healing. In Jakha, the shepherds rub the powder of this on the body of newly born sheep and goats, so that the young mother takes more care to nurse them by milk feeding. The plant is kept in houses and its fumigation after burning on coal is credited with the properties of driving away the evil spirits.

Contd.

Sr. No.	Botanical name	Local name	Use
127.	*S. obvallata* (DC.) Edgew.	Brahmakamal	Roots are used in preparations applied to wounds and cuts. The floral bracts boiled in water are used as fomentation in hydrosele. Flowers are kept between woollen clothes to protect these against insects.
128.	*Senecio chrysanthemoides* DC.	Sersan-mendo	Essential oil obtained from the plant is used as perfumery material. The herb is toxic to animals.
129.	*Sida rhombifolia* Linn.	Biulti	The roots and leaves are dried & powdered, sieved and taken with cold water to cure jaundice.
130.	*Silene conoidea* Linn.	Chiffi-buti	Herb is used as fumigant. Its juice is useful in opthalmia. Contains saponin.
131.	*Skimmia laureola* Sieb. & Zucc.	Shimshar, Patti-dhoop, Kasturi patra	The dried leaves when burnt with desi ghee on red hot coal, emit out highly scented fumes which purify the atmosphere. The orange or purple berries are sweet in taste and are administered to the fever stricken patients to allay the thirst. Infusion of the leaves is given internally to cure ascites.
132.	*Solanum khasianum* Clarke	Kateri	The entire plant is burnt, made into ash and small tablets are made mixing ash with honey and black pepper and administered in whooping cough.

Contd.

Sr. No.	Botanical name	Local name	Use
			The paste prepared from fruits is applied on wounds and cuts for early healing.
133.	*S. surattense* Burm. f.	Kateli	People suffering from asthma smoke it in hooka for relief.
134.	*Tagetus minuta* Linn.	Gaindaghas, Gaindri	Raw paste of the bruised leaves is applied on the naval of the children to expel round-worms.
135.	*Tanacetum longifolium* Wall.	Amritdhara-ghas	Its aroma, especially when found bruised and crushed by sheep is said to cause giddiness. The leaves are made into a round pill of the size of almond nut and swallowed with water in

Contd.

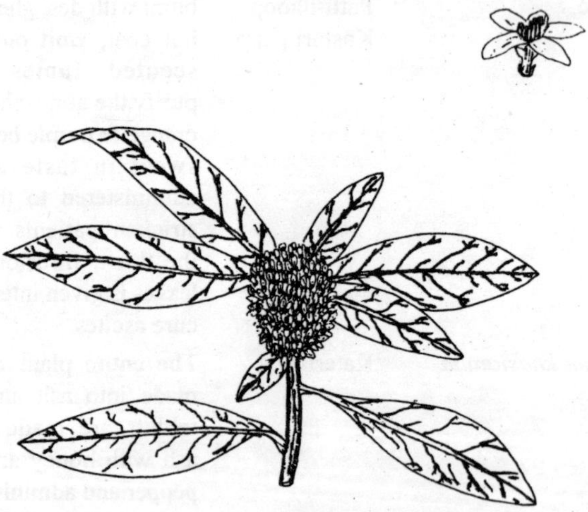

Skimmia laureola Sieb. & Zucc.

Sr. No.	Botanical name	Local name	Use
			stomach pain and indigestion. The expressed juice is useful as an antispasmodic, carminative and antipyretic.
136.	*T. nubigenum* Wall. ex. DC.	Bhurse-khamba	Roots are used as incense to purify houses and to ward off harmful insects. Also yield essential oil known as oil of tansy.
137.	*Taraxacum officinale* Wigg.	Sarshan-mendo, Dandelion, Kanphuli	The sheep and goats browse it as a potent fodder. The whole plant is crushed into a mesh and given internally in snake bites. The paste is applied externally on the wound. Leaves are effectively used for fomentation in swollen parts, boils and sprains.
138.	*Terminalia chebula* Retz.	Harad	The fruits are crushed and powdered and taken with cold water to relieve cough.
139.	*Thalictrum alpinum* Linn.	Atangthang	Roots contain alkaloid taliksimure and fruits contain magnoflorine.
140.	*Thalictrum foliolosum* DC.	Mamiri, Pilijari, Barmoth, Mirchadi	In summers, to get relief from heat-roots are dried, powdered and eaten mixed with butter. The roots are burnt and made into 'Surma' and used for eye problems. The root is also used against snake bite. It is given internally and the paste is applied on the

Contd.

Sr. No.	Botanical name	Local name	Use
			bitten organs to neutralize the poison.
141.	*Thymus serpyllum* Hook. f.	Padumba, Khishikpasa, Banajwain	The decoction of the plant is an effective home remedy for colds, cough, coryza, fever and stomach problems. Infusion of leaves with gur acts as a very good diuretic, increases urine flow and relieves pain. The paste of the herb including a few seeds of black pepper (kali mirch) is said to be efficaceous in alopacia (falling of hair in circles from head, chin and beard etc.). Powdered flowers mixed with gur are given as vermicide in appropriate doses. It is believed that he, who drinks tea medicated with this herb regularly, never suffers from common colds and general stomach ailments.
142.	*Trigonella foenum-graceum* Linn.	Methi	If the child urinates in bed-the powdered methi seeds are given with milk to check this menace.
143.	*Typhonium diversifolium* Schott	Chamush	The tubers are collected during August-September. These are washed, the central hard core removed and then ground into a thick paste of which 2.5-7.5 cm long × 1-2 cm in diameter pieces are

Contd.

Sr. No.	Botanical name	Local name	Use
			moulded. These pieces are macerated with a mixture of garlic and salt and swallowed as such without chewing. Chewing causes severe itching in the throat because of the presence of some chemical substances. This is regarded as an efficaceous remedy for chronic dysentery, complaints of gastro-intestinal tracts and piles. In addition, it forms a staple diet for the poor tribal people and supplements their cereal requirements during these lean months.
144.	*Verbascum thapsus* Linn.	Bhoot-lankra, Jangli tambaku	The crushed leaves are made into a pill and given in constipation and allied stomach pains. Decoction of leaves along with onion is useful for stomach problems. The infusion of plant is given in snake bite. A paste of leaves is also applied locally on the bitten part (wound). The crushed leaves are warmed in mustard oil and used as massage for curing inflamations of the body. The smoke (dhuni) of the plant is utilised to drive away the ghostly instincts from the children. The plant is used by the brahmins and tan-

Contd.

Sr. No.	Botanical name	Local name	Use
			trics in all such worships and oracle rites, where bad spirit is involved and is to be driven out of the place so as to give peace to the inmates.
145.	*Viola serpens* Linn.	Banafsha	The flowers are dried and a decoction is prepared and given for expulsion of phlegm. It is a good cure for sore throat.
146.	*Vitex negundo* Linn.	Bana, Shaephali	The leaves are boiled in cow's urine and when half the quantity is left, a paste is made and applied on wounds and painful body organs to get relief. Boil leaves in water and apply with the help of cloth on swellings to get relief.
147.	*Urtica dioica* Linn.	Kimchhi, Kugeha, Bichhu-buti	The young leaves are cooked as vegetable and regarded as blood purifier. It is said to be a cure for skin diseases. The fine paste of the plant, especially of the root part, is applied in hidden muscular pains/hurt. The shocks of the stings of its twigs two to three times are applied in swollen, rheumatic joints for relief from pain and swelling.
148.	*Woodfordia fruticosa* (L.) Kurz.	Dhai, Dawi	The flowering buds are mixed with 'Mishri' and administered in stomach ache.

Contd.

Sr. No.	Botanical name	Local name	Use
149.	*Zehneria umbellata* Thwaites Syn. *Malothria heterophylla*	Mokri	The ripe fruits are edible for their sweet taste. Seeds are brittle shelled and break easily inside the mouth. It is considered as stomachic and tonic. The root fibers bear spindle shaped tubers at the ends of their tips. This tuberous part of the root is washed, dried, powdered and taken with milk in the dose of about two grams twice a day as a cure for seminal debility, spermatorrhoea and its use improves vitality.

Appendix 6. Demand of state-run Ayurvedic pharmacies (at optimum production capacity)

There are two Ayurvedic pharmacies in government sector under the Department of Ayurveda based at Jogindernagar (Mandi) and Majra (Sirmour) engaged in manufacturing of Ayurvedic medicines for supply to over 1100 Ayurvedic dispensaries/institutions spread all over Himachal Pradesh. Another pharmacy attached to Ayurvedic College Paprola (Kangra) has also become functional recently. The requirement of raw drugs by these pharmacies at full production capacity based on first and second priority (A&B) is as follows:

Sr. No.	Ayurvedic name	Botanical name (official part)	Quantity
		(A)	
1.	Jeevak	*Malaxis muscifera* (Bulb)	50
2.	Rishbhak	*M. acuminata* (pseudobulb)	50
3.	Ridhi	*Habenaria intermedia* (Root)	50
4.	Vridhi	*H. edgeworthii* (Root)	50
5.	Meda	*Polygonatum verticillatum* (Rhizome)	50
6.	Mahameda	*P. cirrifolium* (Rhizome)	50
7.	Kakoli	*Roscoea procera* (Roots)	50
8.	Kshirkakoli	*Paris polyphylla* (Bulb)	50
9.	Kshirkakoli	*Fritillaria roylei*	50
10.	Atis, Ativisha	*Aconitum heterophyllum* (Root)	300
11.	Vatsanabh	*Aconitum ferox/A. deinorrhizum* (Root)	1000
12.	Heritpatri	*Digitalis lanata* (Leaves)	1000
13.	Belladona	*Atropa belladona, A. acuminata* (Leaves, Root)	1000
14.	Somlata	*Ephedra gerardiana* (Twigs)	1500
15.	Kutaki	*Picrorhiza kurrooa* (Rhizome)	2000
16.	Brahmi	*Bacopa monnieri, Centella asiatica* (Leaves, plant)	5000
17.	Chirayata	*Swertia chirayita* (Plant)	1500
18.	Sarpagandha	*Rauvolfia serpentina* (Root)	2000
19.	Shankhpushpi	*Evolvulus alsinoides/Convolvulus pluricaulis*	5000
20.	Ghrit-kumari	*Aloe barbadensis* (Leaves)	5000

Contd.

Sr. No.	Ayurvedic name	Botanical name (official part)	Quantity
21.	Karu	*Gentiana kurroo* (Root)	200
22.	Bidang, Baibidang	*Embelia ribes* (Seeds)	1500
23.	Kuth	*Saussurea costus* (Root)	800
24.	Krishnajirak, Kalajira	*Carum carvi* (Fruit)	600
25.	Khadir (Katha)	*Acacia catechu* (Extract)	9000
26.	Kesar	*Crocus sativus* (Stigma)	100
27.	Ashwagandha	*Withania somnifera* (Root)	1000
28.	Talispatra	*Taxus wallichiana/Rhododendron anthopogon* (Leaves)	5000
29.	Chora, Chorak	*Angelica glauca* (Root)	100
30.	Bankakri Giriparpat	*Podophyllum hexandrum* (Root)	500
31.	Bibhitak, Bahera	*Terminalia bellirica* (Fruit)	10000
32.	Haritaki, Harad	*T. chebula* (Fruit)	10000
33.	Amlaki, Aonla	*Emblica officinalis* (Fruit)	10000
34.	Anardana	*Punica granatum* (Seed, fruit rind)	1500
35.	Karkatshringi Kakarsingi	*Pistacia integerrima* (Galls)	900
36.	Kutaj	*Holarrhena antidysenterica* (Seeds and bark)	10000
37.	Vanaksha, Banaksha	*Viola serpens* (Flowers and leaves)	500
38.	Kantkari (Chhoti)	*Solanum xanthocarpum* (Panchrag)	1500
38A	Kantkari (Brihati)	*S. indicum* (Panchrag)	1500
39.	Shatavari	*Asparagus racemosus* (Root) *A. adscendens* (Roots)	400
40.	Rasanjan, Rasaunt	*Berberis aristata* (Extract)	400
41.	Tulsi	*Ocimum sanctum* (Leafy shoots)	9000
42.	Vansa, Basuti, Adusa	*Justicia adhatoda* (Leaves)	4000

Contd.

Sr. No.	Ayurvedic name	Botanical name	Quantity
43.	Dhatura (Safed)	*Datura stramonium* (Pancharg), (Seeds)	2000
44.	Bhringraj	*Eclipta prostrata* (Plant)	2000
45.	Dhataki pushpa	*Woodfordia fruticosa* (Flowers)	5000
46.	Jatamansi	*Nardostachys grandiflora* (Roots)	400
47.	Bari illaichi	*Amomum subulatum* (Seeds/fruits)	200
48.	Kaunchbeej	*Mucuna pruriens* (Seeds)	500
49.	Tejpatra	*Cinnamomum tamala* (Leaves)	1000
50.	Bhilawa	*Semecarpus anacardium* (Fruits)	1000
51.	Pushkarmool	*Inula racemesa* (Roots)	100
52.	Manjith	*Rubia cordifolia* (Roots)	200
53.	Chameli	*Jasminum officinalis* (Leaves & flowers)	200
54.	Senna, Sannay	*Cassia angustifolia* (Leaves)	500
55.	Arjun	*Terminalia arjuna* (Bark)	1000
56.	Ashok	*Saraca asoka* (Bark)	1000
57.	Bilve	*Aegle marmelos* (Bark & fruit)	400,400
58.	Mamiri	*Thalictrum foliolosum* (Roots)	200
59.	Tajbal	*Zanthoxyllum armatum* (Seeds)	1000
60.	Salampanja	*Dactylorhiza hatagirea* (Root)	200
61.	Malkangni	*Celastrus paniculatus* (Seeds)	500
62.	Katrin	*Cymbopogon* spp. (Leaves)	400
63.	Shalparni	*Desmodium gangeticum* (Plant)	400
64.	Prishniparni	*Uraria picta* (Plant)	400
65.	Hauber	*Juniperus communis* (Berries)	200
66.	Bach, Vacha	*Acorus calamus* (Rhizomes)	1000
67.	Apamarga	*Achyranthes aspera* (Plant)	1000
68.	Suhanjana	*Moringa pterygosperma* (Seeds)	300
69.	Gojihva	*Onosma echioides* (Roots & leaves)	250
70.	Giloye	*Tinospora cordifolia* and *T. malabarica* (Twigs/Sut Giloye)	200,10
71.	Tagar	*Valeriana Jatamansi* (Root)	500

Contd.

Sr. No.	Ayurvedic name	Botanical name	Quantity
		(B)	
1.	Ahiphen, Aphim	*Papaver somniferum* (Latex)	20
2.	Mulhathi, Liquorice	*Glycyrrhiza glabra* (Root)	4000
3.	Pudina	*Mentha* spicata (Leaves)	2000
4.	Karpur	*Cinnamomum camphora,* (Camphor crystals)	400
5.	Anantmool	*Hemidesmus indicus* (Root)	2000
6.	Nirgundi	*Vitex negundo* (Leaves)	2000
7.	Kaiphal	*Myrica esculenta* (Bark)	500
8.	Atibala	*Abutilon indicum* (Shoots, with fruit)	2500
9.	Nagbala	*Grewia tenex* (Twigs)	2000
10.	Bala	*Sida cordifolia* (Plant)	2500
11.	Neem	*Azadirachta indica* (Leaves)	5000
12.	Gandhsrasarini	*Paederia foetida* (Twigs)	1000
13.	Bharangi	*Clerodendrum serratum* (Twigs)	1000
14.	Agnimanth	*Premna integrifolia* (Bark)	400
15.	Gambhari	*Gmelina arborea* (Bark)	400
16.	Akarkara	*Anacyclus pyrethrum* (Seeds/root)	1000
17.	Nishoth	*Ipomoea turpethum* (Root)	200
18.	Priyangu	*Callicarpa macrophylla* (Seeds)	200
19.	Lodhra	*Symplocos racemosus* (Bark)	200
20.	Bijasar	*Pterocarpus marsupium* (Bark/oleo-resin)	200
21.	Irimeda	*Acacia farnesiana* (Bark)	1000
22.	Unnav	*Zizyphus sativa* (Fruits)	5000
23.	Gulab pushpa	*Rosa damascena* (Petals)	500
24.	Chandan Safed	*Santalum album* (Wood)	500
25.	Patha	*Cissampelos pareira* (Root)	100
26.	Chitrak	*Plumbago zeylanica* (Root)	1200
27.	Arlu	*Oroxylum indicum* (Bark)	400
28.	Padhal, Patla	*Stereospermum suaveolens* (Bark)	400
29.	Madanphal, Rada	*Randia dumatorum* (Fruit)	100
30.	Kikar	*Acacia nilotica* (Bark)	300
31.	Sirish	*Albizia lebbek* (Bark)	500

Source: Directorate of Ayurveda (Technical), Jogindernagar, H.P.

Appendix 7. Raw material for dhoop and incense industry

Dhoop and aggarbaties are household requirements in every Hindu family for worshipping deities and gods, performing havanas and yagnas, for purifying home environment and also for repelling mosquitos and insects-pests. The plant materials in large quantities are required to meet the demand of this industry. The selected species utilized for the purpose are given hereunder:

Sr. No.	Source plant (Latin name)	Common name (Parts used)	Availability in Himachal Pradesh
1.	*Jurinea dolomiaea*	Dhoop (Root)	Pangi-Bharmour, Thamsarjot, Dainasar, Kullu, Shimla (Rohru Forest Division)
2.	*Nardostachys grandiflora*	Jatamansi Balchharh, (hairy underground stem)	Manjiban, Tangankhai & Buraskandi in Dodra-kawar Range, Great Himalayan National Park, Kullu
3.	*Valeriana jatamansi*	Tagar, Nihani, Mushkbala (root)	Common in forest shades 1600-3800 m in many parts of the state.
4.	*Skimmia laureola*	Neir, Kasturipatra, Shimshar (Leaves)	Shades of Kharshu Oak between 2600-3200 m.
5.	*Selinum tenuifolium*	Muramansi, Mathosal (Roots)	Temperate and alpine areas between 2700-3800 m.
6.	*S. vaginatum*	Bhutkesi, Mathosla (Hairy roots).	Temperate and alpine areas between 2700-3800 m.
7.	*Cinnamomum camphora*	Kapur (Camphor)	Planted at Jogindernagar, Nauni and Paonta Sahib.
8.	*Taxus wallichiana*	Talishpatra, Thuna, Rakhala (Needles)	Common in association with fir and spruce between 2800-3200 m.
9.	*Juniperus macropoda*	Shugpa, Dhuppi (Twigs with leaves)	Kinnaur and Lahaul-Spiti Divisions of Himachal Pradesh.
10.	*J. communis*	Thekru, Thelu, Chhorshup	Alpine zone all over H.P.
11.	*J. recurva*	Thelu, Theleru, Shingapalu	Alpine zone all over H.P. Often spreading on rock-boulders.

Contd.

Sr. No.	Source plant (Latin name)	Common name (Parts used)	Availability in Himachal Pradesh
12.	*Waldhemia glabra*	Phillu, dhoop (Herb)	Spiti (in Chandratal area) and near Ladakh borders 4800 m.
13.	*Tanacetum longifolium* and allied spp.	Bhurse khamba Tansy (Herb)	Alpine meadows above 3400 m.
14.	*Anaphalis busua* Syn. *A. araneosa*	Dhoop, ruighas (Herb)	1500-2500 m. Common in most parts of H.P.
15.	*Artemisia macrophylla*	Dhoop, ruighas (Herb)	Ladarcha, Kibber to Langza in Spiti, 3500-4800 m.
16.	*Morina longifolia*	Bishkandara, dhoop (Roots)	Chansil, Churdhar, Rohtang, Chandranahan, Dainasar etc. above 3200 m.

Appendix 8. Raw material for manufacturing of phytochemicals for the drug industry

As is evident from the foregiven information, the state produces hundreds of medicinal and aromatic plants utilized by pharmaceutical units in the country, both in Indian systems of medicines as well as in modern pharmaceutical preparations. Some important drug plants yielding such products demanded all over the world are indicated hereunder:

Sr. No.	Products	Source Plant(s)	Common name	Availability in H.P.
1.	Berberis extract (Rasount) and Berberis salts (Berberine hydrochloride & Berberine sulphate) Useful in skin diseases, eye diseases, piles, cholera, menorrhagia, diarrhoea.	*Berberis lycium*	Kashmal	Common between 1200-2800 m
		B. aristata	Daruhaldi	1800-3000 m
		B. asiatica	Chothar	2200-3200 m
		B. chitria	Shimoni	2200-3200 m
		Thalictrum foliolosum	Mamiri Pilijari	Upto 1500 m
		Cissampelos pareira	Patha, Patindu	
2.	Xanthotoxin (used in leucoderma)	*Heracleum candicans*	Padara, Patrala	Sporadic in temperate and sub-alpine areas and common in Spiti division.
3.	Diosgenin and steroids (used in various preparations including birth control pills).	*Dioscorea deltoidea*	Singli-mingli	Between 1600-2800 m
		Trillidium govanianum	–	Chhajpur forest 2700 m
		Costus speciosus	Kemuk	Shiwalik Ranges
		Gloriosa superba	Kalihari	Upto 1300 m
		Solanum indicum	Kantkari	Upto 1300 m
		S. viarum syn. *S. khasianum*	Banbhindi	Common upto 1500 m
		S. laciniatum	Kangaru apple	Planted at Nauni
		S. surratense	Chhoti Kandiari	Shiwalik zone extending Upto 1400 m

Contd.

Sr. No.	Products	Source Plant(s)	Common name	Availability in H.P.
4.	Ephedrine (Useful in asthma)	*Ephedra gerardiana* Wall and allied species	Soma, Chedang, Tutgantha	Lahaul-Spiti, Pooh Division Shali hill.
5.	Santonin (Wormicide)	*Artemisia brevifolia*	Seinski, Nurcha	Pooh, Spiti, Lahaul, Pangi and Kotgarh division.
6.	Hyoscyamine & Atropine (Antidote, sedative, narcotic, anodyne, dilation of pupit)	*Hyoscyamus niger*	Khurasaniajwain, Henebane	Commonly found in dry temperate areas near village. Habitations in Lahaul Division.
		Physochlaina praealta	Laltang	(as above)
		Datura metel	Kala dhatura	Shiwalik areas in Waste lands
		D. stramonium	Safed dhatura	Common upto 2800 m.
		Atropa acuminata	Jalgi, Indian belladona	Sporadic in forest shades 2600-2800 m.
7.	Rutin (Useful in capillary fragidity and haemorrhageal conditions).	*Fagopyrum esculentum*	Ogal,	Cultivated for fruits as coarse grain in tribal areas and
		F. tataricum	Chabru	
		F. cymosum	Phafto	*F. cymosum* occurs in field borders.
		F. dibotrys (D.Don) Hara	Phafra	
8.	Taxol (Useful in breast and uterus cancer)	*Taxus wallichiana*	Talis	2800-3200 m in fir and spruce forest shades.

Steroid saponins can be extracted from *Asparagus, Agave* and *Yucca*. Sources of alkaloids, glycosides, tannins etc., are also reported in the text.

In indigenous system of medicines there is a vast potential of preparing churanas, (Triphala, Darimashtak, Haridrakhandava, Ashwagandha etc.) Asavas (Kumariasava, Drakshasava, Lodhrasava, Kanakasava etc.); Arishtas (Kutajarishta, Draksharishta, Arjunarishta, Khadirarishta etc.); Avleha (Vasavleha, Chyavanprasha (Ashtavargayukta); Kashaya/Kwatha (Tulsiyadi, Manjishthadi, Dashmool); medicated creams, oils, soaps, extractives, tinctures and snuffs. Hundreds of species can be utilised for preparing mother tinctures in Homoeopathic medicines.

Appendix 9. Essential oil-yielding sources

Though over 150 species of aromatic plants have been identified in the state flora, yet the species having commercial importance have been included hereunder:

Sr. No.	Product	Source plants(s)	Common name	Availability in H.P.
1.	Acacia perfume	*Robinia pseudacacia*	Honey-locust	Planted all over the state and now naturalised from Kalka to Kalpa. Also introduced in Spiti.
2.	Angelica oil	*Angelica glauca*	Chora	2200-2800 m, becoming sporadic at many places and endangered in others.
3.	Artemisia oil	*Artemisia* spp.	Charmar, Kubish Nagdona etc.	About a dozen species are growing in abundance in many parts of the state.
4.	Basil camphor	*Ocimum kilimandscharicum*	Kapur-tulsi	Sporadic in shiwalik zone and can be easily cultivated.
5.	Broom or Genet	*Cytissus scoparius*	Broom	Naturalized in various parts of Solan, Shimla & suburbs.
6.	Calamus oil	*Acorus calamus*	Bach, Bare	In marshy areas upto 2800 m.
7.	Cedar wood oil	*Cedrus deodara*	Devdaru	Rich forests occur between 2000-2800 m.
8.	Zinger oil	*Zingiber officinale*	Zinger, Adrak	Commercial cultivation is done in Bilaspur & Sirmour districts.

Contd.

Sr. No.	Product	Source plants(s)	Common name	Availability in H.P.
9.	Hedychium oil	*Hedychium acuminatum*	Kapur-Kachri, Shati	1800-2400 m; more common in Shimla and Chail catchment areas.
10.	Jasmine oil	*Jasminum officinale* & allied species	Chameli, Malti	Upto 2600 m in all parts of the state.
11.	Kuja oil	*Rosa moschata*	Kuja, Kuje-ke-phool	Common upto 2800 m all over H.P.
12.	Kuth oil	*Saussurea costus*	Kuth, Costus	Lahaul, Pangi-Bharmour and Kinnaur.
13.	Hops oil	*Humulus lupulus*	Hops	Pattan valley in Lahaul.
14.	Millfoil/Yarrow	*Achillea millefolium*	Biranjasif	Between 2000-3000 m.
15.	Marjorum oil	*Origanum vulgare*	Sathra	Widely distributed between 1300-3500 m.
16.	Oakmoss oil	*Parmelia* spp.; *Usnea longissima; Ramalina* spp.	Chharila, Sewal, Ujjral	On rocks; stem and branches of high level oaks.
17.	Mint/spearmint oil	*Mentha spicata, M. longifolia*	Pudina, Chachri	Near moist localities also planted in house yards. In moist place even upto 5000 m in Spiti.
18.	Palmarosa oil	*Cymbopogon martinii*, var. 'motia'	Makoraghas	Common in sub-tropical open grasslands.
19.	Scented Pleurospermum	*Pleurospermum brunonis*	Nesar	Alpine rocky slopes between 3400-4200 m.
20.	Taragon oil	*Artemisia dracunculus*	Taragon	Alpine & sub-alpine meadows; dry temperate zone.

Contd.

Sr. No.	Product	Source plants(s)	Common name	Availability in H.P.
21.	Skimmia oil	*Skimmia laureola*	Patti-dhoop	Moist temperate forest (2800-3200 m).
22.	Thyme oil	*Thymus serpyllum*	Banajwain, lepti	Temperate, rocky open slopes and dry alpine areas.
23.	Valerian oil	*Valeriana jatamansi*	Nihani, Tagar, Sugandhbala	Common in temperate forest shades.
24.	Jurinea oil	*Jurinea dolomiaea*	Dhoop	Alpine areas in Pangi Bharmour, Chhota and Bara Bhangal, Mural danda, Chansil pass, Chandranahan and Sanglakanda.
25.	Kapur minjar (Camphorous oil)	*Strobilanthes auriculata*	Kapur minjri	Sporadic in Hamirpur, Bilaspur, Shahpur and Rait area of Kangra.
26.	Vetiver oil	*Vetiveria zizanioides*	Khus	Marshy areas upto 1300 m.
27.	Worm seed oil	*Chenopodium ambrosioides*	Ban bathu, Sugandh vastuk	Common in wastelands upto 2000 m.
28.	Camphor	*Cinnamomum camphora*	Mushki-kapur	Planted in Palampur, Joginder Nagar, Nauni and in several gardens.
29.	Tejpat	*Cinnamomum tamala*	Tejpatra, Gurpatraj	Machhial, Drahal forest of Jogindernagar, and Darang forest near Palampur.
30.	Tejbal	*Zanthoxylum armatum*	Timir	All over H.P. upto 1500 m, sporadic in ravines.

Contd.

Sr. No.	Product	Source plants(s)	Common name	Availability in H.P.
31.	Hyssop	*Hyssopus officinalis*	Juffa, Tengu, Tyangu	Common in Gulling-Burr area in Pin valley. Moderate in Attargu and sporadic in Khab, Kaa loop areas and Kishori bridge near Triloknath.
32.	Pine needle oil	*Pinus roxburghii*	Chir, Saral,	Common all over the state below 1800 m, elevation, goes upto 2300 m in Parvati valley, Kullu.

In addition, *Tagetus* oil from *Tagetes minuta* (Jangligainda) Cedral from *Juniperus macropoda*, Pine oil from the needles of *Pinus wallichiana* (Kail), Tansy from *Tanacetum longifolium* and Ocimum oil from *Ocimum sanctum* (Tulsi) can be harnessed. Species of *Elscholtzia, Nepeta, Salvia, Senecio, Morina, Selinum, Rosemary, Matricaria* etc. which yield valuable essential oil of commerce are being cultivated in the experimental fields for future commercialisation.

Appendix 10. Plants for insecticide and pesticide industry

Importance of conservation of environmental quality and the preservation of natural equilibria in eco-system is now well recognized. Equally well documented are inter-convertible evidences of gross environmental contamination and resultant ecological havoc wrecked by innundation of the biosphere with a veritable flood of synthetic alien molecule, chiefly the conventional synthetic organic insecticides. However, a plethora of drawbacks, disadvantages, undesireable side effects and hazards stemming from the indiscriminate and extensive application of conventional insecticides and pesticides have lead to a reappraisal of their continuing role in insect-pest control programmes. These re-appraisals have resulted into the evolution of pest management from natural resources for increasing the agricultural production without irretrievably damaging the biosphere. The natural resources especially the botanical ones offer unlimited scope towards the realization of this desideratum. Among these, few of such plant species are enlisted hereunder:

Sr. No.	Botanical name (Family)	Common/vern name	Parts used
1.	*Acorus calamus* Linn. (Araceae)	Bach, Bare, Calamus root	Leaves & rhizomes, calamus oil
2.	*Adina cordifolia* Roxb. Benth. & Hook. f. (Rubiaceae)	Haldu	Bark & Juice
3.	*Agava americana* Linn. (Agavaceae)	Kantala, century plant	Leaves & Roots
4.	*Semecarpus anacardium* (Anacardiaceae)	Bhillawa	Juice/Sap
5.	*Arachne cordifolia* syn. *Andrachne cordifolia* Muell. Arg. (Euphorbiaceae)	Gurguli	Leaves
6.	*Arisaema tortuosum* Wall and other spp. (Araceae)	Samp-ki-kumb, Snake maize	Tubers
7.	*Artemisia absinthium* Linn and *A. vulgaris* Linn.	Wormwood, Villaiti afsantin	Flowering tops/ Essential oil
8.	*Azadriachta indica* A. Juss (Meliaceae)	Neem Margosa tree	Leaves, bark & oil
9.	*Berberis aristata* DC and allied spp. (Berberidaceae)	Kashmal, Rasount, Daruhaldi	Extract & salts

Contd.

Sr. No.	Botanical name (Family)	Common/vern name	Parts used
10.	*Boenninghausenia albiflora* (Hook) Reichb.ex Meissn. (Rutaceae)	Pissumar buti	Leaf extract, essential oil
11.	*Butea monosperma* Lamk. Tanbert (Leguminosae)	Dhak, Palash, Flame of the forest	Seeds
12.	*Cedrus deodara* (Roxb.) Loud. (Pinaceae)	Devdar, Himalayan cedar	Cedar wood oil
13.	*Croton tiglium* Linn. (Euphorbiaceae)	Jayapal, Jamalgota, Purging croton	Seeds
14.	*Curcuma longa* Roxb. (Zingiberaceae)	Haldi, Turmeric	Rhizomes
15.	*Cymbopogon nardus* Linn. (Graminae)	Katrin, Citronella, grass	Leaves, oil
16.	*Delphinium vestitum* wall *D. brunonianum* Royle (Ranuculaceae)	Juhi Salyan, Laskar	Roots, leaves
17.	*Diospyros montana* Roxb. (Ebenaceae)	Bistendu, Lohari, Mountain-persimon	Fruits & crushed leaves
18.	*Eucalyptus globulus* Labill. (Myrtaceae)	Kapurmaram, Blue gum	Leaf, decoction and oil.
19.	*Euphorbia royleana* Boiss. (Euphorbiaceae)	Thuhar, Thor, Danda, thor, Shakar Pitam	Latex (milky)
20.	*E. thymifolia* Linn. (Euphorbiaceae)	Chhoti dudhi, Rakta Vinda, Chada	Plant, oil
21.	*Gloriosa superba* Linn. (Liliaceae)	Gigers claws, glory lily, Kalihari, Langli, Agnishikha	Roots (Rhizomes)
22.	*Hedychium acuminatum* Rosc. (Zingiberaceae)	Kapur-kachri, Kachur, Shati, spiked ginger lily	Rhizome, essential oil.

Contd.

Sr. No.	Botanical name (Family)	Common/vern name	Parts used
23.	*Holarrhena antidysenterica* (L.) Wall. ex DC. (Apocyanaceae)	Kutaj, Kurchi Indrajau, Ivory tree	Bark and seeds
24.	*Kalanchoe spathulata* (Poir) DC. (Crassulaceae)	Tatara, Haiza	Leaves
25.	*Madhuca latifolia* (Roxb.) Mc Bride (Sapotaceae)	Mahua	Mahua meal, a residual cake after extraction of oil from the seeds.
26.	*Nicandra physaloides* Gaertn. (Solanaceae)	Apple of Peru	Decoction of leaves
27.	*Nicotiana tabacum* Linn. & *N. rustica* Linn. (Solanaceae)	Tobacco, Tambakhu	Alkaloid Nicotine/ plant extract
28.	*Ocimum* species i.e. *O. gratissimum* Linn. *O. sanctum* Linn. (Labiatae)	Ram-tulsi Tulsi, etc.	Essential oil
29.	*Picrasma quassioides* (D.Don)Benn. (Simaroubaceae)	Charangi, Karwi, Bering, Quassia	Stem wood
30.	*Pieris ovalifolia* D.Don (Syn. *Lyonia ovalifolia*) (Wall.) Drude (Ericaceae)	Anyar	Leaves
31.	*Polygonum hydropiper* Linn. (Polygonaceae)	Jal Marich, Pachir mool, Ghaniri	Leaves and herb
32.	*Ricinius communis* Linn. (Euphorbiaceae)	Erand, Castorbean	Seeds, leaves
33.	*Saussurea costus* Clarke (Compositae)	Kuth	Roots
34.	*Sophora mollis* R.Grah. (Leguminosae)	Buna, Sakina, Arghavan	Seeds
35.	*Trigonella foenum-graecum* Linn. (Papilionaceae)	Methi, Fenugreek	Plant, dried leaves.
36.	*Vitex negundo* Linn. (Verbenaceae)	Bana, Sura, Nirgundi	Leaves
37.	*Zanthoxylum armatum* DC. (Rutaceae)	Tirmir, Tejbal Tumburu	Fruits, branches and thorns

Appendix 11. Tannin-yielding plants

Tans and dyes are of vital importance in leather industry, pharmaceutical as well as cotton industry. Plant resources yielding tans and dyes as available in Himachal Pradesh are enumerated hereunder indicating the part of plant used for extraction of these products e.g. bark, leaf, fruit, root etc.

Sr. No.	Latin name	Local name
Bark tans		
1.	*Acacia nilotica*	Kikar, babul
2.	*A. mollissima*	Wattle
3.	*Bridelia retusa*	Ekdania, khaja
4.	*Cassia auriculata*	Avaram
5.	*Careya arborea*	Kumbhi
6.	*Cassia fistula*	Amaltas, konnai
7.	*Castanopis hystrix*	Indian chestnut
8.	*Casuarina equisetifolia*	Jangli jhau
9.	*Emblica officinalis*	Aonla
10.	*Juglans regia*	Walnut
11.	*Quercus* spp.	Ban, mohru, kharshu etc.
12.	*Lagerstroemia parviflora*	Dhaura, Seja
13.	*L. speciosa*	Jarul, Pride of India
14.	*Shorea robusta*	Sal
15.	*Terminalia arjuna*	Arjun
16.	*T. tomentosa*	Asan
17.	*Tamarix aphylla*	Laljhau, Farash
18.	*T. troupii*	Jhau
Leaf tans		
1.	*Anogeissus latifolia*	Chhal, Dhav
2.	*Carissa spinarum*	Karounda
3.	*Cotinus coggygria*	Tung
4.	*Coriaria nepalense*	Gangar, Bathu
5.	*Emblica officinalis*	Amla
6.	*Elscholtzia polystachya*	Pothi
7.	*Impatiens balsamifera*	Gulmehndi
8.	*Lawsonia inermis*	Mehndi

Sr. No.	Latin name	Local name
Fruit and Gall tans		
1.	*Acacia nilotica*	Babul pods
2.	*Emblica officinalis*	Amla
3.	*Pistacia integerrima*	Kakar-singi; (gall)
4.	*Quercus* spp.	Ban, mohru, kharshu
5.	*Rhus succedenia*	Devil's tree, titre
6.	*Spondias pinnata*	Ambara
7.	*Semecarpus anacardiam*	Bhillawa
8.	*Terminalia bellirica*	Bahera
9.	*T. cheubla*	Harad
10.	*Zizyphus xylopyrus*	Kathber
Root tans		
1.	*Arnebia benthami*	Ratanjot
2.	*A. euchroma*	Ratanjot
3.	*Bergenia ciliata*	Pashanbhed
4.	*B. stracheyii*	Pashanbhed
5.	*Butea monosperma*	Dhak, Palash
6.	*Geranium nepalense*	Laljari
7.	*G. wallichianum*	Laldori
8.	*Juglans regia*	Akhrot
9.	*Onosma hispida* & allied spp.	Gajawan
10.	*Polygonum* spp.	Anjwar, Chhoti-ninai etc.
11.	*Potentilla nepalensis*	Laljari
12.	*Parmelia* spp.	Chharila (Thallus)

Appendix 12. Medicinal and aromatic plants to fit in different cropping systems

Himachal Pradesh has different agro-climatic zones starting from near tropical to semi-arctic in cold desert areas. In each zone and valley areas, various micro-climatic conditions prevail. As such there is a vast scope to cultivate medicinal and aromatic plants in different areas as given below:

Sr. No.	Botanical name	Vern name	Where to cultivate
1.	*Aconitum heterophyllum*	Atis, Patish	Temperate and alpine pastures (2500-4500 m).
2.	*A. deinorrhizum*	Aconite, Vatsnabh	Fringe zone of timber line and alpine pastures.
3.	*Acorus calamus*	Bach, Calamus root	Moist and boggy places near water courses between 1200-2500 m.
4.	*Angelica glauca*	Chora, Angelica root	1800-2800 m in humus rich forest soils.
5.	*Atropa acuminata* and *A. belladona*	Belladona	Between 1800-2800 m.
6.	*Abrus precatorius*	Ratti, Gunja	Field borders shrubberies in Shiwalik zone upto 1400 m.
7.	*Arnebia benthami*	Ratanjot	Alpine slopes/pastures between 3500-4500 m.
8.	*A. euchroma*	Ratanjot	Dry temperate and alpine zones in Lahaul-Spiti and Pooh (Kinnaur).
9.	*A. guttata*	Ratanjot	Dry temperate and alpine zones in Lahaul-Spiti and Pooh (Kinnaur).
10.	*Artemisia brevifolia*	Santonin plant Sainski, Nurcha	Dry alpine and temperate areas in Lahaul-Spiti, Kinnaur and Pangi Bharmour with ease and success.
11.	*A. dracunculus*	Tarragon	Alpine/temperate pastures around 2500-3500 m.
12.	*Aegle marmelos*	Bilve, Bilpatri	Lower areas upto 1300 m.

Contd.

Sr. No.	Botanical name	Vern name	Where to cultivate
13.	*Asparagus racemosus* and other species	Shatavar	Upto 1300 m
14.	*Berberis aristata*	Daruhaldi, kashmal	Village shamlats, as hedge in field borders between 1600-2800 m.
15.	*Bergenia ciliata* and *B. stracheyi*	Pashanbhed	Shady rocks and alpine moist rocky slopes. Can better rhrive in biosphere reserves.
16.	*Betula utilis*	Bhojpatra	Between 2800-4000 m.
17.	*Bryonopsis laciniosa*	Shivlingi	Valley areas upto 1500 m.
18.	*Cassia fistula*	Amaltas	Social forestry plantation/ avenue plantation upto 1500 m.
19.	*Cassia absus*	Chaksu	Sub-tropical areas upto 1400 m.
20.	*Cassia tora*	Chakramarda	Sub-tropical areas upto 1400 m.
21.	*Cassia occidentalis*	Kasmarda, kasondi	Throughout lower areas upto 1500 m.
22.	*Celastrus paniculatus*	Malkangni, Jyotishmati	Sub-Himalayan tract upto 1500 m.
23.	*Centella asiatica*	Mandookparni, Brahmi	In marshy/moist places upto 1500 m.
24.	*Centratherum anthelminticum*	Somraji, Ghorajiri	Foot hills upto 1600 m.
25.	*Chrysanthemum cinerariaefolium*	Pyrethrum	1600-2400 m.
26.	*Cichorium intybus*	Kasni, chicory	1600-2800 m.
27.	*Cimicifuga foetida*	Jiunti	Between 2000-3200 m.
28.	*Cinnamomum camphora*	Camphor tree, Kapur	Foothills upto 1300 m.

Contd.

Sr. No.	Botanical name	Vern name	Where to cultivate
29.	*C. tamala*	Tejpatra	Foothills upto 1300 m.
30.	*Citrullus colocynthis*	Indrayan	Sandy warm soils upto 1000 m.
31.	*Clerodendrum indicum*	Bharangi	Upto 1200 m.
32.	*Colchicum luteum*	Hirantutiya	Chamba (1000-2400 m), Udaipur (Kukunseri) in Lahaul.
33.	*Cordia dichotoma*	Lasura	Can be planted below 1500 m.
34.	*Costus speciosus*	Kemuk, Keu	Foothiils upto 1300 m in moist locations.
35.	*Crataeva nurvala*	Varuna, Barna	Sub-Himalayan tract upto 1200 m.
36.	*Crocus sativus*	Kesar, Saffron	Kinnaur and other places around 1600-2600 m.
37.	*Croton tiglium*	Jamalgota	Below 1000 m.
38.	*Curculigo orchioides*	Kalimushli	Shiwalik region upto 1200 m especially in *Shorea robusta* range.
39.	*Cymbopogon* spp.	Lemon grass, Citronella grass, Palma rosa grass etc.	Shiwalik zone
40.	*Dactylorhiza hatagirea*	Salampanja	Alpine pasture and temperate madows between 2800-4200 m.
41.	*Datura inoxia*	Dhatura	Shiwalik range.
42.	*Delphinium denudatum*	Jadwar, Nirbisi	Between 2600-3800 m.
43.	*Desmodium gangeticum*	Shalparni	Sub-tropical forest shades.
44.	*Didymocarpus pedicellata*	Patharphori, Shilapushpa	Kangra, Chamba & Mandi area upto 1600 m.
45.	*Digitalis lanata* and *D. purpurea*	Hritpatri, Tilpushpi	Temperate zone 1600-2400 m.

Contd.

Sr. No.	Botanical name	Vern name	Where to cultivate
46.	*Dioscorea deltoidea*	Singli-mingli, Ganj	1300-3000 m in shrubberies.
47.	*Dolichos uniflorus*	Kulath	Throughout the region upto 1500 m.
48.	*Eclipta prostrata*	Bhringraj, Bhangra	In moist places upto 1500 m.
49.	*Embelica ribes*	Baibidang	Upto 1500 m.
50.	*Emblica officinalis*	Amla	In Social forestry plantations upto 1400 m.
51.	*Ephedra gerardiana* and *E. intermedia*	Soma	In biosphere reserves in Kinnaur, Spiti, Pangi and Lahaul 2800-5000 m.
52.	*Ferula asafoetida* and *F. jaeschkeana*	Hing Wild hing	Lahaul-Spiti, Kinnaur and Pangi area.
53.	*Foeniculum vulgare*	Mithi saunf	Valley areas upto 1500 m.
54.	*Gentiana kurroo*	Karu	1800-3500 m.
55.	*Gloriosa superba*	Kalihari, Langli	Sub-tropical areas upto 1300 m.
56.	*Glycyrrhiza glabra*	Mulhatti	Lower valley areas upto 1200 m, can be tried in Spiti conditons as it has proved successful in UHF station at Sharboo, Kinnaur.
57.	*Gmelina arborea*	Gambhari	Shiwalik zone
58.	*Hedychium acuminatum*	Shati, Kapur-Kachri	1500-2400 m.
59.	*Helicteres isora*	Marorphali	Upto 1200 m.
60.	*Heracleum candicans*	Padara	2000-3800 m.
61.	*Holarrhena antidysenterica*	Kutaja, Indrajau	Upto 1200 m.
62.	*Humulus lupulus*	Hops	Kinnaur and Lahaul-Spiti.
63.	*Hyoscyamus niger*	Khurasani ajwain	2000-3800 m in dry temperate zone.

Contd.

Sr. No.	Botanical name	Vern name	Where to cultivate
64.	*Hyssopus officinalis*	Zufah-yabis, Zufah	2400-3400 m in Lahaul & Spiti.
65.	*Inula racemosa*	Pushkarmool	1600-3200 m in Lahaul-Spiti.
66.	*Ipomoea hederacea*	Kaladana	Upto 1400 m in lower area of H.P.
67.	*Iris germanica*	Ghurbach, Padmapushkara	2000-2500 m in forest soils.
68.	*Jasminum* spp.	Chamele, Malti	In garden, field borders upto 2800 m.
69.	*Juglans regia*	Akhrot, Walnut	Upto 2800 m in orchards/ border lands.
70.	*Juniperus macropoda*	Dhupi, Shurgu	2600-4500 m especially in cold deserts.
71.	*J. Communis*	Hauber, Hapusha Bitharh	3200-4300 m in temperate and alpine zones.
72.	*Jurinea dolomiaea*	Dhoop	3000-4200 m (*In-situ* measures are required).
73.	*Lepidium sativum*	Chandrashur	Below 1300 m.
74.	*Lilium polyphyllum*	Kshir-kakoli	2000-2800 m
75.	*Madhuca butyracea*	Phalwara, Chiura	Below 1500 m.
76.	*M. indica*	Mahua, Mahwa	Shiwalik zone below 1000 m.
77.	*Matricaria chamomilla*	Babuna	Sub-tropical to temperate regions.
78.	*Melia azadirachta*	Bakain, Mahanim	Upto 1500 m.
79.	*Melissa officinalis*	Bililotan, Balm	Sub-tropical to sub-temperate zone.
80.	*Moringa oleifera*	Sohanjan	Sub-Himalayan tract.
81.	*Mucuna pruriens*	Kaunch	Sub-Himalayan tract.
82.	*Murraya koenigii*	Mithinim	Upto 1500 m.
83.	*Myrica esculenta*	Kaiphal	1200-1800 m.
84.	*Nardostachys grandiflora*	Balchharh, Jatamansi	3500-4200 m in moist temperate areas.

Contd.

Sr. No.	Botanical name	Vern name	Where to cultivate
85.	*Nerium indicum*	Lal kaner	Upto 1600 m.
86.	*Ocimum sanctum* & other spp.	Tulsi	Upto 1500 m.
87.	*Olea europaea*	Olive	1200-1800 m.
88.	*Onosma bracteatum* and *O. echioides*	Gojihva, Shankhahuli	2500-4500 m.
89.	*Operculina turpethum*	Nishoth, turpeth	Below 1100 m.
90.	*Origanum vulgare*	Sathra	1400-3800 m.
91.	*Oroxylum indicum*	Arlu	Below 1200 m.
92.	*Paeonia emodi*	Udsalib	1500-3000 m.
93.	*Peganum harmala*	Harmal	Upto 1600 m in drier parts can be introduced in Spiti cold desert also.
94.	*Physochlaina praealta*	Bajarbang, Nandru	2800-3800 m in Lahaul-Spiti & Kinnaur.
95.	*Phytolacca acinosa*	Matazor, Jalga	1500-2800 m.
96.	*Picrasma quassioides*	Charangi	Upto 2500 m.
97.	*Picrorhiza kurrooa*	Kutaki	2800-4500 m.
98.	*Pistacia khinjuk*	Kakarsingi	Upto 2600 m but galls are produced in sub-tropical regions only.
99.	*Plmbago zeylanica*	Chitrak, Chita	Upto 1500 m.
100.	*Podophyllum hexandrum*	Bankakri, Giriparpat	2800-4200 m.
101.	*Polygonatum cirrihfolium*	Meda	2600-3000 m.
102.	*P. multiflorum*	Meda	1600-3000 m.

Contd.

Sr. No.	Botanical name	Vern name	Where to cultivate
103.	*P. verticillatum*	Mahameda, Salam-misri	2000-2800 m.
104.	*Pongamia pinnata*	Karanj	Below 1300 m.
105.	*Prinsepia utilis*	Bhekhal	1500-2800 m.
106.	*Prunella vulgaris*	Ustakhadus	1500-3500 m.
107.	*Pueraria tuberosa*	Bidarikanda, Siali	Upto 1300 m.
108.	*Punica granatum*	Darim, Anar	1000-1800 m.
109.	*Putranjiva roxburghii*	Jiyapota	Upto 1200 m.
110.	*Randia dumetorum*	Rada, Madanphal	Upto 1300 m.
111.	*Rauvolfia serpentina*	Sarpagandha	Upto 1000 m in Shiwalik zone.
112.	*Rheum australe* and other spp.	Revandchini, Rhubarb	2500-4500 m.
113.	*Ricinius communis*	Castor, Erand	Upto 1500 m.
114.	*Rosmarinus officinalis*	Rosemari	Upto 2000 m.
115.	*Salvia officinalis*	Salvia, safahuss	Upto 2000 m.
116.	*Sapindus mukorossi*	Ritha, Soapnut	Upto 1300 m.
117.	*Saussurea costus*	Kuth	2700-3800 m in drier areas.
118.	*Scilla indica*	Bhuikanda	Upto 1200 m.
119.	*Semecarpus anacardium*	Bhillawa	Upto 1000 m.
120.	*Skimmia laureola*	Neir, Kasturi patra	2200-3200 m.
121.	*Solanum indicum*	Brihati	Upto 1300 m.
122.	*S. laciniatum*	Kangaru apple	Upto 1300 m.
123.	*S. nigrum*	Kakmachi	1300-2600 m.
124.	*S. viarum*	Banbhindi	Upto 1500 m.
125.	*S. surattense*	Kantkari	Upto 1300 m.

Contd.

Sr. No.	Botanical name	Vern name	Where to cultivate
126.	*Sophora mollis* & *S. japonica*	Buna, Sakina	Upto 2200 m.
127.	*Spilanthes acmella*	Akarkara	Upto 1200 m.
128.	*Spondias pinnata*	Ambara	Upto 1000 m.
129.	*Strobilanthes auriculata*	Kapurminjar	Below 1200 m.
130.	*Swertia chirayita*	Chirayata	1300-3200 m.
131.	*S. angustifolia* and *S. purpurascens*	Subst. for Chirata	Upto 1800 and between 1500-3500 m respectively
132.	*Symplocos paniculata*	Lodhra	Upto 2800 m.
133.	*Syringa emodi*	Shafri	2800-3500 m.
134.	*Terminalia arjuna*	Arjun	Upto 1200 m.
135.	*T. bellirica*	Bahera	Upto 1000 m.
136.	*T. chebula*	Harad	Upto 1000 m.
137.	*T. tomentosa*	Asan	Upto 1000 m.
138.	*Thevetia peruviana*	Pila kaner	Upto 1000 m.
139.	*Tinospora cordifolia*	Guduchi, Giloe	Upto 1200 m.
140.	*Trichosanthes cucumerina*	Patol	Upto 1000 m.
141.	*Tylophora indica*	Damabuti	Upto 1000 m.
142.	*Urginea indica*	Jangli piaz, squill	Sub-tropical tract upto 1300 m.
143.	*Valeriana jatamansi*	Nihani, Mushkbala	Upto 3000 m.
144.	*Verbena officinalis*	Karaita, Pamukh	Upto 2200 m.
145.	*Vetiveria zizanioides*	Khaskhas	Moist boggy places upto 1200 m.
146.	*Vinca rosea* (*Catharanthus roseus*)	Sadabahar	Upto 1200 m.

Contd.

Sr. No.	Botanical name	Vern name	Where to cultivate
147.	*Viola odorata* or *V. canescens*	Banaksha	1200-2400 m.
148.	*Vitex negundo*	Nirgundi	Upto 1500 m.
149.	*Viscum album*	Banda, Mistletoe	1200-2600 m.
150.	*Withania coagulens* and *W. somnifera*	Ashwagandha	Upto 1300 m.
151.	*Woodfordia fruticosa*	Dhataki pushpa	Upto 1300 m.
152.	*Zanthoxylum armatum*	Tejbal	Upto 1300 m.
153.	*Zingiber officinale*	Zinger, Sunthi	Upto 1300 m.

Appendix 13. Domestication hints for medicinal and aromatic plants

During the past about 3 decades, attempts have been made to bring various species under domestication in herbal gardens. More than 230 species have been planted in herbal gardens in the state under various schemes. Observations recorded during the process about their artificial regeneration are tabulated.

Sr. No.	Botanica name	Local name	Medicinal/ Aromatic	Methods of domestication
1.	*Abelomchus moschatus*	Lata-kasturi	Aromatic & medicinal	Seeds
2.	*Abrus precatorius*	Ratti, Gunja	Medicinal	Through seeds
3.	*Achyranthes aspera*	Puthkanda, Apamarga	Medicinal	Seeds and splits of root.
4.	*Acorus calamus*	Bach, Bare	Medicinal & aromatic	Rhizomes
5.	*Aegle marmelos*	Bilve	Medicinal	Seeds
6.	*Agave americana*	Ramban, Kantala	Medicinal	Bulbils
7.	*Ajuga bracteosa*	Neel-Kanthi	Medicinal	Seeds
8.	*Aloe barbadensis*	Ghee-kuwar	Medicinal	Bulbils
9.	*Ammomum subulatum*	Bari-ilaichi	Medicinal	Seeds, slips
10.	*Arctium lappa*	Jangli-kuth	Medicinal	Seeds/Roots
11.	*Artemisia annua*	Malaria plant	Aromatic/ medicinal	Seeds
12.	*Artemisia vulgaris*	Chharmar	Medicinal	Seeds
13.	*Asclepias curassavica*	Kaktundi	Medicinal	Seeds
14.	*Asparagus adscendens*	Sahasrapal	Medicinal	Seeds/suckers/ sprouts
15.	*A. filicinus*	Shatrawari	Medicinal	Seeds/suckers/ sprouts
16.	*A. racemosus*	Shatavar	Medicinal	Seeds/suckers/ sprouts
17.	*Atropa belladona*	Belladona	Medicinal	Seeds/stem/ root cuttings
18.	*Bauhinia variegata*	Kachnar	Medicinal	Seeds
19.	*Berberis lycium*	Kashmal	Medicinal	Seeds/cuttings

Contd.

Sr. No.	Botanica name	Local name	Medicinal/ Aromatic	Methods of domestication
20.	*Boerhaevia diffusa*	Punernava	Medicinal	Seeds/stem cuttings with root portion intact.
21.	*Bryonopsis lacinosa*	Shivlingi	Medicinal	Seeds
22.	*Calotropis procera*	Ak, Madar	Medicinal	Seeds
23.	*Cardiosperum halicaccabum*	Kanphuti	Medicinal	Seeds
24.	*Cassia absus*	Chaksu	Medicinal	Seeds
25.	*C. occidentalis*	Kasundi, Kasmard	Medicinal	Seeds
26.	*C. sophera*	Kasmard	Medicinal	Seeds
27.	*C. tora*	Chakramard	Medicinal	Seeds
28.	*Catharanthus roseus*	Sadabahar, Sadasuhagan	Medicinal	Seeds/cuttings
29.	*Centella asiatica*	Mandook-parni	Medicinal	Runners
30.	*Cinnamomum camphora*	Camphor	Aromatic & medicinal	Seeds
31.	*Cissampelos pareira*	Patha	Medicinal	Seeds/cutting
32.	*Costus speciosus*	Kemuk	Medicinal	Rhizomes
33.	*Cryptolepis buchanani*	Dudhibel	Medicinal	Seeds/cuttings
34.	*Cymbopogon nardus*	Citronella grass	Aromatic/ medicinal	Seeds, slips
35.	*C. flexuosus*	Malabai grass	Aromatic/ medicinal	Seeds, slips
36.	*Datura metel*	Kala Dhatura	Medicinal	Seeds
37.	*D. stramonium*	Safed Dhatura	Medicinal	Seeds
38.	*Digitalis lanata*	Foxglove, Hritpatri	Medicinal	Seeds
39.	*Dioscorea bulbifera*	Varahikand	Medicinal	Seeds/tubers/ bulbils
40.	*D. deltoidea*	Singli-mingli	Medicinal	Rhizomes

Contd.

Sr. No.	Botanica name	Local name	Medicinal/ Aromatic	Methods of domestication
41.	*Eclipta prostrata*	Bhringraj	Medicinal	Seeds
42.	*Emblica officinalis*	Amla	Medicinal	Seeds
43.	*Ephedra gerardiana*	Somlata	Medicinal	Seeds/stem divisions with root part.
44.	*Erythrina variegata*	Pangra	Medicinal	Seeds
45.	*Fagopyrum esculentum*	Buck-wheat	Medicinal	Seeds
46.	*Foeniculum vulgare*	Mithi-saunf	Aromatic/ medicinal	Seeds
47.	*Glaucium flavum*	Yellow horned poppy	Medicinal	Seeds
48.	*Gloriosa superba*	Kalihari, Langli	Medicinal	Seeds, Tubers
49.	*Heracleum candicans*	Patrala	Medicinal	Seeds, Stem cuttings/root splits
50.	*Hyoscyamus niger*	Khurasani ajwain	Medicinal	Seeds
51.	*Jasminum officinale*	Chameli	Aromatic/ medicinal	Seeds & cuttings
52.	*Justicia adhatoda*	Basuti, Vasa	Medicinal	Seeds
53.	*Matricaria chamomilla*	Babuna	Medicinal/ aromatic	Seeds
54.	*Melissa officinlais*	Balm	Aromatic/ medicinal	Seeds/Suckers.
55.	*Mentha arvensis*	Kala Pudina	Aromatic	Seeds & Runners/ Suckers.
56.	*M. longifolia*	Safed Pudina	Aromatic	Seeds & Runners/ Suckers.
57.	*M. piperita*	Peppermint	Aromatic	Seeds & Runners/ Suckers.

Contd.

Sr. No.	Botanica name	Local name	Medicinal/ Aromatic	Methods of domestication
58.	*M. spicata*	Pahari Pudina	Aromatic	Seeds & Runners/ Suckers.
59.	*Mimosa pudica*	Lajwanti	Medicinal	Seeds
60.	*Mucuna prurlens*	Kaunch	Medicinal	Seeds
61.	*Murraya koenigii*	Mithanim, Curry leaf	Aromatic/ medicinal	Seeds
62.	*Neirum indicum*	Lal Kaner	Medicinal	Seeds/cuttings (more easy).
63.	*Ocimum basilicum*	Sweet basil	Medicinal/ aromatic	Seeds
64.	*O. kilimandscharium*	Kapur tulsi	Medicinal/ aromatic	Seeds
65.	*O. sanctum*	Tulsi	Aromatic/ medicinal	Seeds
66.	*Origanum vulgare*	Ban Marua	Medicinal/ aromatic	Seeds/Suckers
67.	*Oroxylum indicum*	Arlu, Tatpalanga	Medicinal	Seeds
68.	*Plantago major*	Jangli Isabgol	Medicinal	Seeds
69.	*Plumbago zeylanica*	Chitrak, Chhichha	Medicinal	Seeds/cuttings
70.	*Punica granatum*	Daru, Anardana	Medicinal	Seeds/cuttings
71.	*Putranjiva roxburghii*	Jiapota	Medicinal	Seeds
72.	*Rheum australe*	Revandchini	Medicinal	Seeds/root splits with stem
73.	*Ricinus communis*	Erand, castor bean	Medicinal	Seeds
74.	*Rosmarinus officinalis*	Rosemari	Aromatic	Seeds, cuttings/ layering
75.	*Salvia bisor*	Sage plant	Medicinal/ aromatic	Seeds

Contd.

Sr. No.	Botanica name	Local name	Medicinal/ Aromatic	Methods of domestication
76.	*S. moorcroftiana*	Thuth	Medicinal/ aromatic	Seeds
77.	*S. lantata*	Chhota thuth	Medicinal/ aromatic	Seeds
78.	*S. sclarea*	Clary Sage	Aromatic	Seeds
79.	*Solanum laciniatum*	Kangaru apple	Medicinal	Seeds
80.	*S. nigrum*	Makoi	Medicinal	Seeds
81.	*S. sissymbrofolium*	Kandiari	Medicinal	Seeds
82.	*S. surratense*	Chhoti kantkari	Medicinal	Seeds
83.	*S. viarum* (*S. kharianum*)	Ban bhindi	Medicinal	Seeds
84.	*Terminalia arjuna*	Arjun	Medicinal	Seeds
85.	*T. bellirica*	Bahera	Medicinal	Seeds/ grafting
86.	*T. chebula*	Harad	Medicinal	Seeds/grafting
87.	*Thalictrum foliolosum*	Mamiri	Medicinal	Seeds
88.	*Thevetia peruviana*	Pila-kaner	Medicinal	Seeds/Cuttings
89.	*Tinospora cordifolia*	Giloye	Medicinal	Branch cuttings
90.	*Tylophora indica*	Damabuti	Medicinal	Cuttings
91.	*Withania somnifera*	Ashwagandha	Medicinal	Seeds
92.	*Woodfordia fruticosa*	Dhai, Dhatki	Medicinal	Seeds & Cuttings
93.	*Zanthoxylum armatum*	Tirmir, Tejbal	Medicinal	Seeds

The production technology for *Solanum khasianum* (Kateri, Banbhindi), *Catharanthus roseus*, (Sadabahar), *Withania somnifera* (Ashwagandha), *Salvia officinalis, Cymbopogon* spp., *Tagetus minuta*, (Jangli Gainda), *Ocimum* spp. (Tulsi), *Pimpenella anisum* (Anise), *Plantago ovata* (Isabgol), *Cassia angustifolia* (Senna), *Papavar somniferum* (Ahiphena), *Psoralea corylifolia* (Babchi), *Chlorophytum arundinaceum* (Safed Musli), *Urginea indica* (Jangli Piaz), *Lavendula officinalis* (Lavender), *Vetiveria zizanioides* (Khas Grass), *Rosa demascena* (Bulgarian Rose), *Chryranthemum cinerariaefolium* (Pyrethrum), *Pelargonium graveolens* (Geranium) etc. have been developed by CIMAP, ICAR coordinated centres, RRLs of CSIR, IIHR, Bangalore etc. CIMAP, Lucknow has developed superior varieties in Metha, Cymbopogon, Vetiver, Coriander and Hyoscyamus. Dr. Y.S. Parmar University of Horticulture and Forestry has also developed agrotechnology for over one dozen species of medicinal & aromatic plants.

Appendix 14. Aromatic plants responsible for health hazards

Though, the original richness of the flora of alpine hills and passes like Sach pass (Pangi) Mani-Mahesh (Bharmour), Rohtang pass (Lahaul-spiti), Thamsar jot (Bara Bhangal-Kangra), Dainasar (Chhota Bhangal-Kangra), Shikari Devi (Mandi), Chur Chandni (Sirmour), Chansal and Mural Danda (Shimla), Manjeban (bordering H.P. and U.P. near Dodra Kawar), Kinner Kailash of Kinnaur, Mudh-Bhaba, Pin-Parvati, Mane-Manirang, Ladarcha, Parangla, Demul pass, Langza, Kibber, Kunzum, Burr peak and Rongtonk peak all in Spiti have been disturbed by excessive grazing and human exploitation, yet the peaks and passes under reference possess quite a good number of rare and highly aromatic plants. The plants while in full growth during August-Sept. emit out different aroma and scent, which often causes nausea and giddiness—even vomiting to the sensitive persons allergic to strong scents. Some of the plants identified are listed hereunder:

Sr. No.	Botanical name	Local name
1.	*Pleurospermum brunonis* Benth.	Nesar, Losar
2.	*Rhododendron anthopogon* D. Don	Talish, Talli, Talsar
3.	*R. lepidotum* Wall.	Simris
4.	*Tanacetum longifolium* Wall.	Tansy, Bhurse-Khamba
5.	*Heracleum candicans* Wall.	Radara, Patrala
6.	*Selinum tenuifolium* Wall.	Muramansi, Mathosal
7.	*S. vaginatum* C.B. Clarke.	Bhutkesi, Mathosla
8.	*Morina longifolia* and *M. persica* Linn.	Bikh-akhwar, Bishkandara, dhoop
9.	*Origanum vulgare* Linn.	Sathra
10.	*Valeriana pyrolaefolia* Decaisne	Nihani, Tagar
11.	*Angelica glauca* Edgw.	Chora, Chamchorai
12.	*Artemisia dracunculus* Linn.	Chambro, Barma, Taragon
13.	*Jurinea macrocephala* Benth	Dhoop
14.	*Senecio amlexicaulis* Wall.	Surjee
15.	*Artemisia brevifolia* Wall. ex DC.	Narcha, Seuski, Bhurse
16.	*A. macrocephala* Jacq. ex Bess.	Dhoop
17.	*Waldhemia glabra* Decne	Phillu
18.	*Thymus serpyllum* Linn.	Banajwain, Padumba
19.	*Primula macrophylla* D. Don	Nili-ka-phool

Contd.

Sr. No.	Botanical name	Local name
20.	*Saussurea obvallata* (DC.) Edges	Brahmakamal
21.	*Aquilegia fragrans* Benth.	Lamo, Lundud dorje
22.	*Ferula jaeschkeana* Vatke	Kait
23.	*Nepeta podostachys* Benth.	Ribhuksu
24.	*Dracocephalum heterophyllum* Benth.	Ribhuksu
25.	*Hysoppus officinalis* Linn.	Tengu, Tyangu

Plants like bhang, dhatura, aconites (mohra) rati (seeds) etc. are narcotic and poisonous. Deaths from aconite poisoning have been many and exudate of Rhus (devils' tree), *Sapium insigne*, thuhar, aak cause blisters and skin eruptions. Pollens of *Parthenium, Chenopodium* and pines cause allergic problems.

Appendix 15. Potent substitutes for indigenous and exotic drug materials

Sr. No.	Ayurvedic name	Original material	Substitute
1.	Agnimanth	*Clerodendron phlomidis* Linn. f	i) *Premna integrifolia* Linn ii) *P. mucronata* Roxb.
2.	Aftimun Vilayati	*Cuscuta epithymum* L.	*Cuscuta reflexa* Roxb.
3.	Amlavetasa	*Garcinia pedunculata* Roxb.	*Rheum australe* & *R. moorcroftianum*
4.	Ashwakarna	*Dipterocarpus alatus* Roxb.	*Terminalia tomentosa* W&A
5.	Anise	*Pimpenella anisum* L.	i) *Anethum soya* Kurz. ii) *Carum carvi* L. iii) *Foeniculum vulgare* Mill
6.	Bans	*Bambusa bambos* Druce	*Curcuma angustifolia* Roxb.
7.	Bidang	*Embelia ribes* Burn. f.	*Embellia robusta* CB Clark (*Myrsine africana* is used as an adulterant for bidang).
8.	Daruhaldi	*Berberis aristata* DC.	i) *Berberis asiatica* Roxb. ex. DC. ii) *B. lycium* Royle
9.	Dhatura	*Datura metel* Linn.	i) *Datura innoxia* Mill ii) *D. stramonium* Linn.
10.	Dugdhika	*Euphorbia thymifolia* Linn.	*Euphorbia prostrata* Ait
11.	Dioscorea	*Dioscorea deltoidea*	i) *Solanum laciniatum* ii) *S. khasianum* C.B. Clarke
12.	Jeevak	*Malaxis muscifera* Ridley	*Pueraria tuberosa* DC.
13.	Kakoli	*Lilium polyphyllum* D. Don.	*Withania somnifera* Dunal
14.	Kshirkakoli	*Frilillaria roylei* Hook	*Withania somnifera* Dunal
15.	Lakuch	*Artocarpus lakoocha* Roxb.	*Solanum surattense* Burm f.

Contd.

Sr. No.	Ayurvedic name	Original material	Substitute
16.	Lamajika	*Cymbopogon jwarancusa*	*Vetiveria zizanioides* (Linn.) Nash.
17.	Madyanti	*Lawsonia inermis* Linn.	*Jasminum sambac* Ait.
18.	Mahameda	*Polygonatum verticillatum* Allioni	*Asparagus racemosus* Willd.
19.	Male fern	*Dryopteris flix-mas* (Linn.) Schott.	i) *Dryopteris barbigera* (Moore) Kuntze ii) *D. blandfordii*
20.	Meda	*Polygonatum cirrhifolium* Royle	*Asperagus racemosus* Willd.
21.	Muramanse	*Selinum tenuifolium* Wall.	*Nardostachys grandiflora* DC.
22.	Moorva	*Marsdenia tenacissima* W. & Arn.	*Choenomorpha macrophylla* (Roxb.) G. Don
23.	Mustak	*Cyperus rotundus* Linn.	i) *Cyperus scariosus* R.Br ii) *C. arundinaceum* Baker
24.	Pashanbhed	*Bergenia ciliata* Wall.	i) *Bergenia strecheyii* Hook. f. ii) *Aerva lanata* Juss iii) *Didymocarpus pedicellata* R. Br
25.	Rakta-punarnava	*Boerhaavia diffusa* Linn.	*Boerhaavia rependa* Willd.
26.	Ridhi	*Habenaria intermedia* D.Don	*Dioscorea bulbifera* L.
27.	Rishbhak	*Malaxis wallichii* Lindl.	*Pueraria tuberosa* DC.
28.	Sal	*Shorea robusta* Gaertn. f.	*Vateria indica* Linn.
29.	Sarpagandha	*Rauvolfia serpentina* (L.) Benth	*Rauvolfia canescens* L.
30.	Somvalli	*Sarcostemma brevistigma* W&A	*Ephedra gerardiana* Wall.
31.	Swarnakshiri	*Euphorbia thomsoniana* Boiss	*Argemone mexicana* Linn.
32.	Belladona	*Atropa belladona* L.	*Atropa acuminata* Royle ex. Lindley

Contd.

Sr. No.	Ayurvedic name	Original material	Substitute
33.	Squill	*Urginea scilla*	*Urginia indica* (Roxb) Kunth.
34.	Aconite	*Aconitum napellus* Linn.	i) *Aconitum chasmanthum* stapf ex. Holmes ii) *A. demorrhizum* Stapf
35.	Psyllium	*Plantago psyllium*	*Plantago ovata* Forsk.
36.	Valerian	*Valeriana officinalis* L.	i) *Valeriana jatamansi* Jones ii) *V. hardwickii* Wall.
37.	Ephedra	*Ephedra sinica*	*Ephedra gerardiana* Wall.
38.	Santonin plant	*Artemisia cina*	*Artemisia brevifolia* Wall.
39.	Nala, Lobelia	*Lobelia inflata* L.	*Lobelia nicotianaefolia* Roth
40.	Gum tragacanth	*Astragalus gummifera*	*Sterculia urens* Roxb.
41.	Turpeth	*Ipomoea purga*	*Ipomoea turpethum* (L) R.Br
42.	Gentian	*Gentiana lutea*	i) *Gentiana kurroo* Royle ii) *Picrorhiza kurrooa* Royle ex Benth.

Source: The Ay. Pharmacopoeia of India, Part-I (1978): 241-259, Mehra, P.N., 1986, 52-59; Fazal, Ummul, 1986, 103-108.

Bibliography

Ahluwalia, K.S. 1952. Medicinal Plants of Kangra Valley. *Indian Forester* Vol. 78: 181-194.

Ahuja, B.S. 1965, Medicinal Plants of Saharanpur. Survey of Medicinal Plants, Central Council of Ayurvedic Research Gurukula Kangri Vishwavidyalaya, Hardwar.

Aiyer, K., Narayana and Kolammal, M. 1966. Pharmacognosy of Ayurvedic Drugs (Kerala). Deptt. of Pharmaconosy, University of Kerala, Trivandrum, Series I., No. 1-10.

Ali, Usman, S., Pilladi, Gopal Krishna, Nair, K. Vasudevan and Cheliadurai, V. 1981. Contribution to the exact identity of Brahmi and Mandookparni; Bull. Med.-Ethnobot. Res. Vol. 2 No. I, pp. 23-20.

Allport, Noel, L. 1943. Chemistry and Pharmacy of Vegetables drugs; George Newnes Ltd. Town House, Southampton Street Strand, W.C. 2, London.

Altschul, S.V.R. 1973. Drugs and food from little known plants-Harward University, Cambridge, Massachusetts.

Anonymous, (1945-1976), Wealth of India, Raw Materials, Vol. I-XI, Publication and Information Directorate, C.S.I.R., New Delhi.

Anonymous, 1948. Report of the committee on Indigenous Systems of medicine, Vol. I, Ministry of Health, Govt. of India, New Delhi.

Anonymous, 1968. British Pharmaceutical Codex. The Pharmaceutical Press, London.

Anonymous, 1977. Himachal Mein Ayurved. Directorate of Ayurveda, Govt. of H.P.

Anonymous, 1978. The Ayurvedic Pharmacopoeia of India Part-I, Govt. of India Ministry of Health.

Anonymous, 1978-79, 1980. Annual reports, Regional Research Laboratory (CSIR), Jammu-Tawi.

Anonymous, (1980). British Pharmacopoea. Vol. I., Her Majestry's Stationery Office, London.

Anonymous, 1986, The useful plants of India. Publications and Information Directorate, CSIR, New Delhi.

Anonymous, 1990, Phytochemical Investigation of Certain Medicinal Plants used in Ayurveda. Central Council for Research in Ayurveda and Sidha, New Delhi.

Anonymous, (1991). Drug research Perspective: A CDRI Series Bibliography of CDRI Contributions 1951-1990. Central Drug research Institute, Lunknow.

Arber, Agnes, 1953. Herbals-their origin and evolution—a chapter in the History of Botany (1470-1670). 2nd Ed. Cambridge University Press.

Aswal, B.S. and Mehrotra, B.S. 1994. Fora of Lahaul-Spiti (A cold desert in North West Himalayas). Bishen Singh Mahendra Pal Singh, Dehradun.

Atal, C.K. and Kapur, B.M. 1977. Cultivation and Utilization of medicinal and Aromatic Plants, Regional Research Laboratory, Jammu-Tawi.

Atkinson, E.T. (Reprint), 1980. Flora of the Himalayas (with special reference to Kumaon, Garhwal and parts of Tibet; Cosmo Publication, New Delhi.

Bahadur, N. (Kr.) 1971. *Quercus dilatata is now O. himalayan.* Ind. For. 9: 423-425.

Bailey Liberty Hyde and Bailey, Ethel Zoe. 1977. Hortus Third, A concise Dictionary of Plants cultivated in United States and Canada, Macmilk Publishing Co., Inc. New York, Collier Macmillen Publishers, London.

Bal, S.N. 1932. Catalogue of Medicinal Plants Exhibits in the Industrial Section of the Indian Museum, Govt. of India, Central Publication Branch, Calcutta.

Banerjee, S.K., Gupta, B.D., Sheldrivk, W.S. and Holfe, G. 1980. Diametric ligusticum lactose from *Angelica glauca*; International Research Congress on Natural products and Medicinal Agents; Strasbourg, July, 6-11.

Bedi, Ramesh 1970. Garments of Shiva's Attendants Nagarjun, Vol. XIII 5: 11-14.

Bentley, Robert and Trimen, Henry, 1980. Medicinal Plants Vol. I-IV J. and A. Churchill, New Bulington Street, London.

Bhakuni *et al.* 1969.

Billore, K.V., Uniyal, M.R. 1973. Study of the group Mansitraya in Ayurveda; Abst. First Scientific Seminar, Drug Research, CCRIMM, New Delhi, Feb. 10-13, pp. 21.

Biwas, K. 1956. Common Medicinal Plants of Darjeeling and Sikkim, Himalayas; Govt. Printing Press, West Bengal.

Bor, N.L. (Reprint), 1980. Manual of India Forest Botany, J.J. Prints, Delhi.

Bor, N.L. and Raizada, M.B. (1954). Some beautiful Indian climbers and shrubs. The Bombay Natural History Society, Bombay.

Brahmavarchas, 1982. Jari-Bution Dwara Swasthya Swarakshan; Vedmate Gayitri Trust, Shantikunj, Haridwar.

Brandis, D. 1874. Forest Flora of North West and Central India; M.H. Allen & Co., London.

Chakraberty, Chandra (Reprint) 1983. An interpretation of Ancient Hindu Medicine; Neeraj Publishing House, Delhi, pp. 429-490.

Chakrapani, 1941 Ed. Charakasamhita, Nirnaysagar Press, Bombay.

Chakravarty, H.L. 1959. Monograph on Indian Cucurbitaceae Records of Botanical Survey of India, Vol. XVII, No. 1.

Champion, H.G. and Seth, S.K. 1968. A revised Survey of the Forest Types of India. Govt. India Publication, Delhi.

Chandler, R.F., Hooper, S.N., Harvey, M.J. 1982. Ethnobotany and Phytochemistry of Yarro, *Achillea millefolium*, Compositae; Eco. Bot. Vol. XVII. No. 36(2), April-June, pp. 203.

Chatterjee, P.N., Singh, Pratap and Sivaramakrishna, V.R. 1968. Efficacy of various insecticides including indigenous vegetable products, II; Preliminary note on Testing the Flea Repellant properties of an essential oil from Boenninghausenia albiflora (Hook.) Breichb. Ind. For., 94(4): 324-325.

Chauhan, N.S., 1974. *Ainslaea aptera* DC. A hitherto unknown herb used in acute gastritis; Sachitra Ayurveda, Feb. 1974. 479-99.

Chauhan, N.S., 1974, Medicinal plants of Una forest division, Una District, H.P. Nagarjun: 31-39.

Chauhan, N.S., Uniyal, M.R. and Sannd, B.N. 1979. A Preliminary study of the Indigenous Drugs used at the Tibetan Medical Centre, Dharamsala (H.P.), Nagarjun, April, 1979. 8: 190-193.

Chauhan, N.S., 1983, Some rare and interesting medicinal plants of H.P. All India Summer Institute on Agroforestry in the Western Himalaya, July 22 to August 7.

Chauhan, N.S., 1984. Medicinal Wealth of Pabbar valley in Himachal Pradesh. M.Sc. Thesis, HPKVV, Palampur.

Chauhan, N.S., 1987(a), Himachal Pradesh mein aushdhiya podhon ki kasht ki sambhavana. Ban Sandesh 20(4):18-27, Shimla.

Chauhan, N.S., 1987(b), Himachal Pradesh ke Jan Jatiya Kshetra Chhota Bhengal ki banoushadhyan, Audhyaniki evam Vaniki 2(½): 9-19. UH&F, Solan.

Chauhan, N.S., 1987(c), Collection, identification, cultivation and utilization of medicinal and aromatic plants in H.P. Status paper presented to the sub-committee on Medicinal plants Development in H.P. under the state Science, Technology and Environment Council, April 4, Shimla.

Chauhan, N.S., 1988a, Exploitation and Utilization of Herbal Resources of H.P. Invited lecture delivered "Workshop on Oppurtunities identification in Agro-based and Food-Products pp. 23-32, STEP, Holiday Home, Shimla, November 4-5.

Chauhan, N.S., 1989a, Herbal Wealth of Cold Desert area of Himachal Pradesh. Proceedings, National Seminar-cum-Workshop on Cold Desert Area in India, 65-68, Govt. Press, H.P. Shimla.

Chauhan, N.S., 1988c, Indigenous plants as insecticides. All India Workshop-cum-Seminar on Plant Protection in Horticulture. In NK Sharma and Ashok Verma (Ed): 399-412, DEE, UH & F, Solan.

Chauhan, N.S., 1988d, Endangered Ayurvedic Pharmacopoeial plant resources of H.P. in P. Kaushik (Ed.) Indigenous Medicinal Plants. Including Microbes and Fungi: 199-205. M/s Today & Tomorrow Printes and Publishers, New Delhi.

Chauhan, N.S., 1988e, Ethnobotanical study of medicinal plants of Himachal Pradesh. In P. Kaushik (Ed.) Indigenous medicinal plants:187-198. M/s Today and Tomorrows Printers & Publishers, New Delhi.

Chauhan, N.S., 1988f, Exploitation Potential of Medicinal and Aromatic Plants in H.P. Invited paper presented on the seminar, "Development Mix and Environmental Consequences" H.P. Vigyan Manch, HPU, Summer Hills, Shimla, May 18-21.

Chauhan, N.S. and Sharma, G.K., 1988, Banoushadhiyon ka Himachal Pradesh ki Samajik Evam arthik Sthiti mein meahatva. Audhyaniki evam Vaniki 1(1):32-37, DEE, UH&F, Solan.

Chauhan, N.S., and Khosla, P.K., 1988, Commercially important medicinal plants of H.P. In P.K. Khosla (Ed.) Trends in Tree Sciences: 81-89. ISTS. Publication.

Chauhan, N.S. and Negi, Y.S., 1988, Production potential of medicinal and Aromatic plants in Himachal Pradesh. In P.K. Khosla (ed.) Trends in Tree Sciences: 81-89. ISTS Publication.

Chauhan, N.S., 1989b, Medicinal and Aromatic Plants of Himachal Pradesh and their role in Social Forestry programme IIIrd Forestry Conference (proceedings) Vol. II:89(1) 89(9):FRI Dehradun.

Chauhan, N.S., 1989c, Potential of Aromatic Plants Flora in Himachal Pradesh, Indian Perfumer 33(2):118-122.

Chauhan, N.S., 1990, Medicinal orchids of Himachal Pradesh. Orchid Society of India (special number). J. Orchid Soc. India, 4 (1,2):99-105.

Chauhan, N.S., Kaith, B.S. and Mann, S.K. 1992. Anti-inflammatory activity of *Rheum asutrale* roots. Int. J. Pharmacognosy 30(2): 93-97.

Chauhan, N.S., 1994. Herbal Wealth of Lahaul-Spiti. Souvenir, Keylong Fair, Lahaul & Spiti.

Chauhan, N.S., and Sanjeev Thakur, 1995. Conservation of Forest Genetic Resources of HP. In press with Indian Jour. For. Res. Dehradun.

Chauhan, N.S., 1996a. Plant Resources of Economic Use in Himachal Pradesh. DEE-UH&F, Solan: 77 pages.

Chauhan, N.S., 1996b. Business potential of medicinal and aromatic plants.

Science Tech Entrepreneur, NSTEDB, DST. GOI, Vol. 2:34-38.

Chauhan, N.S., 1997. Aushdiya Evam Sugandhit Poudhon ki Krishi ki Awashyakta Kyon? Udhyamita, Bhopal (MP), Vol. 5(10): 15-19.

Chawla, J.S., Puri, S.C. and Anand, S.M., 1980. Vanillin from Deodar wood lignin; National Symposium on Utilization of Agriculture and Forest Residues, Punjab University, Chandigarh; Sept., 26-28.

Chopra, R.N., Chopra, I.C., Handa, K.L. and Kapur, L.D., 1958, Chopras Indigenous Drugs of India. IInd Ed. U.N. Dhar and Sons, Pvt. Ltd. 15. Bankin Chatterjee Street, Calcutta-12.

Chopra, R.N., Nayar, S.L. and Chopra, I.C., 1974, Supplement to the Glossary of Indian Medicinal Plants CSIR Publication, New Delhi.

Chopra, I.C. and Handa, K.L., 1961, Review of Research on Indian Medicinal and Allied Plants, ICAR, New Delhi.

Chopra, I.C. and Kapoor, L.D., 1968. Steroid Sapogenin bearing plants of India. Ind. For. 94(8): 620-629.

Chopra, R.N., Chopra, I.C., Handa, K.L. and Kapur, L.D., 1958. Chopras Indigenous Drugs of India. Iind Ed. U.N. Dhar and Sons, Pvt. Ltd., 15, Bankin Chatterjee Street, Calcutta-12.

Chopra, R.N., Nayar, S.L. and Chopra, I.C., 1956. Glossary of Indian Medicinal Plants, CSIR, New Delhi.

Chowdhery, H.J. and Wadhwa, B.M., 1984. Flora of Himachal Pradesh, Analysis. Vols. 1-3. Botanical Survey of India, Calcutta.

Collet, H., 1902. Flora Simlensis. Thacker Spink and Co. Calcutta and Shimla, Reprinted 1971. Bishen Singh Mahendra Pal Singh, Dehradun.

Chopra, R.N., Badhwar, R.L. and Gosh, S., 1965. Poisonous Plants of India, 2 Vols., ICAR, New Delhi.

Dastur, J.F., 1951, Useful plants of India and Pakistan, D.B. Taraporewala Sons & Co., Ltd., Bombay.

Dastur, J.F., 1962, Medicinal Plants of India and Pakistan, D.B. Taraporevala Sons and Co., Pvt. Ltd., Bombay.

Dery, A.C., 1964. Folk-lore or Medicinal Plants of Bhagirathi Valley (Himalayas)—in proceeding of the International Symposium on Medicinal Plants, Kandy, Ceylon, pp. 7-20.

Dey, A.C., 1980, Indian Medicinal Plants used in Ayurvedic Preparations. Bishen Singh Mohendra Pal Singh, Dehradun.

Dhar, S.K., Gupta, S. and Chandokhi, N., 1968, Antifertility activity of some indigenous plants. Indian pharmacological Society, XI Annual Conference, New Delhi.

Dorfier Hans-Peter and Rosett Gerhard, 1989, The Dictionery of Healing Plants. Blandford Press. London, New York, Sydney.

Drury, C.H., 1873. The useful plants of India. International Book Distributors. Dehradun (Reprint: 1978).

Dymock, William; Warden, C.J.H., Hooper, David (1889-1891). Pharmacographica Indica; Thacker Sprink and Co., Calcutta.

Gaitonde, B.B., Joglekar, S.N., Kulkarni, H.J. and Nahar, S.D., 1977. Anti-inflammatory, Analgesic and Anti-pyretic activity of Indigenous Medicinal Plants. *Oxalis corniculata* Linn. Jour. Res. Ind. Med., Yoga and Homeopathy, 12(2): 12-17.

Gamble, J.S. (Reprint) 1972. Manual of Indian Timbers, Sampson Law, Marston & Co. Ltd. London.

Garg, M.P., 1986, Export Import of medicinal plants. Proceeding of Regional Seminar on Medicinal Plants (Central Region) held at Nainital (Uttar Pradesh) 26-28th December, 1986:187-189.

Goswami, P.C., 1982. Agroforestry practices and prospects as a combined land use system; Ind. For. 108: 385-387.

Grieve, M. (Mrs.), 1989, A Modern Herbal, Vol I and II Hafner Publishing Co., New York.

Guenther, E., 1948-52. The essential oil (6 Vols.) D. Van Nostrand Co. Inc., New York.

Gulati, B.C., Shawi, A.S., Garn, S.N. and Channa, G.N., 1980, Essential oil of *Juniperus recurva* var *squamata* and other oils of Juniperus spp., Indian Perfumer, Vol. XXIV. 2: 57-65.

Gupta, M.P., 1988, Present status and Future Prospects for the cultivation and collection of Medicinal Plants of Himachal Pradesh. Ind. For. 114(1):19-25.

Gupta, R., 1964. Survey record of medicinal and aromatic plants of Chamba Forest Division; Himachal Pradesh. *Indian* Forester 90: 445-468.

Gupta, R., 1977. Medicinal plants component of Indian Forests; Indian farming, Feb. 1977: 96-98.

Gupta, Rajendra 1977. Cultivation of *Digitalis lanata* Ehrh. In Chamba Hills, Himachal Pradesh; Indian Forester, 93(1): 33-40.

Gupta, R.K. 1968. Flora Nainitalensis Navyug Traders, Desh Bandhu Gupta Road, New Delhi-5.

Guru, L.V., 1979. Researches in Indian Medicine; Nagarjun, April 1979. 22(8): 187-189.

Hedyatullah, S., 1964. Studies in the cultivation of some rare Indigenous Drug Plants. Proceed. Int. Symp. On Med. Plants, Dec., 1964, Kandy, Ceylon; pp. 47-50.

Hooker, Sir, J.D., 1872-1897 (Reprint 1961). Flora of British India, Vol. 1-7, Reeve and Co. Ltd. The Cast House, Brook NR, Ashford, Kent, England.

Hutchinson, J. (Reprint 1960). The families of flowering plants, Vol. I & II, 2nd Edition, Orford, University Press, London.

Issar, R.K., 1977. Pharmacognostical study of the Yunani Drug, Mamiran-

I-Chini (*Thalictrum foliolosum* DC.). Jour. Res. Ind. Med. Yoga and Homo., 12(1): 34-40.

Jain, S.K., 1968. Medicinal Plants. National Book Trust, India, New Delhi.

Kanji Lal, Upendernath, 1969, Forest Flora of Chakrata, Dehradun and Saharanpur Forest Divisions, Uttar Pradesh. Manager of Publications, Delhi.

Kapur, L.D., 1968. The Indigenous Vegetable Drugs and their role in Modern Medicines—Jour. Res. Ind. Med., 2(2): 244-245.

Karnick, C.R., 1981. Phytochemical screening of some Medicinal Plants used in Ayurvedic System of Medicine. Bull. Medico-ethnobotanical Res., CCRAS, 2: 3 pp.

Karnick, C.R., 1982. Observation on the effect of use of Medicinal Plants on different Nakshatras as per older Ayurvedic texts. Nagarjun, 26(4): 79-84.

Karnick, M.G. and Sharma, O.P., 1968. Note on Composition of *Origanum vulgare* seed cake. Ind. For. 94: 344-345.

Kharo, A.K., Feroz, Hasen and Sharma, M.K., 1983. Anti-inflammatory drugs from the plants; Jour. Scientific Res. Plants and Medicines. 41(1): 4-11.

Khory, Rustomjee Naserwamjee and Katrak, Nanabhai Nawroji (Reprint 1984). Meteria Medica of India and their therapeutics. Neeraj Publishing House, Delhi-52.

Khoshoo, T.N., 1983. Energy from Biomass; Indian Farming, June, 1983, pp. 10-13.

Kirtikar, K.R. and Basu, B.D., 1975 (Reprint). Indian Medicinal Plants. Vol. I-IV Lalit Mohan Basu, 49, Leader Road, Allahabad, India.

Langer, J.G., Gupta, O.P. and Atal, C.K., 1980. Chemical trials on Picrorhiza kurroa as on immunoregulator; Proceedings. XIIIth Annual Conference, Ind. Pharmacological Society, RRL. Jammu, Sept., 1980.

Lindley, John 1839. Flora Medica. A.K. Distributors at Delux Offset Printers, New Delhi-35 (Reprint, 1981).

Maheshwari, P. and Singh, Umrao, 1981, Dictionary of Economic Plants in India, ICAR, New Delhi.

Mamgain, M.D., 1971. Himachal Pradesh Distt. Gazetteer (Kinnaur); Quick Printers, Kuldip Nagar, Ambala Cantt.

Manandhar, N.P., 1980, Medicinal Plants of Nepal Himalaya, Ratna Pustak Bhandar, Kathmandu, Nepal.

Martindale, 1977, The extra Pharmacopoiea (27th Ed.) Edited by Ainley Wade, London Pharmaceutical Press 1-Lambeth, High Street, SE I.

Matthias, Hermann 1973. Marvellous world of Herbs and Medicinal Flowers, Abbey Library, London.

Mehra, P.N., Bhatnagar, J.K. and Handa, S.S., 1969. Researches in

Pharmaconosy in India. Res. Bull (N.S.). Punjab University, Chandigarh. 20(3-4): 261-337.

Mehra, P.N., Puri H.S., 1973, Key to the identification of well known Aconite tubers of India, Abst. 1st Scientific Seminar, Drug Res. CCRIMH, New Delhi. Feb., 11 & 12, p. 30.

Mehra, P.N., 1986, Prospects of research on medicinal plants. Proceedings of Regional Seminar on Medicinal Plants (Northern Region) held at Manali (H.P.), Ministry of Health and Family Welfare, Govt. of India, New Delhi: 54-59.

Mehra, P.N., 1969. Pharmacognosy of Indian drugs NSS Bulletin No. 4, Chandigarh.

Menon, A.K., 1960. Indian Essential oil (A review) CSIR, New Delhi.

Mishra, S.S., 1986, A note on regional seminar on medicinal plants (Central Region) Proceedings of Regional Seminar on Medicinal Plants (Central Region) Held at Nainital (U.P.) 26-28 Dec. 1986:55-57.

Mitra, Jyotir 1970. Medicinal Plants of Vedic Antiquity as mentioned by the Western Indologists; Nagarjun, 13(12): 9-21.

Mitra, R., Kapur, S.L., 1975. Herbs that cure-I, Sachitra Ayurveda, Sept. 1975, pp. 198-201.

Mitra, R., Prasad, S., 1971. Further studies on the Pharmacognosy of *Picrorhiza kurrooa* Linn. (Kutki). J. Res. Ind. Med. 6(3): 300-307.

Moor. N.S., Vayaskara, 1953. Ayurvedic Flora Medica; Fasciculus-I; Vaidyasarathi, Kottayam.

Mukerji, B., 1953, Indigenous Drugs, Indian Pharmaceutical Codex, CSPR (Delhi).

Nadkarni, K.M., 1927. The Indian Materia Medica, 2 Vols. (3rd revised edition); Popular Book Depot, Bombay.

Nair, N.C., 1977, Flora of Bushahr Himalaya, International Bio-Science Publishers, Hissar, Madras.

Nautiyal, B.P. and Nautiyal, Subhash, 1983, Some medicinally important tree species of U.P. Himalaya: Relevance in Regional Development and Ecological Security Jour. Sci. Res. P and Medicine, 4(1):14-22.

Nautiyal, Prakash Chandra 1983. Berberis—A threatened plant of Himalayan Region; Science Reporter (Science Spectrum), June 1983; pp. 349-350.

Negi, J.P., 1997. Policy document on propagation and processing of herbal/ medicinal plants in Himachal Pradesh. Govt. of Himachal Pradesh, Deptt. of Indian Systems of Medicine and Homeopathy: 1-19. (under consideration of Govt.).

Oza, G.M., 1972. On the Botanical Identity of Bhang, Charas, Judiciary and Society; Ind. For. 98(6): 349-355.

Pandey, Gangasahai and Chunekar, K.C., 1969. Bhav Prakash Nighantu (Hindi) Chowkhamba vidya Bhavan, Varanasi-I.

Pandey, M.P., J. Prasad and L.P. Awasthi, 1988, Antiviral effect of the essential oils from Lemon Grass (*Cymbopogon flexuosus*), Mentha (*Mentha arvensis*) and Vetiver (*Vetiveria zizanoides*). In P. Kaushik (ed.) indigenous medicinal plants:47-50. M/S Today and Tomorrow's Printers and Publishers, New Delhi.

Pandey, V.N., 1986, The problems, potentialities and prospects of Ayurvedic medicinal plants in Himalayan region. Proceedings, of Regional Seminar on Medicinal Plants (Central Region) held at Nainital (U.P.) 26-28th December, 1986:35-43.

Parmar (Dr.) Y.S., 1977. Trees as Monuments in Himachal Pradesh. Indian Farming, Feb. 1977, pp. 5-7.

Pathak, N.N. Karnick, C.R., 1981. Newer observations on the crude drugs termed as Tikta from Ladakh (Himalaya) region. Nagarju, 25(1): 4-6.

Pillai, A.P.G., 1986, Investigations on the botanical sources of soma. Proceedings of Regional Seminar on Medicinal Plants (Central Region), held at Nainital (Uttar Pradesh) 26-28th December: 128-140.

Polunin, Oleg and Staintion, Adam, 1984, Flowers of the Himalaya Oxford University Press, Delhi, Calcutta, Bombay, Madras.

Prakash Om 1961. Food and Drinks in Ancient India; Munshi Ram, Manohar Lal, Nai Sarak, Delhi.

Prasad Durga and Kaul, P.L., 1965. Working plan for the Pabbar valley forests. Albion Press Kashmere Gate, Delhi-6.

Prasad, K., 1986, Medicinal Plants in U.P. Proceedings of Regional Seminar on Medicinal Plants (Central Region) held at Nainital (U.P.): 98-101.

Raghunathan K. and Mitra, Roma, 1982, Pharmacognosy of Indigenous Drugs. Vol. I-II Central Council for Research in Ayurveda and Sidha, New Delhi.

Raizada, M.B., 1968. Some name changes in the flora of India. Ind. For. 94(6): 432-462.

Rajasekharen, S. and Srivastava, T.N., 1977. Ethnobotanical study on Vacha and a preliminary clinical trial on Bronchial asthma; J. Res. Ind. Med. Yoga and Homoeopathy. 12(4): 92-95.

Rao, P.G., Zutsi, U., Kaul, J.L. and Atal, C.K., 1978. Pyrrolizidine alkaloids from *Senecio jacquemontiana* and *Cynoglossum wallichii*; Indian Pharmaceutical Congress.

Rao, T. Ananda 1961. A botanical tour in Kashmir State, Records of Botanical Survey of India, Vol. XVIII, No. 2.

Rathore, Y.K., Singh, Haqqi, Tansir Ul-haq, Lal Chhote, Bhadoria, A.K., Singh, Sharma, Rajeev Kumar, 1986, An appraisal of medicinal plants with reference to Ayurvedic system of medicine. Proceedings of Regional Seminar on Medicinal Plants (Central Region) held at Nainital (U.P.):26-28th Dec. 1986:170-186.

Ratra, Parminder, S. and Misra, K.C., 1970. Seasonal variation in chemical components of *Achyranthes aspera* Linn. and *A. bidentata* Bl. Ind. For. 96(5): 372-375.

Roxburgh, William (Reprint) 1971. Description of Indian Plants—Today and Tomorrow's Printers and Publishers, New Delhi-5.

Sahni, K.C., 1981. Western Himalayan Flora. In J.S. Lall (Ed.). The Himalaya, Aspects of change; Oxford University Press, New Delhi.

Sanecki, Kay, N., 1956. Wild and Garden Herbs, W.H. and L. Collingridge Ltd. London.

Santapau, H., 1958, Floristic studies in India, Memoir, Ind. Bot. Society, 1:117-121.

Sarin, Y.K., 1967. A survey of vegetable raw material Resources of Lahaul. Ind. For., Vol. 93, No. 7: 489-499.

Schery, Robert W., 1954. Plants for man, George Allen and Unwin Ltd., Ruskin House, Museum Street, London.

Shabnam, S.R., 1964. Medicinal Plants of Chamba. Indian For., 90: 50-63.

Sharma, Durga Pradas, 1974. Banoushadhi Adharit Udhyog; Ek Sakshatkar; Sachitra Ayurveda (Hindi). Feb. 1974, pp. 471-472.

Sharma, Davinder nath and Chauhan, N.S., 1976. Himachal Pradesh Mein Prapya Ayatit Pratinidhi Banoushadh Dravya (Hindi); ban Sandesh, Oct. Dec., 1976, pp. 27-28.

Sharma, Davinder, Misra, D.D., 1972. Scope of phytochemical industry in Himachal Pradesh: Nagarjun, July, 1972. 15(11): 26-27.

Shiva, M.P., Singh, N.P. and Giri, R.V., 1986, Biological calendar for cultivation of temperate medicinal plants grown at Chakrata Proceeding of Regional Seminar on Medicinal Plants (Central Region) held at Nainital (U.P.):106-111.

Singh, B. Dubey, S.D., 1981. Standardization, Nagarjun, 25(4): 80-81.

Singh, Gurcharan and Kachru, P., 1976. Forest Flora of Srinagar and Plants of Neighbourhood; Bishan Singh, Mahendra Pal Singh, 23-A, Connaught Place, Dehradun.

Singh, Janardhan 1982. Anthelmintic activity of *Urtica dioica* Linn. (Bichhu-buti)-Nagarjun, 26(2): 130-135.

Singh, R.V., 1982, Fodder Trees of India, Oxford and IBH Publishing Co., New Delhi.

Singh, Ram Chandra Pal 1961. District Handbook-4 (Mahasu); Census, 1961, Himachal Pradesh.

Sood, Meenu, 1992, Screening of *Valeriana jatamansi* Jones collections for valepotriates content and development of cultivation technique for promising collection, Ph.D. Thesis, Dr. Y.S.P. University of Horticulture and Forestry, Solan.

Sood, R.P., Kalia, N.K, Brada, B.L., 1978. Scope for commercial exploitation of various aromatic plants in Palampur (HP). Ind. Perf. 22(2): 136-37.

Sood, R.P., Kalia, N.K. and Sobti, S.N., 1982, Scope for development of phytochemical and Essential oil industry in Palampur area (Kangra-valley). Nagarjun: 30-34 pp.

Sood, V.K., Karnic, M.G., Sagar, V. and Sood, K.G., 1966. Preliminary investigations on the essential oil of *Boenninghausenia albiflora* (Hook) Reich., Ind. For., 92, pp. 295.

Srinivasan, E., 1974. Industrial Pharmacy of Indigenous medicines (Ayurveda, Sidha and Yunani) with particular reference to the innate problems pertaining thereto; Nagarjun: Oct. 1974, pp. 24-28.

Tewari, P.V., Misra, D.N., Chaturvedi, C., 1982. Medicinal preparations used for obstetric and Gynaecological disorders in Medieval India, Nagarjun, 25(6): 124-129.

Thakur, Chandrashekhar, G., 1965. Introduction to Ayurveda (Basic Indian medicine). The times of India Press, Bombay, pp. 105.

Trease, G.E., 1952. Text Book of Pharmacognosy; 6th Ed., London.

Trivedi, V.P., Dixit, R.S. and Lal, V.K., 1980. Orchids in the Drug markets of Bareilly, Kanpur and nearby Districts; Nagarjun: April, 1980, pp. 157-163.

UNESCO, 1960, Medicinal Plants of the Arid Zones, United Nations Educational, Scientific and Cultural Organization, Paris.

Uniyal, M.R. and Chauhan, N.S., 1971, Medicinal Plants of Uhal valley in Kangra Forest Division, H.P. Jour. Res. Ind. Med. VI (3): 287-299.

Uniyal, M.R. and Chauhan, N.S., 1972, Commercially important Medicinal Plants of Kullu Forest Division, H.P. Nagarjun 4 (2): 28-32.

Uniyal, M.R. and Chauhan, N.S., 1973, Traditionally important medicinal plants of Kullu Forest Division, H.P. *Jour. Res. Ind. Med.* VII (1): 76-85.

Uniyal, M.R. and Issar, R.K., 1966. Botanical studies of Ashtavarga, Nagarjun; 9(8): 276-284.

Verma, P.K., Punia, M.S., and Sharma, G.D., 1988, Cultivation prospects of medicinal plants in Haryana in P. Kaushik (Ed.) Indigenous Medicinal Plants (Including Microbes and Fungi). Today and Tomorrows Printers and Publishers, New Delhi.

Wadhwa, B.M., and Hajra, R.K., 1986, Medicinal Plant resources in North-Western and Central Himalayas and their Conservation. Proceeding of seminar on medicinal Plants (Central Region) held at Nainital (U.P.) 26-28th December, 1986: 112-120.

Wallis, T.E., 1951. Text Book of Pharmacognosy 2nd Ed. J. and A Churchill Ltd., London.

Watt, George (Reprint 1972), A Dictionary of the Economic products of India; Periodical Experts, Delhi.

Willis, J.C., 1960. A Dictionary of the Flowering Plants and Ferns, 6th Ed. University Press, Cambridge.

Wren, R.C., 1956. Potter's New Cyclopaedia of Botanical Drugs and Preparations, Sir Issac Pitman and Sons, Ltd., London.

Youngken, Herbere W., 1951. Pharmaceutical Botany; Blakiston Company, Philladelphia and Toronto.

INDEX OF BOTANICAL NAMES

Abies pindrow 479, 507
Abies spectabilis 47, 48
Abies webbiana 452
Abrus precatorius 49, 51, 470, 479, 510, 579
Abutilon indicum 479
Acacia arabica 56
Acacia catechu 52, 54, 479
Acacia farnesiana 565
Acacia mollissima 577
Acacia nilotica 56, 57, 479, 510, 565
Achillea millefolium 59, 60, 479, 510, 571
Achyranthes aspera 62, 64, 470, 479, 510, 516
Achyranthes bidentata 479, 517
Aconitum chasmanthum 479, 500, 510
Aconitum deinorrhizum 65, 67, 479, 500, 510, 517, 579
Aconitum ferox 562
Aconitum heterophyllum 69, 71, 452, 470, 479, 499, 500, 579
Aconitum napellus 598
Aconitum violaceum 452, 479, 499, 497, 514
Aconogonum rumicifolium 518
Acorus calamus 72, 74, 452, 470, 479, 498, 518, 569, 574, 579
Adhatoda vasica 259, 513
Adiantum lunulatum 452, 479, 510
Adiantum pedatum 479
Adiantum venustum 76, 518
Adina cordifolia 574
Aegle marmelos 77, 79, 480, 579
Aerva lanata 597
Aesculus indica 80, 81, 480, 510, 519
Ageratum conyzoides 36
Agrimonia eupatorium 480
Ailanthus altissima 480
Ainslaea aptera 452, 519
Ajuga bracteosa 83, 84, 480, 520
Ajuga parviflora 36
Albizia lebbek 85, 57, 480, 510
Albizia odoratissima 480
Albizia procera 480, 510
Aleurites montana 480
Allium carolinianum 480, 510, 520
Allium cepa 45
Allium govanianum 480
Allium jaquemontii 480, 510
Allium sativum 88, 480
Allium stracheyi 500
Alnus nitida 37, 38
Aloe barbadensis 90, 470, 480, 510
Althaea officinalis 480
Amaranthus caudatus 480
Amaranthus spinosus 480
Ammi majus 461
Amomum subulatum 480
Amphicome emodi 481
Anagalis arvensis 481
Anaphalis busua 567
Andrachne cordifolia 574
Androsace rotundifolia 521
Anethum soya 596
Angelica glauca 91, 93, 452, 470, 481, 498, 500, 570, 579
Anogeissus latifolia 481
Aquilegia fragrans 500, 521
Aquilegia pulbiflora 502

Arachne cordifolia 574
Arctium lappa 94, 481, 510, 522
Ardisia solanacea 481
Arenaria serpyllifolia 481
Argemone mexicana 481
Arisaema tortuosum 574
Arnebia benthamii 96, 97, 98, 99, 470, 481, 502, 510, 579
Arnebia euchroma 481, 510, 522, 579
Arnebia guttata 481, 522, 579
Artemisia absinthium 574
Artemisia brevifolia 100, 101, 452, 470, 481, 502, 510, 569, 579
Artemisia cina 598
Artemisia dracunculus 481, 511, 522, 571, 579
Artemisia scoparia 481, 511, 523
Artemisia sieversiana 502
Artemisia vulgaris 481, 511
Artocarpus lakoocha 481, 511
Asclepias curassavica 481
Asparagus abscendens 102, 481, 511, 523
Asparagus racemosus 580
Astragalus condolleanus 502
Astragalus strobiliferus 481
Atropa acuminata 104, 452, 482, 502, 569, 579
Atropa belladonna 106, 470, 579
Azadirachta indica 107, 108, 482, 511, 574
Bacopa monnieri 482
Baliospermum montanum 110, 482
Bambusa bambos 596
Barleria cristata 482
Bauhinia purpurea 482
Bauhinia variegata 111, 112, 482, 511, 523
Bauhinia vehlii 482, 511, 523
Berberis aristata 114, 116, 452, 482, 502, 568, 574, 580
Berberis chitria 482, 502, 568
Berberis lycium 117, 118, 452, 482, 498, 568
Berberis pseudoumbellata 502
Bergenia ciliata 120, 121, 482, 524, 580
Bergenia stracheyi 123, 124, 482
Betula alnoides 37
Betula utilis 125, 126, 482, 498, 503, 511, 579
Bidens pilosa 482, 524
Boenninghausenia albiflora 575
Boerhavia diffusa 482, 511
Bombax ceiba 128, 130, 482
Brassica campestris 482
Brassica juncea 483
Bridelia retusa 483
Bryonopsis laciniosa 470, 483, 511, 580
Buchanania lanzan 483
Bunium persicum 132, 452, 470, 483, 499, 503
Bupleurum falcatum 525
Butea monosperma 133, 135, 483, 511, 575
Buxus wallichaiana 483
Caesalpinia crista 483
Calendula officinalis 483
Callicarpa macrophylla 483, 511
Calotropis gigantea 137, 138, 483
Calotropis procera 483, 511
Caltha palustris 483, 525
Cannabis sativa 139, 141, 483
Capparis spinosa 470, 483
Capsicum annuum 143
Careya arborea 483
Carica papaya 483
Carissa opaca 36
Carpinus viminea 37
Carum bulbocastanum 132
Carum carvi 144, 145, 470, 483, 499, 503
Casearia tomentosa 36
Cassia absus 147, 483, 580
Cassia fistula 148, 149, 484, 511, 580
Cassia occidentalis 150, 484, 511, 579
Cassia tora 151, 484, 511, 580
Cassiope fastigiata 525

Cedrus deodara 153, 484, 511, 570, 575
Celastrus paniculatus 484, 511, 580
Celtis caucasica 37, 38
Centella asiatica 157, 473, 484, 512, 560
Centratherum antheliminticum 481, 512, 525, 580
Chaerophyllum villosum 499
Chenopodium album 484, 425
Chenopodium ambrosioides 156, 484, 572
Chrysanthemum cinerariaefolium 470, 580
Cichorium intybus 157, 159, 484, 580
Cimicifuga foetida 580
Cinnamomum camphora 484, 572, 580
Cinnamomum tamala 160, 161, 484, 499, 503, 572, 581
Cissampelos pareira 162, 163, 484, 528, 568
Citrullus colocynthis 484, 512, 581
Clerodendrum indicum 581
Clerodendrum phlomidis 484
Clerodendrum serratum 164, 484
Clerodendrum viscosum 37
Codonopsis ovata 484, 529
Colchicum luteum 484, 503, 581
Colutea nepalensis 485, 530
Convolvulus arvensis 485
Cordia dichotoma 485, 581
Coriandrum sativum 165, 485
Corydalis govaniana 167, 168, 485, 512
Corylus colurna 485
Costus speciosus 169, 170, 470, 485, 503, 568, 581
Cotinus coggygria 530
Cotoneaster bacillaris 37
Crataeva nurvala 485, 581
Crocus sativus 172, 470, 485, 503, 581
Croton tiglium 575, 581
Cryptolepis buchanani 485, 512
Curculigo orchioides 485, 503, 581
Curcuma longa 485, 575
Cuscuta europaea 485
Cuscuta reflexa 174, 175, 530
Cydonia oblogna 485
Cymbopogon jwarancusa 485
Cymbopogon martinii 485, 511, 571
Cymbopogon nardus 575
Cynodon dactylon 176, 485, 530
Cyperus rotundus 177, 178, 485, 512
Cyperus scariosus 486
Dactylorhiza hatagirea 180, 181, 452, 470, 486, 499, 503, 581
Daphne cannabina 37
Datura fastuosa 183, 486
Datura inoxia 581
Datura metel 184, 186, 485, 512, 569
Datura stramonium 187, 188, 486, 512, 569
Delphinium denudatum 485, 581
Delphinium vestitum 531, 575
Dendrocalamus strictus 37
Desmodium gangeticum 486, 512, 581
Desmodium tiliaefolium 37
Didymocarpus pedicellata 486, 581
Digitalis lanata 470, 486, 581
Dioscorea bulbifera 486, 512
Dioscorea deltoidea 190, 192, 452, 470, 486, 498, 503, 568, 582
Diospyros montana 575
Diospyros tomentosa 37
Diplazium esculentum 38
Dodonaea viscosa 36
Dolichos uniflorus 194, 485, 582
Drocera peltata 195, 485
Duchesnea indica 532
Eclipta alba 486
Eclipta prostrata 196, 197, 470, 512, 582
Embelia ribes 279, 563, 596
Embelia robusta 596
Embelia tsjeriam-cottom 486
Emblica officinalis 199, 200, 486, 582

Ephedra gerardiana 202, 203, 452, 486, 499, 503, 569, 582
Ephedra intermedia 486, 582
Ephedra major 487
Ephedra sinica 463, 598
Equisetum arvensis 487, 533
Eremurus himalaicus 533
Eucalyptus citriodora 487
Eucalyptus globulus 575
Eulaliopsis binata 36
Eulophia campestris 487, 503
Euonymus tingens 37
Euphorbia hirta 487
Euphorbia pilosa 487, 533
Euphorbia prostrata 596
Euphorbia royleana 487, 535, 575
Euphorbia thymifolia 487, 575
Evolvulus alsinoides 487
Fagopyrum cymosum 487, 569
Fagopyrum dibotrys 487, 569
Fagopyrum esculentum 205, 487, 569
Fagopyrum tataricum 487, 569
Feronia limonia 37
Ferula jaeschkeana 487, 506
Ficus benghalensis 487
Ficus racemosa 487
Ficus religiosa 487
Flacourtia indica 36
Foeniculum vulgare 487, 582
Fragaria vesca 37
Fraxinus xanthoxyloides 487, 535
Fritillaria oxypetalum 487, 506
Fritillaria roylei 487, 506
Fumaria parviflora 206, 488
Gallium rotundifolium 536
Gentiana Kurroo 207, 208, 488, 499, 506, 582
Gentiana lutea 598
Gentiana tenella 488
Gentiana tianschanica 488, 506, 536
Geranium nepalense 210, 488, 512
Geranium wallichianum 488, 537
Geum elatum 488, 537
Ginkgo biloba 211, 506
Girardinia diversifolia 488, 537
Girardinia heterophylla 452
Glaucium flavum 45, 461, 590
Gloriosa superba 212, 214, 470, 488, 506, 512, 568, 575, 582
Glycyrrhiza glabra 582
Gmelina arborea 488, 582
Grewia optiva 37
Habenaria edgeworthii 488, 506
Habenaria intermedia 488, 506
Hedychium acuminatum 215, 217, 452, 488, 512, 571, 575, 582
Hedysarum cashmirianum 506
Helicteres isora 218, 219, 488, 512, 582
Heliotropium indicum 488
Hemidesmus indicus 488, 512
Heracleum candicans 221, 452, 470, 488, 506, 568, 582
Hippophae rhamnoides 223, 470, 488
Hippophae salicifolia 225, 470, 488
Holarrhena antidysenterica 225, 227, 470, 489, 512, 576, 582
Humulus lupulus 229, 230, 486, 538, 571, 582
Hydrocotyle rotundifolia 489
Hyoscyamus niger 231, 232, 452, 470, 489, 513, 569, 582
Hypericum perforatum 234, 235, 489
Hyssopus officinalis 237, 238, 470, 489, 513, 573, 583
Ilex dipyrena 37
Indigofera pulchella 37
Inula racemosa 240, 241, 489, 507, 513, 583
Ipomoea hederacea 583
Ipomoea purga 598
Iris germanica 489, 513, 583
Jasminum humile 242, 244, 489, 513
Jasminum officinale 245, 489, 513, 571
Jasminum sambac 597
Jatropha curcas 489
Juglans regia 246, 247, 489, 513, 583

Juniperus communis 250, 251, 489, 513, 583
Juniperus macropoda 253, 254, 489, 583
Juniperus recurva 255, 489
Jurienea dolomiaea 256, 257, 453, 470, 489, 498, 507, 572, 583
Justicia adhatoda 259, 489, 513
Kalanchoe pinnata 489, 540
Kalanchoe spathulata 540, 576
Kydia calycina 37
Lactuca sativa 489
Lagerstroemia parviflora 577
Lamium album 540
Lannea coromandelica 490
Lantana camara 36
Lawsonia inermis 490
Lepidium sativum 490, 583
Leptodermis lanceolata 36
Leucas cephalotes 260, 261, 490, 513
Leucas lanata 540
Lilium polyphyllum 490, 507, 513, 583
Linum usitatissimum 490
Lippia nodiflora 490
Litsea chinensis 490, 540
Litsea umbrosa 490, 513
Lobelia inflata 598
Lobelia nicotianaefolia 463, 598
Lonicera spinosa 40
Lyonia ovalifolia 36, 576
Macrotomia benthami 96
Madhuca indica 583
Madhuca latifolia 576
Malaxis acuminata 490, 507
Malaxis muscifera 490, 507
Malaxis wallichii 458, 597
Mallotus phillippinensis 262, 264, 490, 513, 541
Malva rotundifolia 490, 541
Malva verticillata 490, 541
Malvastrum coromandelianum 490
Mangifera indica 490, 541
Marsdenia tenacissima 597
Martynia annua 490, 513
Matricaria chamomilla 470, 490, 583
Melia azadirachta 583
Melissa officinalis 490, 583
Mentha longifolia 265, 267, 491, 513, 543
Mentha piperita 491
Mentha spicata 491, 571
Meriandra strobilifera 543
Micromeria biflora 544
Milletia auriculata 491, 513
Mimosa pudica 491
Mirabilis jalapa 491
Moghania fruticosa 36
Monochoria vaginalis 491
Morchella esculenta 268, 267, 491, 498
Morina coulteriana 491
Morina longifolia 491
Moringa oleifera 269, 271, 491, 583
Moringa pterygosperma 564
Mucuna pruriens 272, 491, 513, 583
Mucuna prurita 272
Murraya koenigii 274, 275, 491, 513, 583
Myrica esculenta 276, 277, 491, 513, 583
Myricaria germanica 39
Myrsine africana 279, 491
Nardostachys grandiflora 280, 281, 453, 491, 507, 513, 544, 583
Nepeta ciliaris 544
Nepeta podostachys 545
Nerium indicum 283, 284, 491, 584
Nerium odorum 491
Nicandra physaloides 576
Nicotiana arbor-tristis 285, 287
Ocimum basilicum 492
Ocimum gratissimum 576
Ocimum kilimandscharicum 570
Ocimum sanctum 288, 289, 470, 492, 576, 584
Olea cuspidata 461
Olea europaea 492, 584

Olea ferruginea 492, 514
Onosma bracteatum 291, 292, 492, 514, 584
Onosma echioides 514
Onosma hispidum 492, 507
Operculina turpethum 492, 584
Orchis latifolia 180
Origanum vulgare 293, 294, 492, 545, 571, 584
Oroxylum indicum 296, 297, 492, 514, 584
Ougeinia oojeinensis 37
Oxalis corniculata 492, 545
Oxyria digyna 546
Paederia foetida 565
Paeonia emodi 298, 492, 584
Papaver somniferum 565
Paris polyphylla 492, 514
Parmelia perlata 300, 492
Parrotiopsis jaquemontiana 30
Parthenium hysterophorus 36
Peganum harmala 584
Pennisetum orientale 36
Persea gamblei 37
Peucedanum graveolens 492
Physochilana praealta 301, 302, 507, 546, 569, 584
Phytolacca acinosa 304, 492, 584
Picea smithiana 37, 38, 403
Picrasma quassioides 576, 584
Picrorhiza kurrooa 305, 306, 453, 470, 492, 499, 507, 546, 584
Pieris ovalifolia 576
Pinus gerardiana 492, 499, 507
Pinus roxburghii 308, 573
Pinus wallichiana 309, 453, 546
Piper cubeba 448
Pistacia integerrima 453, 492, 499, 547
Pistacia khinjuk 311, 312, 514, 584
Plantago major 492, 547
Plantago ovata 460, 598, 592
Plantago psyllium 598
Pleurospermum brunonis 548, 571
Plumbago zeylanica 314, 315, 493, 514, 548
Podophyllum hexandrum 317, 318, 453, 470, 493, 498, 507, 514, 584
Polianthes tuberosa 493, 514
Polygonatum multiflorum 508, 584
Polygonatum verticilatum 453, 493, 508, 514, 585
Polygonum aviculare 320, 321, 493, 514
Polygonum hydropiper 548, 576
Polygonum recumbens 549
Pongamia pinnata 323, 493, 585
Populus ciliata 38, 40
Potentilla fulgens 324, 493, 514
Potentilla nepalensis 453, 493, 498, 549
Premna integrifolia 565, 596
Premna latifolia 493
Primula macrophylla 594
Prinsepia utilis 325, 326, 493, 551, 585
Prunella vulgaris 327, 328, 493, 585
Prunus amygdalus 493
Prunus cerasoides 329, 331, 493, 552
Psoralea corylifolia 470, 493
Pterocarpus marsupium 565
Pteridium aequilinum 38
Pueraria tuberosa 332, 493, 585
Punica granatum 333, 335, 493, 514, 552, 585
Purerarria tuberosa 514
Putranjiva roxburghii 337, 494, 585
Pyracantha crenulata 338
Pyrus pashia 36, 37
Quercus baloot 38
Quercus floribunda 37
Quercus leucotrichophora 36
Quercus semecarpifolia 39, 301, 403
Randia dumetorum 340, 341, 490, 514, 585
Randia tetrasperma 36
Raphanus sativus 494
Rauvolfia canescens 597

Rauvolfia serpentina 342, 344, 470, 494, 508, 514, 585
Rhamnus triquetra 553
Rheum australe 349, 350, 456, 473, 494, 498, 508, 514, 585
Rheum moorcroftianum 348, 349, 494, 508, 514
Rhododendron anthopogon 351, 352, 494, 508, 514, 553
Rhododendron arboreum 353, 354, 553
Rhododendron campanulatum 355, 356, 494, 514
Rhododendron lepidotum 508
Rhus parviflora 37
Rhus succedanea 38, 578
Ribes orientale 40
Ricinius communis 357, 359, 494, 514, 576, 585
Robinia pseudacacia 570
Rosa damascena 494
Rosa macrophylla 39
Rosa moschata 571
Roscoea alpina 360, 361, 494
Roscoea procera 508
Rosmarinus officinalis 470, 494, 585
Rubia cordifolia 553
Rubia manjith 366, 367, 494, 514
Rubus ellipticus 36, 37
Salix viminalis 38, 40
Salvia bisor 288
Salvia glutinosa 37
Salvia moorcroftiana 366, 367, 453
Salvia officinalis 470
Salvia sclarea 45
Santalum album 565
Sapindus mukorossi 368, 369, 494, 514, 585
Sapium insigne 554
Sarcococca pruniformis 37
Sarcostemma bervistigma 597
Saussurea bracteata 508
Saussurea costus 370, 371, 453, 494, 499, 508, 571, 575, 585
Saussurea gossypiphora 508, 554
Saussurea jacea 40
Saussurea obvallata 509, 555
Saussurea sacra 509
Scilla indica 585
Selinum tenuifolium 456, 494, 498
Selinum vaginatum 373, 453, 494, 498
Selinum wallichianum 374
Semecarpus anacardium 494, 509, 574, 585
Senecio amplexicaule 44
Sesamum indicum 376, 494
Sesbania grandiflora 494
Shorea robusta 495
Sida acuta 495
Sida cordifolia 377, 378, 495
Sida rhombifolia 380, 495, 555
Sida veronicaefolia 495
Skimmia laureola 470, 495, 509, 555, 572, 585
Solanum khasianum 381, 555
Solanum indicum 495, 568, 588
Solanum laciniatum 470, 568, 585
Solanum nigrum 383, 384, 495, 515, 585
Solanum surattense 386, 387, 495, 515, 556, 568, 585
Solanum verbascifolium 37
Solanum viarum 495, 515, 568
Solanum xanthocarpum 563
Sophora japonia 586
Sophora mollis 576, 586
Sphaeranthus indicus 495
Spilanthes acmella 338, 495, 586
Spondias pinnata 495, 515, 586
Staphylea emodi 509
Staphyllococcus aureus 309
Stephania glabra 495
Sterculia urens 598
Stereospermum suaveolens 495
Strobilanthes auriculata 572, 585
Swertia angustifolia 390, 393, 495, 515
Swertia chirayita 389, 391, 453, 495, 509, 515, 586

Swertia purpurascens 495, 515
Symplocos paniculata 394, 395, 495, 515
Symplocos paniculatus 394
Syringa emodi 586
Syzygium cumini 496
Tagetes minuta 396, 556
Tamarix aphylla 577
Tanacetum longifolium 496, 556
Taraxacum officinale 398, 399, 496, 557
Taxus baccata 401
Taxus wallichiana 401, 402, 453, 496, 498, 509, 569
Teramnus labialis 496
Terminalia alata 496
Terminalia arjuna 404, 405, 496, 515, 586
Terminalia bellirica 407, 408, 496, 586
Terminalia chebula 410, 411, 470, 496, 557, 586
Terminalia tementosa 496, 515, 586
Thalictrum foliolosum 412, 413, 453, 496, 515, 557, 568
Thermopsis inflata 509
Thevetia peruviana 496, 515, 586
Thevetia neriifolia 415
Thymus serpyllum 416, 417, 453, 496, 558, 572
Tinospora cordifolia 418, 515, 586
Tinospora malabarica 496, 515
Tinospora sinensis 456
Trachyspermum ammi 496
Tribulus terrestris 496
Trichosanthes cucumerina 496, 515, 586
Trichosanthes dioica 496
Trigonella foenum-graecum 496, 558, 576
Trillidium govanianum 420, 421, 568
Tylophora indica 496, 586
Typhonium diversifolium 558
Ulmus wallichiana 37
Uraria lagopodioides 496, 515
Uraria picta 422, 497, 515
Urginea indica 423, 497, 509, 586
Urginea scilla 598
Valeriana hardwickii 425, 453, 497, 499, 509, 515
Valeriana jatamansi 426, 427, 453, 470, 497, 499, 572, 586
Valeriana officinalis 598
Valeriana pyrolaefolia 497, 509, 515
Vanda roxburghii 497
Vanda tessellata 428, 509
Vateria indica 597
Verbascum thapsus 559
Verbena officinalis 497, 586
Vernonia cinerea 497
Vetiveria zizanioides 430, 497, 572, 585
Viburnum nervosum 497, 515
Viola canescens 497, 587
Viola odorata 587
Viola serpens 431, 432, 453, 497, 498, 560
Viscum album 434, 435, 497, 587
Vitex negundo 436, 437, 497, 560, 576, 587
Vitis vinifera 497
Waldhemia glabra 594
Withania coagulens 587
Withania somnifera 439, 440, 470, 497, 515, 587
Woodfordia fruticosa 442, 443, 497, 515, 560, 587
Wrightia tomentosa 497
Xanthium strumarium 444, 497
Zanthoxylum alatum 572
Zanthoxylum armatum 446, 447, 497, 509, 576, 587
Zeheneria umbellata 561
Zingiber officinale 570, 587
Zizyphus mauritiana 497, 515
Zizyphus sativa 565
Zizyphus xylopyra 578

Index of Local Names

Aak 137, 483
Aali-ra-tat 148
Aconite 579
Adhasisi 444
Adrak 570
Adusa 259
Aftimun 596
Agast ke pushpa 494
Agnimanth 484
Agnishikha 212
Ajaka 288
Ajwain, 496
Akalbir 486
Akalkada 388
Akanda 137, 483
Akarkara 388, 495, 586
Akashbel 530
Akasvela 174
Akhrot 246, 489, 583
All-heal 327
Allioni 453, 493, 514, 597
Aloe 90
Alsi 490
Am ki guthli 490
Amalaki 199
Amaltas 148, 484, 580
Amarvel 485
Amarvela 174
Ambara 495, 586
Ambashtha 162
American wormseed 156
Amla 199, 582, 590
Amlaki 563
Amlavetasa 596
Amli 225, 488
Amrit dhara ghas 556
Amrita 418, 487
Anantmool 488
Anar 585
Anardana 591
Angelica 32, 594
Angelica root 579
Angur 497
Anise 596
Anjir 487
Anjubar 320, 493
Anjwar 578
Anyar 576
Aonla 199, 486
Apamarga 479, 588
Aphim 565
Apple of Peru 576
Aragwadha 148
Arandi 357
Archa 346, 348
Arishta 368
Arjan 404
Arjun 404, 496, 586, 592
Arjuna 404
Arka 137
Arlu 492, 584, 591
Arni 493
Asan 496, 586
Asgandh 442
Ashok 564
Ashvaha 415
Ashvagandha 439
Ashwath 487
Asmania 202
Ashwagandha 497, 587, 592

Atibala/Atribala 479
Atirasa 428
Atis 479, 500, 579
Atish 69
Atvisha 69
Austakhaddus 327
Avaram 577
Avartani 218
Babchi 493
Babul 56, 479
Babul pods 578
Babuna 490, 583, 590
Bach 72, 479, 498, 518, 579, 588
Badam 496
Badi Kateri 495
Bael 77
Baheda 496
Bahera 407, 496, 586, 592
Bahira 407
Bai vidang 486
Bajarbang 301, 584
Bajardant 480
Bajra-danti 324
Bajradanti 493
Bakain 583
Bakar 493
Bal kunwari 211
Bala 377
Balameshika 363
Balchhharh 280, 507, 583
Balm 583, 590
Ban 577
Bana 436, 560
Banafsha 498, 560
Banfshah 431
Banajwain 416, 496, 558
Banaksha 431, 497, 587
Banbakri 164
Banbalang 479
Banbathu 156
Banbhindi 381, 585, 592
Banda 434, 497, 509, 587
Banhaldi 488
Bankakri 317, 498, 507, 584
Bankhor 80
Bannalia 320
Banokra 444
Ban-palandu 423
Ban-piyazi 423
Bans 596
Banshte 259
Bansulpha 206
Bantulsi 293
Bara bichhua 488
Bara elwan 150
Bara-gokhru 497
Barakotha 260
Baranda 288
Barau 398
Bare 72, 479, 498, 518, 588
Barela 377
Bargad 487
Bari ilaichi 480, 588
Bari kantkari 495, 515
Bariara 495
Bariyar 377
Barma 522
Barna 581
Barsanga 274
Bartang 492
Baryan 72
Basant 489
Bashti 259
Basloghas 545
Bassant 234
Basuti 259, 489, 590
Bathu 484, 526
Batindu 162, 528
Bel 77, 480
Belladona 579, 588
Ber 497
Berang 279
Berberis roots 498
Bergenia 39, 40, 43, 120, 123, 482
Bering 576
Bethar 250, 489
Betharh 255
Bhang 139, 483

Bhangolu 139
Bhangra 196, 582
Bhangru 196
Bharangi 164, 581
Bharda 332
Bhares 205
Bhargi 164
Bhedara 255
Bhekhal 325, 493, 551, 585
Bhekhla 325
Bhenkal 325
Bhilava 494, 509
Bhojpatra 125, 482, 498, 503, 580
Bhooj 125
Bhotia-badam 485
Bhringraj 486, 582, 590
Bhringraja 196
Bhuikanda 585
Bhuj 125
Bhurja granthi 125
Bhurja patra 125
Bhurse 594
Bhurse khamba 557
Bhutbishi 167
Bhutjata 167, 485
Bhutkes 167
Bhutkeshi 373
Bhutkesi 494, 498
Bibhitak 407
Bichhu buti 537, 560
Bidang 279, 491
Bidari 332
Bidari-kanda 585
Bihidana 485
Bikh-akhwar 594
Bidarikanda 585
Billi Lotan 490
Bilpatri 77, 579
Bilve 480, 579, 588
Birnjasif 59
Birch 125
Birmi 401, 498
Biroja 308
Bish kandara 491
Bishkaral 412
Bishkhapra 495
Bishloghas 502
Bistendu 575
Bitharh 250, 583
Blue gum 575
Blue pine 309
Box-myrtle 276
Braa 485, 530
Brahm kamal 509
Brahmakamal 555
Brahmi 154, 482, 484, 580
Bras 353
Brihati 585
Broom 570
Buckwheat 205, 590
Budagur 202
Buna 586
Burans 553
Burdock 94, 481, 522
Burweed 444
Byaghrerand 489
Calamus root 579
Calotropis 137
Camphor 580, 589
Caper 483
Caraway 503
Carway 144
Cassia pod 148
Castor 585
Castor-bean 357, 591
Catechu 52
Cedarwood 153
Cedral 573
Chabru 487
Chachri 265, 543
Chaitra 412
Chakramard 150, 151, 484, 580, 589
Chaksu 147, 483, 580, 589
Chakunda 150, 151
Chalai 253
Chambro 594
Chamchora 91
Chameli 489, 539, 590

Chamush 558
Chandan safed 565
Chandrashoor 490
Chandrashur 583
Chandrika 342
Charangi 584
Charas 139
Chaulai 480
Chedang 569
Chhal 481
Chhambra 481
Chharila 300, 492
Chharma 223, 488
Chharmar 588
Chhe 202
Chhedang 202
Chhershi 250
Chhershup 250
Chhota dhatura 444
Chhota doda 508
Chhoti dudhi 487
Chhoti kandiari 568
Chhoti Kateri 495
Chhoti-Ninai 578
Chhuchha 225, 544
Chiayun 268
Chikory 157
Chikori 484
Chil 309
Chilgoza 492, 507
Chir pine 308
Chirata 389
Chirayata 495, 509, 586
Chirayita 389
Chirh 308
Chirindi 490
Chironji 483
Chita 314, 584
Chitra 314
Chitrak 314, 493, 584, 591
Chitrakmula 314
Chitraparni 422
Chora 91, 481, 498, 500, 579
Chorak 500
Choraka 91
Chothar 114, 482, 502
Chuchi 346, 348
Chuk 225, 488
Chukri 346, 348, 498
Chulu-ka-banda 434
Chura 91
Citronella grass 581, 589
Cocklebur 444
Coriander 165
Costus 370
Country mallow 377
Cowhage 272
Crabseye 49
Curry leaf 274, 491, 591
Curry patta 274
Dabra 422
Damabuti 496, 586, 592
Damtuli 76
Danda 535
Dandasa 246
Dandelion 398, 557
Danti 110
Dantimool 482
Daran 333
Darim Giloe 333, 493, 585
Darmu 333
Daru 333, 552, 591
Daruhaldi 114, 117, 502, 580
Daru-haridra 114, 482
Daya 483
Deadly night shade 104
Devdar 153
Devdaru 153, 484
Dhai 442, 497, 560, 592
Dhai-ki-phool 442
Dhak 133
Dhamani 291
Dhaniya 165, 485
Dhataki 442, 497
Dhataki pushpa 442, 587
Dhatriphala 199
Dhatura 184, 187, 581
Dhatura (Safed) 187, 486, 512, 564

Dhaura 577
Dhava 481
Dhoklamb 120
Dhoop 255, 256, 351, 489, 491, 498, 507, 583
Dhouturo 187
Dhup 253
Dhupi 583
Dhuppi 566
Dhurlughas 540
Dhurpi sag 260
Dhustura 184
Dimo 98
Dioscorea 190, 498, 503
Dipya 231
Donojhan 481
Doob 176, 530
Doob-grass 176
Doorva 176
Dori 456
Dorighas, 498
Draksha 497
Dronpushpi 260, 490
Drougalbel 272
Dru 144
Drub 176
Dudhi 487
Dudhi mohra 479, 517
Dudhli 398
Dugdhapheni 398
Dugdhika 596
Dulal 398
Dungtuli 452
Duno 480
Durwa 485
East Indian screw tree 218
Ekdania 577
Elwan 151
Emblic 199
Emblica 199
Ephedra 40, 598
Erand 357, 494, 585, 591
Eranda 357
Farash 577
Farna 480, 500
Fenugreek 576
Flame of the forest 133
Fossil tree 211, 506
Foxglove 486, 589
Gaggar 355
Gaindri 396, 556
Gajavan 291
Gajawan 578
Gajwanbhed 507
Galgal 484
Gambhari 488, 582
Gandhela 274
Gandhini 346
Gandhloghas 396
Gangar 577
Ganj 190, 503, 582
Gaojavan 291
Garlic 88
Garmezu 481
Gatikpa 123
Gauj 491
Gentian 598
Ghee-Kuwar 90, 588
Ghorajiri 580
Ghrita-Kumari 90
Ghrit-kumari 480
Ghuggi Badshah 508, 554
Ghurbach 583
Gigers claws 575
Giloe 418, 586
Giloye 496, 592
Gindaru 495
Gingilly 376
Giriparpat 317, 584
Glory lily 212
Gojihva 291, 492, 584
Gokhru, 496
Goma 260
Goncha 272
Gorakhmundi 495
Guchhi 268, 491, 498
Guduchi 418, 496, 586
Gulab pushpa 565

Gulabans 491
Gulancha 418
Gule banafsha 497
Gule gulab 494
Gulebanaksha 431
Gulegaojavan 291
Gulje 418
Gulmehndi 577
Gulmundi 495
Gumabuti 260
Gunja 49, 479, 579, 588
Gurguli 574
Gurhbheli 481
Gurpatraj 160
Haiza 576
Haldi 485
Haldu 574
Hallo 490
Halu 366
Hanspadi 76
Hansraj 76, 479
Hapusha 250, 583
Harad 410, 496, 557, 586, 592
Harde 410
Harhjori 487
Haritaki 410, 496
Harmal 584
Harsinghar 285
Hasha 416
Hashish 139
Hathi-sundi 488
Hathpanja 180
Hauber 250, 489, 583
Hawthorn 338, 485
Hemp 139
Hempushpika 242
Henebane 569
Heritpatri 562
Hieunsheli 525
Himalayan cedar 575
Himalayan Foxtail-lily 506
Himalayan high level silver fir 47
Hiamalayan Paeony 298
Himalayan yew 401
Hing 582
Hiranpadi 485
Hirantutiya 484, 581
Holy 288
Honey combed mushroom 268
Hops 229, 489, 538, 582
Horse-radish tree 269
Horse tail 487
Horse-chestnut 480
Horsegram 194
Horsemint 265
Hritpatri 486, 581, 589
Hunraj 320
Hunter's tea 320
Hyssop 237
Indian aconite 65
Indian belladona 104, 502
Indian chenopodium 156
Indian chestnut 577
Indian Junipers 253
Indian Kapok tree 128
Indian Laburmum 148
Indian liquorice 49
Indian Podophyllum 317
Indian Squill 423
Indian Valerian 426
Indian horse-chestnut 80
Indrajau 225, 489, 497, 582
Indrayan 484, 581
Irimeda 565
Jadwar 581
Jafran 172
Jaitum 492
Jakhmbuti 293
Jal marich 576
Jalga 304, 492, 584
Jalgi 104, 502
Jalneem 482
Jalpipli 490
Jamalgota 575, 581
Jamiri 484
Jamun guthli 496
Jangli gainda 396
Jangli hing 487

Jangli jhau 577
Jangli Kuth 94, 481, 522, 588
Jangli Lahsun 480, 520
Jangli piaz 423, 497, 509, 586
Jangli pudina 491
Jangligainda 573
Janpapro 300
Jari dhoop 256
Jarul 577
Jatamansi 280, 491, 507, 583
Jayapal 575
Jeewak 490, 507
Jharka 104, 304, 492, 502
Jhau 481, 523
Jhingan 490
Jhinti 482
Jiaputa 337
Jiunti 580
Jiyapota 493, 585
Jogi badshah 509
Jonkaphal 218
Jonkmari 481
Joob 530
Jufah 489
Juffa 237
Juhi 575
Jungli podina 265
Juti 337
Jyotishmati 580
Kabra 480
Kachnar 111, 482, 523, 588
Kachur 215
Kahbung 434
Kahi 276
Kahu 492
Kail 309, 546
Kail cones 456
Kaiphal 276, 491, 583
Kait 506
Kaizaban 351
Kakamachi 383
Kakar 311
Kakarsinghi 499
Kakarsingi 584, 584
Kakhum 279
Kakmachi 585
Kaknasa 490
Kakoli 360, 487, 490, 494, 507, 508
Kakrasingi 311
Kakru 311
Kaktundi 481, 588
Kala dhatura 486, 589
Kalabis 225
Kaladana 583
Kala-jirah 132
Kalajira 483, 499, 503
Kalazira 483
Kali sariva 485
Kaliar 482
Kaligewar 525
Kalihari 212, 488, 506, 582, 590
Kalijari 366
Kalijhant 479
Kalimushli 581
Kalimusli 485, 503
Kamala 262
Kamila 262, 490, 541
Kampillak 262
Kanak 184
Kanakchampa 242
Kanchnar 111
Kanda-ri-mokri 317
Kandiari 381, 592
Kaner 283, 491
Kanermool (Rakta) 491
Kangaru apple 585, 592
Kanor 80, 519
Kanphul 398, 496, 557
Kanta-chaulai 480
Kantala 480, 588
Kantkari 381, 386, 585
Kantkari (Brihati) 495, 515
Kantkari (Chhoti) 592
Kaphal 276, 532
Kapoor kachri 452
Kapur 484, 580
Kapurkachri 215, 488, 582
Kapurmaram 575

Kapurminjar 586
Kapur-Tulsi 570
Karaita 497, 586
Karal 488, 537
Karanj 493, 585
Karanja patra 323
Karkatsringi 311
Karoo 453
Karounda 577
Karpur 565
Karru 305
Karu 207, 488, 499, 506, 536, 553, 582
Karveer 283
Karvi 305
Karvi Patish 499
Karvi Buti 519
Karvir 491
Karwi 576
Kasani 157
Kashini 157
Kashmal 114, 117, 482, 502, 580, 588
Kashmiri Patha 355, 494
Kasmard 484, 580, 589
Kasni 157, 484, 580
Kasondi 580
Kasturi patra 495, 509, 555, 585
Kateli 386, 556
Katha 479, 563
Katha tree 52
Kathber 578
Kathu 205
Kat-karanj 483
Katphal 458, 513
Katrin 564, 575
Kaunch 272, 491, 583, 591
Kaur 207, 305, 481, 506
Kelon 153
Kemuk 169, 485, 503, 581, 589
Kesar 172, 485, 503, 581
Kesaraja 196
Keshara 172
Keu 169, 485, 581
Kewli 153
Khadir 52
Khadir 563
Khair 52, 479
Khairwal 482
Khaja 483
Khamet 98, 522
Khanda 202
Khanor 80, 480, 519
Kharenti 377, 495
Kharshu 577, 578
Khas 430, 497
Khaskhas 586
Khor 246
Khubbasi 490
Khurasani ajwain 231, 489, 582, 590
Khus 572
Khuskhus grass 430
Kikar 56, 479
Kinnauri phool 296
Kins 190
Kiratatikta 389
Kirmala 502
Kiwanch 272
Knotgrass 320
Koche 480, 521
Kolroi 47
Konnai 577
Kooth 370
Kovidara 111
Krishmadhastura 183
Krishnajeerak 132
Kshri-kakoli 492, 506, 583
Kubish 570
Kuja 571
Kuje-ke-phool 571
Kulath 194, 582
Kulthi 194, 486
Kumari 90
Kumbhi 483
Kumbro 524, 531
Kura 225, 489
Kurchi 225
Kuri 285, 531
Kuruvira 283

Kushmol 114, 502
Kushtha 169, 370
Kustehindi 370
Kutaja 225, 582
Kutaki 584
Kuth 370, 494, 499, 508, 585
Kuthakhar 80
Kutki 492, 499, 507, 546
Kware 90
Kwar patha 90
Laghu kantkari 386
Lahsun 88
Lahsunia salep 487, 506
Lajwanti 491, 591
Lakuch 458, 511
Lakucha 481
Lal kaner 283, 584, 591
Laldori 210
Laljari 96, 210, 488, 493, 522
Laljhau 577
Lalmirch 143
Laltang 301, 507, 546
Lamajika 597
Lamjak 485
Lamo 509, 521
Langli 212, 488, 582, 590
Lanka mirch 143
Laot 480, 520
Lasun 88
Lasurha 485
Leadwort 314
Lechu 348
Leechu 508
Leelkounthe 83
Lemon grass 581
Lepte 416
Lepti 572
Lepto 416
Leur 253
Lilkounthe 83
Liquorice 565
Lobelia 598
Lodar 394
Lodh 394
Lodhra 394, 495, 509, 585
Lojh 394
Ludud-dorje 500
Lundud dorje 595
Machhi 280, 544
Machoti 320
Madanphal 340, 493, 585
Madar 137, 483, 589
Madyanti 597
Maghz badam 493
Mahameda 493, 508, 585
Mahanim 583
Mahaphala 77
Maharukh 480
Mahua 583
Mahwa 583
Maiden hair 479
Maiden hair fern 76
Maiden hair tree 211
Mainphal 340
Makoi 383, 495, 592
Makora ghas 485
Male fern 597
Maljhan 482
Malkangni 484, 580
Malti 583
Mamira 412
Mamiri 483, 496, 557, 592
Mandukparni 154, 489
Mandookparni 484, 580, 589
Manjistha 363, 494, 553
Manjit 363
Manjith 363, 536
Mansi 280, 544
Margosa 107
Marorphali 218, 488, 582
Masho 416
Mashparni 496
Masreen 96, 488, 537
Matazor 304, 584
Mathosal 374, 498
Mathosla 373
Mawo 144
Meda 493, 508, 584

Meda-lakri 490
Meenachari 360
Mehndi 300, 490
Menphal 340
Methi 496, 558
Methonni 212
Millfoil 59
Mindha 340
Mistletoe 434, 497, 587
Mitha-Indrajau, 497
Mitha neem 274
Mithinim 491, 583
Mitha patish 479, 499, 500
Mithi saunf 582, 590
Mithpatra 160
Mohra 500, 517
Mohra-bish 479, 500
Mohru 577, 578
Mokri 561
Mooli 494
Moorva 597
Morua 351
Motha 177
Motia 512
Motininai 346
Mridupatra 490
Mrigashinga 218
Mukhajali 195, 486
Mulhathi 565
Mulhatti 582
Munire 525
Muramanse 597
Muramansi 374, 494
Musabbar 90
Musha-ro-khor 444
Mushkbala 426, 453, 497, 499, 586
Mushki-Kapur 572
Mustak 177
Myrobalan 450
Nagarmotha 177, 485, 486
Nagbala 495
Nagdaun 509
Nala 598
Nandru 301, 584
Neelkanthi 83, 588
Neem 107, 482
Neir 585
Neoza 499
Nepali chirayata 495
Ner 495
Nera 351
Nesar 548
Nesrawlo 374
Nihani 425, 426, 497, 499, 509, 586
Nilabansa 482
Nili-Ka-Phool 549
Nilkanthi 480
Nilkonthe 520
Nim 107
Nimba 107
Nimbu 484, 528
Nirbisi 486, 581
Nirgandi 436, 497
Nirodhak buti 521
Niragundi 436
Nishoth 492, 584
Nismoli 320
Nurcha 100, 579
Nut grass 177
Oakmoss 571
Ogla 205
Oleander 283
Olive 492, 584
Pachak 370
Padara 221, 488, 506, 582
Paddam 329
Padhal 495
Padmakashta 329
Padmakh 329
Padmapushkara 583
Padumba 416, 558
Pajja 329, 493
Pakhanbed 120, 123
Palah 133
Palash 133
Pamukh 586
Pangduma 416
Panja 180

Panwarh 151
Papita 483
Papra 317, 366
Papri 483
Parasikayavani 231
Parijata 285
Pashanbhed 482, 580
Pashanbheda 120, 123
Patha 162, 484, 589
Pathar-chat 120
Patharphori 486, 581
Patindu 162, 528
Patis 479
Patish 69, 500, 579
Patla 495
Patmakh 493
Patol 496, 586
Patol patra 496
Patraj 160
Patrala 221, 488, 506, 590
Patti-dhoop 555
Phafra 487
Phafto 569
Pharna 480, 521
Phenita 368
Phillu 255
Phumbasa 502
Picrorhiza 305
Pila kaner 415, 496, 586, 592
Pili-chameli 489, 539, 590
Pilijari 412, 496, 557
Pipal 487
Pipermint 491
Pipjul 535
Pipling-tsa 525
Pissu buti 544
Pissumar buti 575
Pitakharenti 380
Pitkaner 415
Pitmalti 489
Pitpatra 488
Plash 483
Podina 265
Pokarmul 388
Pongam 323
Poshkar 240
Poskarmul 240
Pothi 577
Poutijia 337
Prayer beads 49
Priangu 483
Pride of India 577
Prishtparni 422
Prisniparni/Prishniparni 422, 496, 497
Priyangu 511, 565
Psyllium 598
Pudina 491, 543
Punarnava 482
Pushkarmool 240, 489, 507, 583
Puthkanda 62, 479, 516, 588
Puti karanj 323
Putranjiva 337
Pyrethrum 580
Rachi 202
Rada 340, 493, 585
Rai 483
Rajagiri 480
Rajnigandha 493
Rakhala 401
Rakhyala 401
Rakta 491
Rakta-Punarnava 597
Rasanjan 114
Rasaunt 452
Rasna 428, 497
Rasona 88
Rasont 114
Rasount 568, 574
Ratanjot 96, 98, 210, 291, 481, 492, 502, 522, 579
Rati 595
Ratpacho 83, 520
Rattanbat 351
Ratti 49, 479, 579, 588
Red chillies 143
Red bead vine 49
Reetha 368
Regmalpatta 285

Reli 151
Resin 450, 483, 487, 495, 522, 545
Revandchini 346, 348, 494, 508, 585, 591
Rhodo flowers 353
Rhubarb 346, 348, 494, 585
Rhus 595
Ribhuksu 545, 595
Ridhi 488, 506
Rikhpat 317
Ringworm plant 151
Rishbhak 490, 507
Ritha 368, 494, 585
Rohini 262
Rosemari 494, 585, 591
Rudanti 506
Rudravanti 502
Ruighas 567
Sadabahar 586, 589
Sadah 486
Safahuss 585
Safed bach 489
Safed dhatura 187, 486, 589
Safed Manjith 536
Safed musali 360, 481, 508, 538
Safed pudina 265, 590
Safed rall (resin) 495
Safed siris 480
Safed musli 494
Safeda 487
Saffron 172, 581
Sag-angur 482
Sahasrapal 102, 588
Sahdevi 495, 497
Sahinjan 269, 491
Sainski 579
Sainsarbuti 102, 523
Sakina 586
Sal 503, 577
Salad 489
Salam mishri 508, 585
Salampanja 180, 486, 499, 503, 581
Salep 180, 503
Salmali 128
Salyan 531
Sambhalu 436, 497
Sambhalu beg 436
Samp-ki-kumb 574
Sanjana 269
Sannay 564
Santonin 481, 569
Santonin plant 100, 481, 579
Saona 296
Saptala 487
Saral 308
Sardanda 484, 529
Sariva 488
Sarla 223, 488
Sarpagandha 342, 494, 508, 585
Sarphar 412
Sarson 482
Sathjalari 452
Sathra 293, 492, 584
Satla 487
Sea-onion 423
Seinski 100, 481
Seja 577
Self-heal 327
Semul 128
Semulmusli 128
Senna 530, 564, 592
Senski 452
Sesame 376
Sewal 571
Shafri 586
Shaileya 300
Shalparni 486, 581
Shamshad 279
Shankhahuli 291, 584
Shankhpushpi 487
Shargar 355
Shatamuli 102
Shatavar 102, 580, 588
Shatawari 102, 588
Shati 215, 582
Shatpushpa 492
Shephali 285
Shephalika 285

Shia-Zira 144
Shilapushpa 581
Shimshar 555
Shingapalu 566
Shingpillu 255
Shingu jira 144
Shinkhrao 300
Shirish 85
Shivlingi 483, 580, 589
Shobri 366
Sholar 301
Shugpa 566
Shurgu 583
Shyonak 492
Shyonaka 296
Siahzira 132
Siali 585, 585
Silk cotton tree 128
Simul 128
Simrath 355
Simris 594
Simrung 355
Sindur 262
Singli-mingli 190, 486, 498, 503, 582, 589
Sinsim 376
Siras ki chhal 480
Sirinh 85
Siris 85, 480
Sirish 85
Snake maize 574
Soa 492
Soanjana 269
Soapnut 368, 585
Sohanjan 583
Soma 202, 487, 582
Somlata 202, 486, 499, 503, 590
Somvalli 597
Sonapatta 296
Spearmint 571
Spikenard 280
Squill 586
St. John's Wort 234
Sthoneyak 401
Stone flower 300
Sugandhbala 426
Suhanjana 564
Sundew 195
Sunthi 587
Sura 436
Suranjan 503
Surjee 594
Svetbarela 380
Swaran khshiri 481
Sweet belladona 304
Sweetflag 72
Swet-barela 495
Tagar 425, 426, 497, 509
Tagetus oil 573
Talispatra 47, 351, 494, 498, 508
Talisa 351
Talish 594
Talispatri 479, 496, 509
Talli 594
Talsar 351, 508, 553
Tamalpatra 160
Tambakhu 576
Tansy 496
Taragon 571, 594
Tatara 576
Taur 482
Teeta 488
Tejbal 446, 497, 509, 587, 592
Tejpat 572
Tejpatra 484, 499, 503, 581
Tejphal 446
Tengu 237, 489
Teragon 481
Tesu 133
Thangi 401
Thekru 566
Thelu 255
Thoohar 487
Thor 575
Thorn apple 187
Thuhar 575
Thuma 487, 535
Thuna 566

Thuner 401
Thuno 401
Thuth 366, 592
Tikta 305, 488, 506, 536
Til 376
Til tail 494
Tilani 497
Timbri 446
Tirku 223
Tirmir 446, 497, 509, 592
Titre 499, 578
Tosh 47, 479
Totoli 482
Trayamana 207
Tuberose 493
Tukhlang 231
Tukhm khatmi 480
Tukshu 346
Tulsi 288, 492, 584, 591
Tumbru 446
Tumburu 576
Tung 530
Tungnut 480
Turpeth 584
Tushusho 363
Tutgantha 202
Tyangu 237
Udsalap 298
Udsalib 298, 492, 584
Udumbar 487
Ujjral 434
Unnav 565
Ushare-revand 346, 348
Ushir 430
Ustakhadus 327, 585
Vacha 72
Vakuchi beej 493
Valerian 454, 598
Vanaksha 563
Van-haldi 215
Vansa 563
Varahikand 486, 589
Varidhi 488
Varun 485
Varuna 581
Vasa 259, 590
Vasaka 259, 489
Vat 487
Vatsnabh 65, 479, 500, 579
Vettiver 430
Vibhitaki 407
Vidarikand 493
Vijaya 139, 483
Vishalyakarni 549
Vridhi 506
Walnut 85, 246, 583
Wattle 577
Wetyar 255
Wild Endive 157
Wild Hing 506, 582
Wild marjornum 293
Wild olive 492
Wild thyme 416
Wild-egg-plant 386
Woodfordia 442
Wormwood 574
Yarrow 59
Yellow jasmine 242
Zakhmehayat 489
Zergul 483
Zinger 587
Zufah 583
Zufah-yabis 583